Transducers—the Electricity Producers

Thermocouple

A thermocouple changes heat energy into electrical energy. It gives only millionths of a watt.

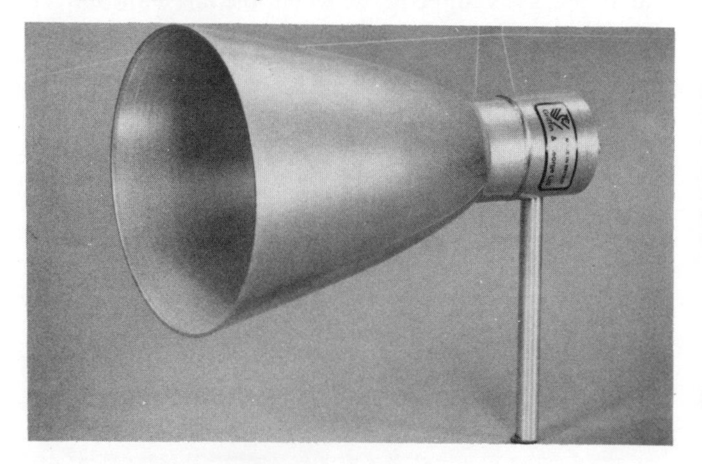

Microphone

A microphone changes sound energy into electrical energy. It gives only millionths of a watt.

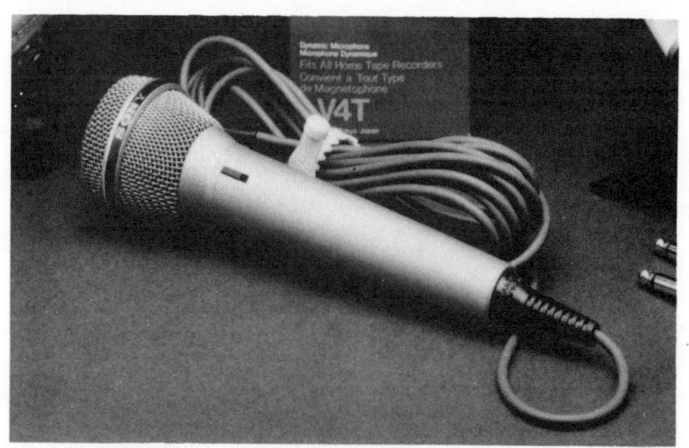

Hi-fi Cartridges

Hi-fi cartridges change movement energy into electrical energy. They give millionths of a watt.

Photocell

A photocell changes light energy into electrical energy. It gives millionths of a watt.

Aerial

An aerial changes the energy of radio waves into electrical energy. Outputs are usually very small.

The photographs show some of the transducers available with the power they can provide. There are, of course, many others but you can see that most of them produce only small amounts of power. These transducers are used at the input to an electronic circuit and the electricity they produce controls the action of the circuit.

Transducers — the Electricity Users

Loudspeaker

A loudspeaker changes electrical energy into sound energy. A typical power value is thirty watts.

Motor and Relay

Motors and relays change electrical energy into movement energy. Can be made to handle large amounts of power.

Light Bulb

A light bulb changes electrical energy into light energy. Some can handle hundreds of watts.

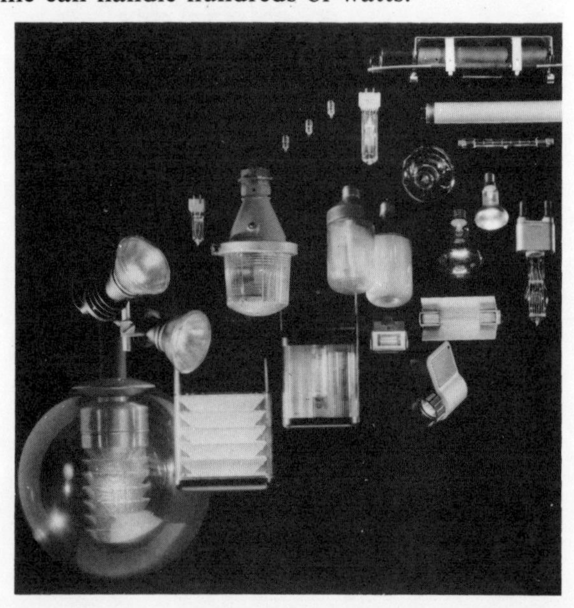

Transmitting Aerial

A transmitting aerial changes electrical energy into the energy of radio waves. The largest aerials handle millions of watts.

These are the transducers that *take in* electrical energy. They handle the *output* from electronic circuits and you will notice that they handle much larger quantities of power than the input transducers. This is of special importance since we have to be able to detect this output with our own senses, our own eyes, ears, touch and so on. (The exception to this, in the examples given above, is the radio transmission which is meant to be detected by a radio receiver.) Thus these transducers have been designed to handle the power levels which we need in our everyday lives.

ELEMENTARY ELECTRONICS

MEL SLADDIN

Cottenham Village College
Cottenham

HODDER AND STOUGHTON
LONDON SYDNEY AUCKLAND TORONTO

Preface

This book arose out of the experience of teaching electronics in a London comprehensive school without the aid of a suitable textbook. Whilst the organisation of practical work is well advanced in many schools and, indeed, the possibilities are almost endless, the author found a distinct shortage of written material which is necessary to raise the status of the subject to be on a par with the other science subjects on the curriculum. This book is an attempt to fill that gap and provide the electronics teacher with a practical workbook from which pupils can obtain enough detail to assemble working circuits. It also aims to provide the theoretical background material to enable pupils to enter the public examinations in the subject.

It is further hoped that the layout and content of the book will encourage teachers and students with no previous experience of electronics to embark on this fascinating subject. To that end, it has been arranged that each page covers one topic, that the topics form a comprehensive basic electronics course in themselves and that the reader need not refer to any other material to gain a first hand working grasp of electronics.

The author is indebted to that core of teachers who have worked so hard to make electronics a going classroom concern and many of the circuits in the book stem from their experience over many years.

M. Sladdin
December 1981

British Library Cataloguing in Publication Data

Sladdin, M.
Elementary electronics.
1. Electronics
I. Title
537.5 TK7815

ISBN 0 340 24643 X

First published 1983

Reprinted 1980, 1981, 1983, 1984

Typeset in 11/12 pt Times (Monophoto) by Macmillan India Ltd. Bangalore

Printed in Hong Kong
Hodder and Stoughton Educational,
a division of Hodder and Stoughton Ltd.,
Mill Road, Dunton Green, Sevenoaks, Kent.
by Colorcraft Ltd.

Contents

Acknowledgments

The author and publishers thank the following for giving permission to reproduce photographs in this book: Barr & Stroud Ltd., Binatone International Ltd., Black and Decker Ltd., British Broadcasting Corporation, Ferranti Ltd., Griffin and George Ltd., Grundig (Great Britain) Ltd., IBM United Kingdom Ltd., Mullard Ltd., Philips Electronic and Associated Industries Ltd., Plessey Ltd., Pye Telecommunications Ltd., Pyser Ltd., Rank Wharfedale Ltd., Sony (UK) Ltd., Timex Corporation.

The publishers would like to thank L. Goldsmith and R. Ward of the Physics Department, The Skinner's Companies School, Tunbridge Wells, for their assistance.

1 INTRODUCTION
The Electronics Age

We are now all so familiar with the large number of electronic devices in our lives that it is easy to forget what a new industry this is. Although radio, television and even computers pre-date the Second World War, our present industry really came to life with the production of the transistor in the early 1950s. Since then, growth has been explosive in its speed and quite remarkable in the variety of products now obtainable.

It seems that, in every aspect of our lives, an electronic device has been made to help us. Some of these we are well aware of. Pocket calculators have become more common than slide rules. Electronic watches are gradually replacing mechanical ones. Both of these electronic devices can be made to operate more accurately than their mechanical equivalents and at less cost.

A pocket calculator

Digital watch

Probably the most outstanding achievement of the industry that has made all this progress possible is miniaturisation. Although some products, for instance computers, could in theory be used whatever their physical size, many products depend for their use on being manufactured a convenient size. The photographs show three such examples.

How has the miniaturisation of these products been achieved? The answer is the production of components called integrated circuits and microprocessors.

Examine some simple components, transistors and resistors say, and note that a typical transistor may have a body size roughly that of a small pea. Now examine the photograph of an integrated circuit. The one shown passes through the eye of a sewing needle and yet contains 120 components. The first microprocessor produced in Europe is roughly the same size and contains over 9000 components! With miniaturisation on this scale it is nearly impossible to imagine the technology needed to make the components but clearly the possibilities for use are tremendous.

A pocket television

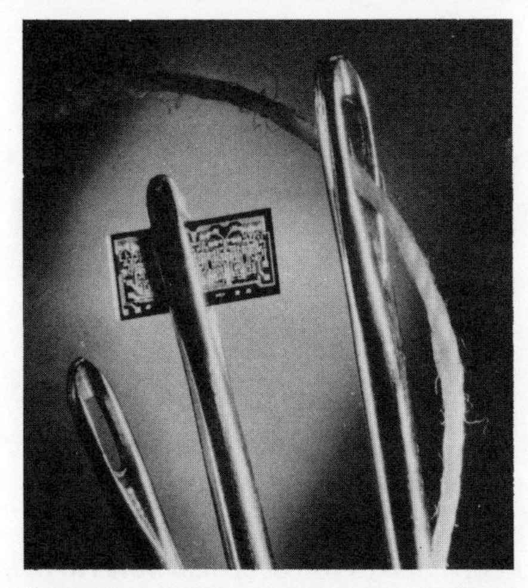
Internal construction of an integrated circuit

The sheer speed of progress in the field of electronics gives rise to many problems for people who work in the industry. Not the least of these is the constant problem of trying to stay up to date with all the latest developments and techniques. A great deal of specialisation is needed to enable the engineer to stay ahead in his subject.

Another problem which has to be solved is that of servicing. Increasingly, devices are made from separate units which can be plugged together very easily. If a device breaks down, the faulty unit is replaced without the actual cause of breakdown being found. The two-way radio shown below, although a very complicated design, is made from seven easily replaceable units, called modules. Should the radio ever fail, replacement of a module is very easy.

Computers are made from boards of components which can be plugged in as easily as a household plug. This enables the service engineer to do his job without having to be as knowledgeable as the person who designed the computer.

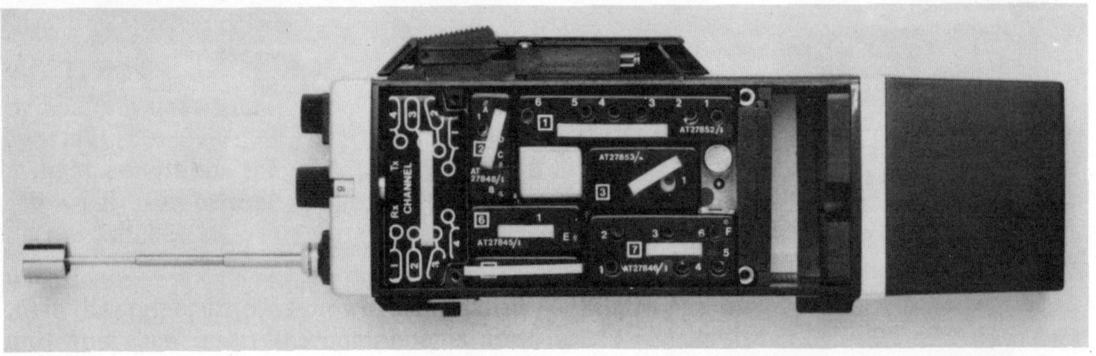

Inside view of a two-way radio

It is usual that expanding industries have a very beneficial effect on other industries, especially the materials industry. This is the case with the electronics industry and a good example is the development of fibre optics. Very narrow rods of glass can now be made and light is sent down these rods, without any loss from the sides of the rod. These beams of light can be modified to carry messages and the whole system could be used to replace our present cable system for the telephone. This new method can carry far more messages in the same space than metal cables but the system depended on the development of very high quality glass. The glass now in use is so clear that a window a kilometre thick would be as clear as an ordinary house window.

As development follows development, so more industries come to use and rely on electronic equipment and, in so doing, make advances which were previously impossible. Space travel, medical research and, sadly, armaments are all fields in which electronics play an important part. Again, man is faced with the problem of putting his discoveries to beneficial use and resisting the temptation to put others in peril for his own gain.

The photographs below show the laser being put to peaceful and non-peaceful uses.

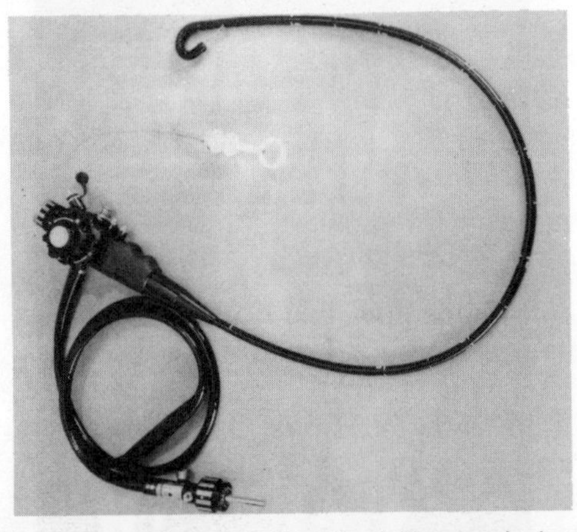

This endoscope makes use of optical fibres to help surgeons see inside the body.

This tank gunsight makes use of a laser for finding the image of its target

Electronic Systems

Here are the major jobs performed by electronic systems.

Alternating Current Amplification

Producing the volume you need is the job of an alternating current amplifier.

Direct Current Amplification

The sensor does not produce sufficient power to be recorded. The direct current amplifier gives an output which can be seen.

Computation

Within these neat cabinets, the computer is carrying out its operations at the rate of millions of operations per second.

Information Storage

The tiny magnetic discs in this unit can store nearly 500 billion digits (numbers 0 and 1)

Telecommunications

Communication without wires. A 2-way radio in operation.

Systems Control

Furnaces at British Steel plants are monitored and controlled by these computers.

Transducers

In this course you are going to study the behaviour of some of the most important components and some of the circuits in which they are used. Whilst these will not reach the complexity of the advanced devices seen in the first two pages you will soon be able to use the basic ideas and language that led to these developments.

To control the electronic circuits we use, we must be able to give them instructions and information in electrical signals. The circuit will then perform a service for us and this will give an end product which is also electrical.

A circuit can be considered thus:

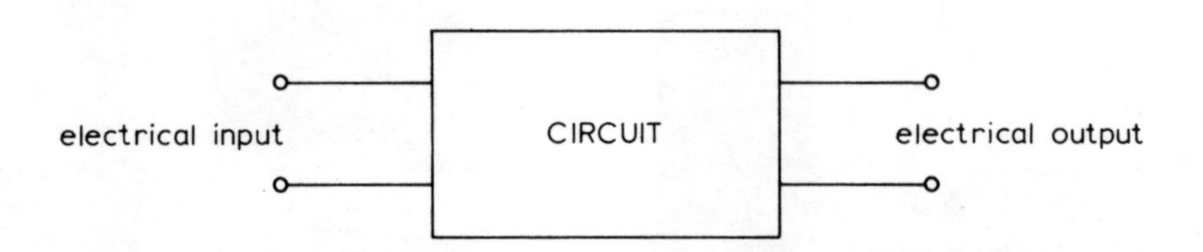

In the example below, we want our electronic circuit somehow to produce a louder copy of the sound for us. To achieve this, it must receive electrical signals and it will then output larger electrical signals.

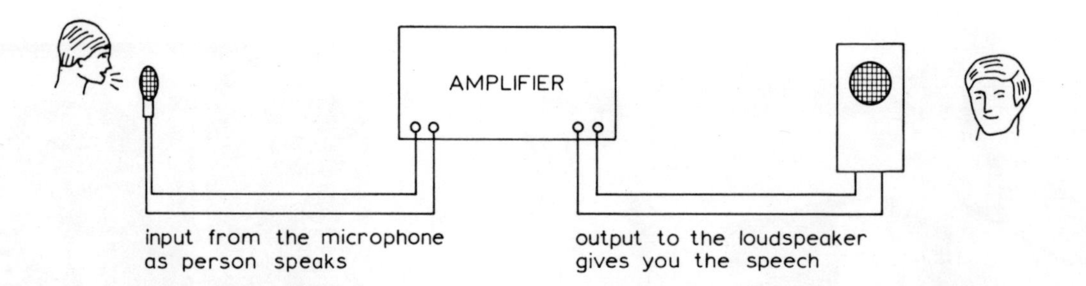

In this example, the microphone and loudspeaker are your contact with the amplifier circuit. It is no use just shouting at your amplifier—without the microphone it can't 'hear' you. Without the loudspeaker you can't 'hear' the amplifier output. The microphone and loudspeaker are *transducers*. They are our contact with the electronic circuitry.

*The **microphone** changes **sound** into **electricity**.*
*The **loudspeaker** changes **electricity** into **sound**.*
} Two important transducers.

A transducer changes one form of energy into another.

In electronics, we are usually interested only in transducers that convert to and from electrical energy. The purpose of this section is to show you which transducers are available and how much electrical power is involved. As a guide to the amount of power involved, remember that the single bar of an electric fire is probably converting about 1000 watts of electrical power to heat and light. The figures given are very approximate since the power will often depend on how the transducer is used. However, you will notice that many of the transducers handle only a very small amount of power.

Using Electronic Components

For your constructional projects, you will have to learn to recognise and handle the different components you need. In particular, you will find it vital to be able to:

1. Learn the symbols
2. Recognise components by their appearance
3. Read the value of a component
4. Learn the correct method of assembly.

To help you with these four problems:

1. The symbols of the components are given in the table below and on page 8.
2. Photographs on the following pages show you the appearance of the common types of each component.
3. There are three ways you can find the value of a component:
   (*a*) *A number* stamped on the case. Most batteries, for instance, will have a number stamped on their case, giving the size of their voltage. Some components need more than one number to describe the job they do and you should look out for them all.
   (*b*) *A colour code* stamped on the case. Bands of colours on the case represent numbers and you will learn this code shortly.
   (*c*) *A type number* stamped on the case. Some components need quite a few numbers to tell you how they work and in this case they are given one type number so that you can then look it up in the manufacturer's handbook, for the details.
4. There are three main problems in using a component in a circuit:
   (*a*) Some components with two leads have a positive and negative lead and are said to be polar. You have to be sure that you put these components in the circuit the right way round. Components with three leads usually have to have the three leads identified and then correctly connected in the circuit.
   (*b*) Some components can be damaged by the heat from the soldering iron and so they must be protected during soldering. Some components can be damaged by heat whilst they are operating and are protected by a heat sink.
   (*c*) Some components are mechanically delicate and must not be roughly handled. In particular, care is often needed when you are bending their leads.

Now memorise the information about components in the following tables and refer to the tables when you do any construction work.

Continued on p. 8

Name	Symbol	Number of leads	Value	Notes on method of assembly
Cell		2	Number	Polar
Battery	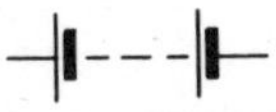	2	Number	Polar
Resistor	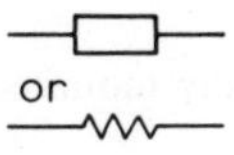 or	2	Colour code (usually)	
Variable resistor	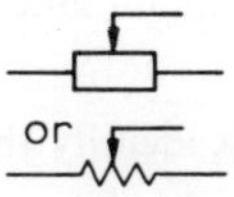 or	2	Number or colour	Two outside terminals identical, middle terminal different
Capacitor		2	Colour or number	
Capacitor	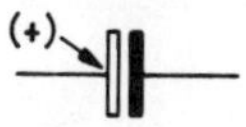 (+)	2	Colour or number	Polar—can be large and require mounting clips
Variable capacitor		2	Number	Panel mounted

Name	Symbol	Number of leads	Value	Notes on method of assembly
Inductor		2	Number	Can be heavy, needs bolting in place
Diode		2	Type number	Polar, heat sensitive
Transistor	or	3	Type number	Three leads, emitter (e), base (b) and collector (c) must be identified, heat sensitive
Transformer		Varies	Number	Usually heavy, needs bolting in place, must identify input and output, must insulate high voltage leads
Ammeter	A	2	Maximum reading on dial important	Polar
Voltmeter	V	2	Maximum reading on dial important	Polar
Light-dependent resistor		2	Type number	Can be delicate
Light-emitting diode		2	Type number	Polar, heat sensitive
Loudspeaker		2	Number	
Microphone		2	Number	
Fuse		2	Number	
Switch		2	Usually the maximum	Usually mounted on a panel
Push-to-make switch		2	permitted voltage and	Usually mounted on a panel
Push-to-break switch		2	current values	Usually mounted on a panel
Change over switch		3	are given	Must identify the 3 leads
Conductors crossing				
Conductors joining				
Photo-voltaic cell		2		Polar
Earth				

Identification

Resistors

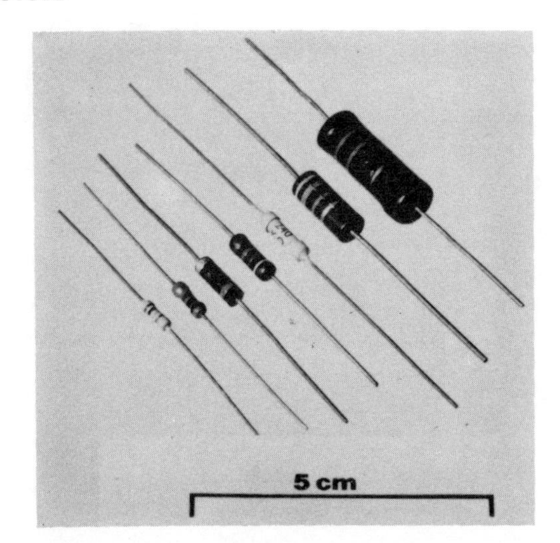

Variable Resistors

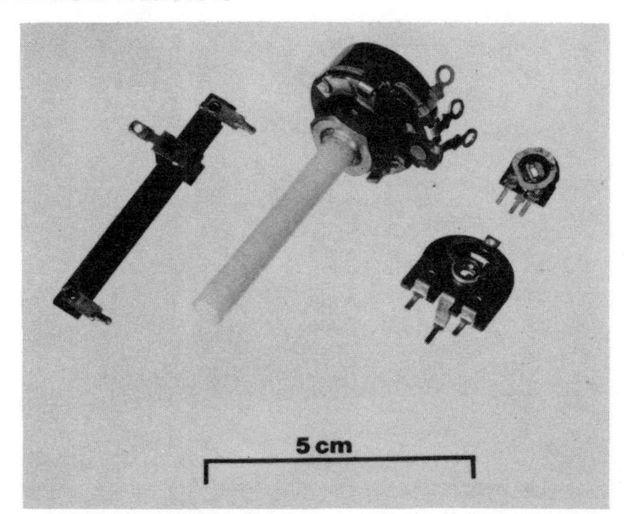

Capacitors

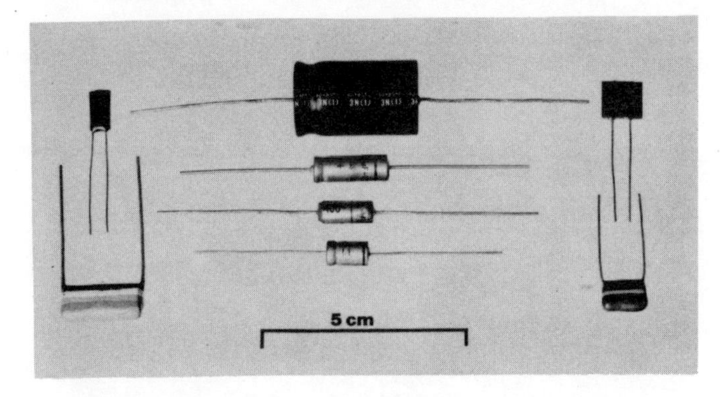

Variable Capacitors

Diodes

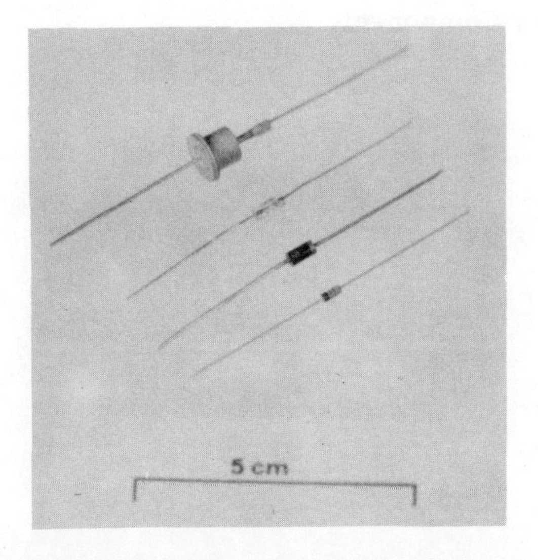

Inductors

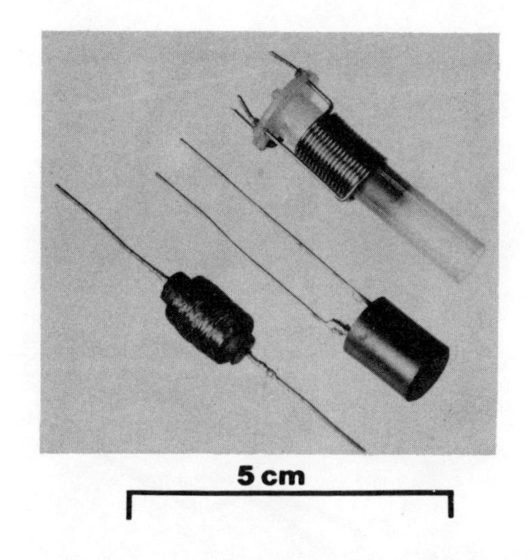

Continued on p. 10

Transistors

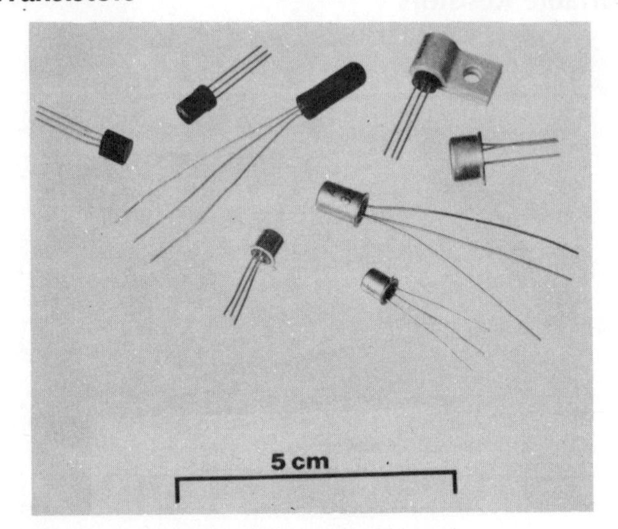

Power Transistors

Meters

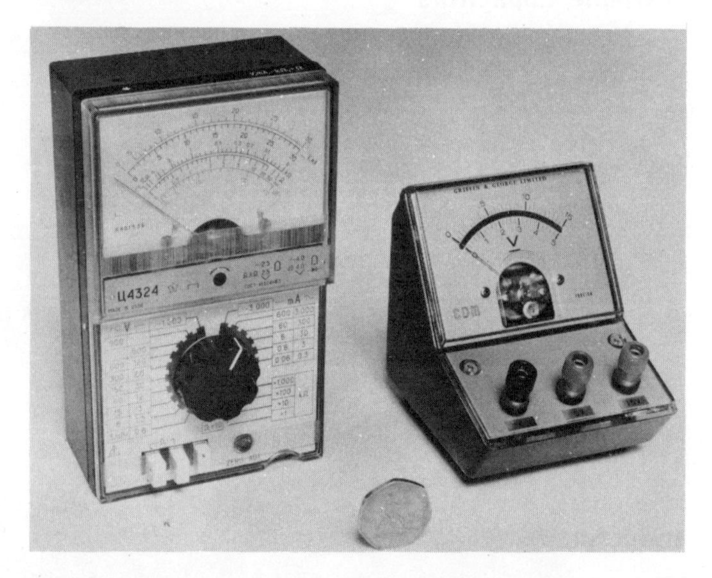

Three More Components

(a) *Light dependent resistor* (b) *Light emitting diode*
(c) *Thermistor*

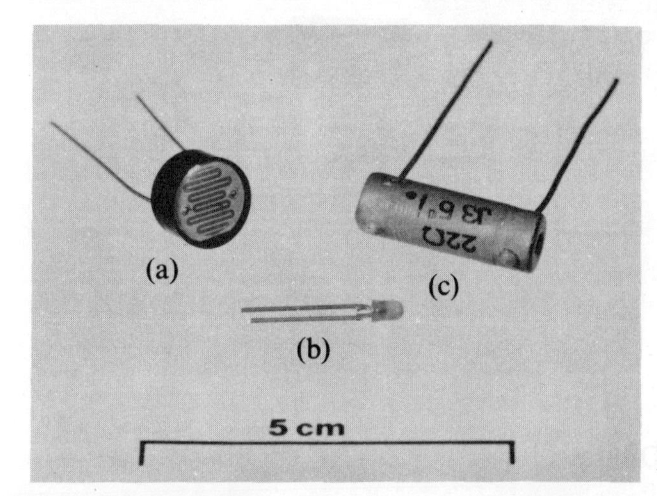

Tools

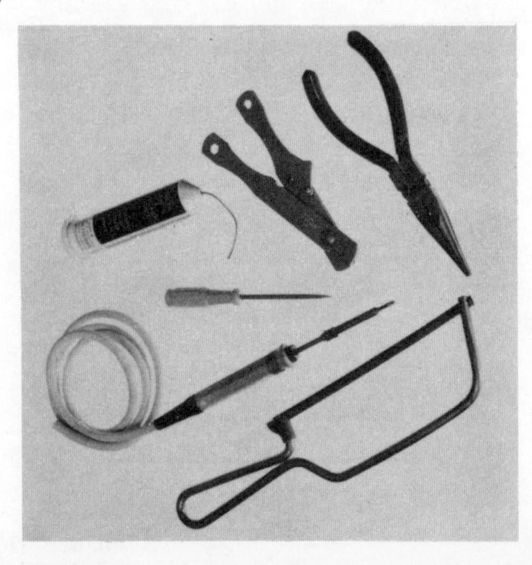

Other Components

A Rapid Construction Method—Using the S-DeC

The S-DeC will be used for many of your teaching circuits and you will find it very useful for two reasons:
1. Circuits can be assembled very rapidly.
2. Components can be used over and over again.

For your permanent circuits, you may use one of the circuit boards on page 14.

Nature

The photograph shows the essential nature of the S-DeC. Components are plugged into holes in the board and the underboard connections provide most of wiring. Any additional wiring is provided by wire links plugged into the holes in the board. No soldering is needed at all, so circuits can be assembled in a few minutes.

Top of S-DeC. The holes in each row are connected under the board. Brass strips form an underboard connection.

Care

Any damage to the underboard connections will cause much annoyance since your circuits will give unexpected results and you will find it very difficult to trace the cause of the trouble. It is absolutely essential to avoid any damage to these underboard connections, so, for successful use of S-DeC, the following instructions must be strictly followed.

1. All leads must be perfectly straight and free of kinks.
2. Any lead which proves difficult to remove must not be forced out.
3. Put only one lead in each hole.
4. Any connecting lead used must be single strand.
5. Do not push leads in too far, or they may touch others under the board.
6. Do not use any lead which has previously been soldered.

Practice Circuit—A Wailer

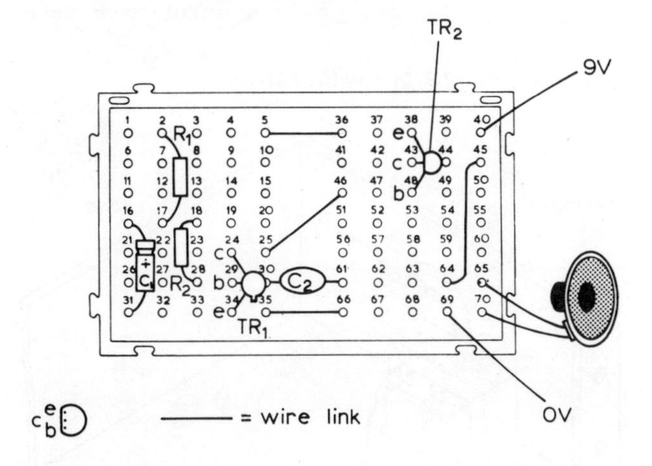

R1 33kΩ
R2 33kΩ
C2 0.1μF
C1 47μF
TR1
TR2
8OΩ
TR1 BC108 TR2 2N3702

Instructions

Like all circuits in this book, the diagram shows which holes are to be used for every component. Take care to identify the leads of your transistor and electrolytic capacitor, and then assemble the circuit. The 47 μF capacitor is an electrolytic type and you should look for the + marking on its case. The transistors need their leads lengthening by soldering an extra 2 cm of bare wire to each. Then you must seek help in identifying the three leads or look ahead to Chapter 7.

When the battery is connected (last of all) a note should be heard which gradually rises in pitch. This tells you that your wailer circuit is working correctly, and that your assembly technique is successful. Now carefully remove the components and hand them in for storage.

Construction Work

You will be given full instructions on how to construct the electronic parts of your projects, but for help with the mechanical construction (building cases, etc.) you should try to get specialist advice. The next two pages contain hints on some of the important stages.

Remember that, with a little practice, you will have a high success rate in electronic construction, but this is only half the story. Your work needs to be well presented to be of practical use. It must be carefully boxed up, have good controls, clear labelling and a neat appearance so that it compares well with bought equipment.

Cases

The diagram shows one of the stages in making a case from sheet aluminium. You will need full details of how to make neat bends, using a vice, how to finish off the edges and how to keep a good surface finish. Notice that all the necessary holes have been drilled before bending starts.

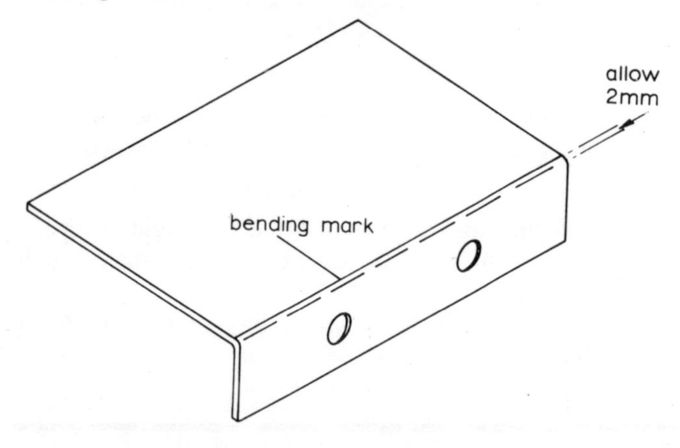

Allowing for the bend when marking out a metal chassis

You will probably need to be able to drill thin metal sheeting and this needs special care. You will need special drills, used at the correct speeds, firm clamping and backing materials to avoid damage.

Most projects need some controls to be mounted on the front panel of the case. Here you see a variable resistor held in place by a nut and locking washer.

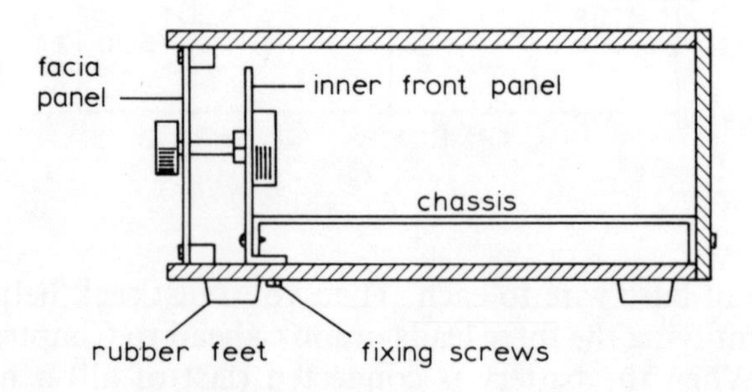

The terminal block gives a secure way of connecting two cables. No loose wires and insulating tape here! The neatly curled cables enable some slack to be kept in them and yet still keep a neat layout.

This diagram shows one of the ways of achieving a good finish with a home made plywood case. The covering is an adhesive-backed 'Fablon' type, with a wood-grain finish. You will need to know how to lap corners, how to avoid creasing and what to do about protruding screw heads.

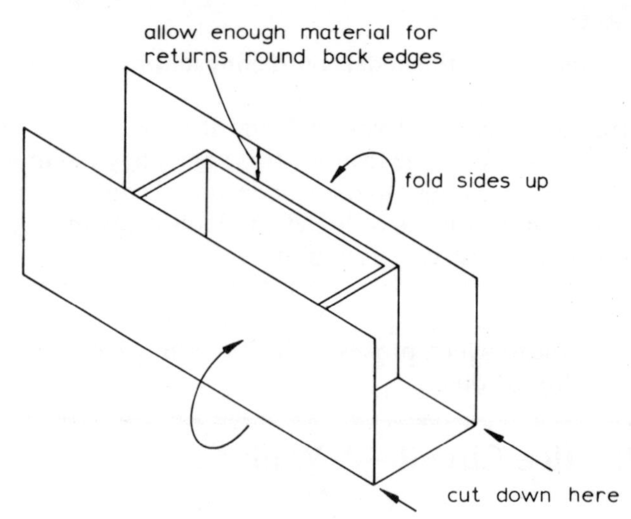

Stages in covering a box with 'Fablon'

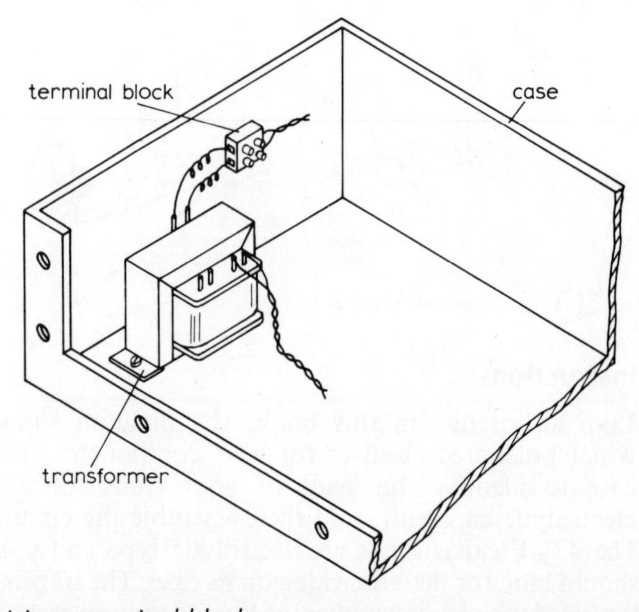

Using a terminal block

Safe Working

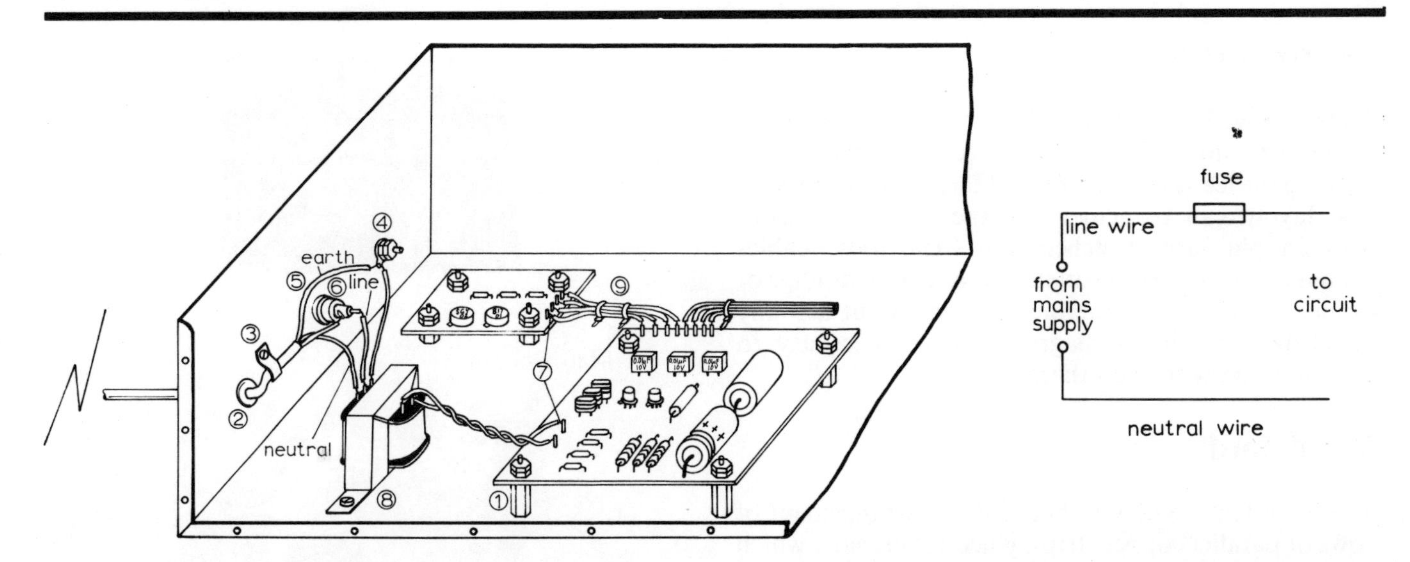

The diagram shows the essential features of construction of mains-operated equipment. Special care is needed for any apparatus which is plugged into the mains supply because the high voltage can be lethal. ***Never operate any mains apparatus which you have built until it has been tested by your teacher.*** Examine the following features carefully.

1. *The circuit board* is fixed in place by bolts and locking nuts, to keep a large enough gap between it and the case. This makes sure that none of the wiring of the board can touch the casing. If the casing is plastic, then the problem of shorting through the case doesn't arise, but it is a still good practice to keep your circuit board firmly fixed in place.
2. *Grommets* are used to protect the cables where they enter the case, from possible damage by the sharp edges of the drill holes. You have to match the internal diameter of the grommet to the cable and drill the case so that the grommet will be held firmly in place.
3. *The cable clip* makes sure that the cable cannot be jerked out accidentally and expose bare wires.
4. *The earth lead* must be connected to the casing if it is metal. Good electrical contact is needed here so any paint or insulation must be completely removed before the nut and bolt are inserted. Note the use of the solder tag for reliable connection of the cable to the nut and bolt assembly.
5. *Protection* from possible electrical shocks is given by the earth lead. If the high voltage wire (the line) should accidentally touch the case for any reason, the earth wire provides an easy route for the current to flow into the ground and this should also cause the fuse to blow.
6. *The fuse* is placed in the line wire immediately it enters the case. Its purpose is to cut off the supply if the current rises above a certain value. Choose a fuse with a current rating just above that at which the circuit operates. If, for any reason, the current reaches this fuse rating, then the fuse burns out and the supply is cut off, so protecting the equipment and the user. The fuse in the plug must be of similar value.
7. Always strip off the minimum amount of insulation from any cable. Have just enough bare wire to make your soldered joint, but no more.
8. Any heavy component, such as a transformer, should be bolted in place, preferably to the case itself.
9. *Cables* which follow the same route through the equipment should be clipped or taped together. Circuits with neat wiring are much easier to check.
10. *The mains plug* is connected as shown. Notice that the outer sheath extends inside the plug and is firmly gripped by the clamp. Wrap the cable clockwise around the terminal screw.

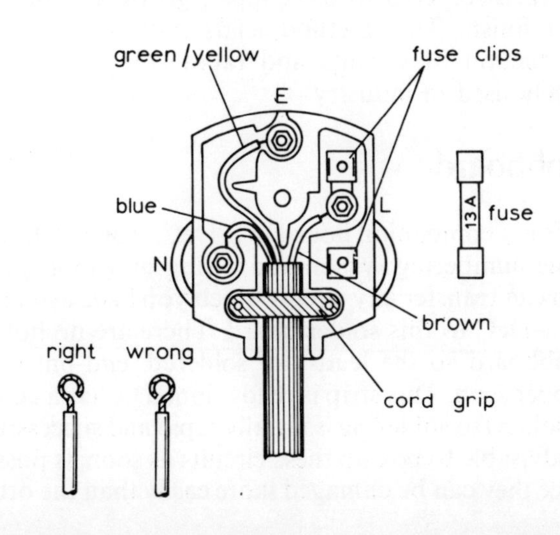

Construction Boards

Matrix Board

Matrix board is a pre-drilled board of insulating material. Components are firmly secured to the pins by bending the leads and are then soldered in place. A very neat layout can be achieved if the connecting wires between pins are stretched taut. Any loose cables should of course be of insulated type. The advantage of the matrix board is that the finished circuit actually looks like the circuit diagram and so it is easy to compare one with the other.

5 cm

Matrix board

Veroboard

The main feature of Veroboard is the arrangement of rows of parallel copper strips, glued to one side, which form the majority of the wiring. All components and any additional wiring are placed on the non-copper side and their leads pushed through the holes until the components are flush with the board. Leads are soldered to the copper strips and then trimmed. It is sometimes necessary to have breaks in the copper strips and these are achieved with the aid of a drill bit, turned by hand. This gives a very fast construction method, but it is often difficult to relate the position of components on the board to the circuit diagram.

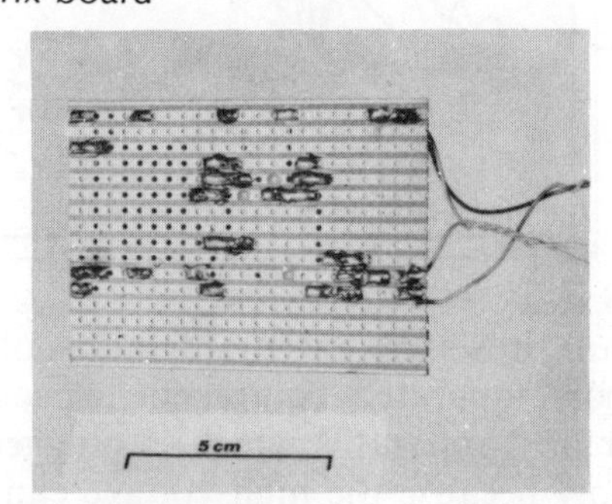

Copper side of Veroboard

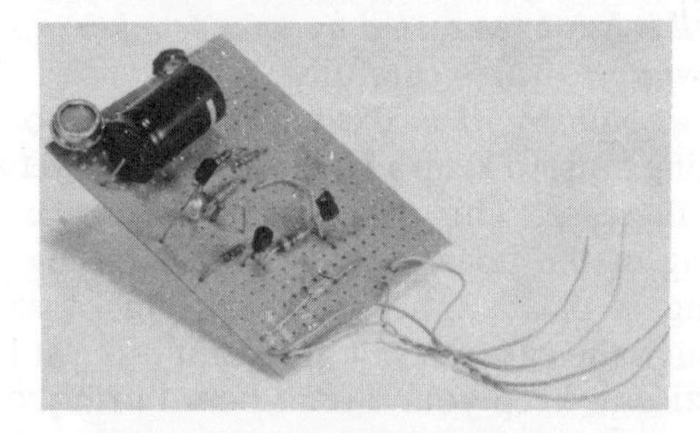

Component side of Veroboard

Printed Circuit Board

A printed circuit board (p.c.b.) usually starts out life as a sheet of insulating material with a layer of copper glued to one side. A plan is then drawn on the copper, showing which areas are to remain, to act as the wiring, and all the unwanted copper is removed by a corrosive chemical. Holes are then drilled in the board and the components are pushed flush with the non-copper side of the board, their leads through the holes. The leads are then soldered to the copper, giving a strong and neat finish. This method lends itself to large-scale production of circuits and rapid soldering so it is widely used in industry.

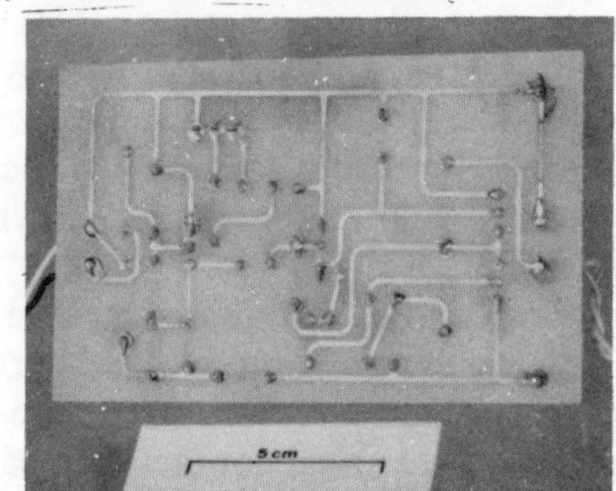

Printed circuit board

Blobboard

This is a companion board to S-DeC. You will find the same numbering system and wiring on Blobboard so you can transfer any circuit which you have assembled on S-DeC to this solder board. There are no holes in Blobboard so the leads are soldered 'end-on' to the copper strip. This strip is ready tinned (it has a coating of solder) so soldering is usually rapid and successful. It is advisable to box up these circuits as soon as possible since they can be damaged more easily than the others.

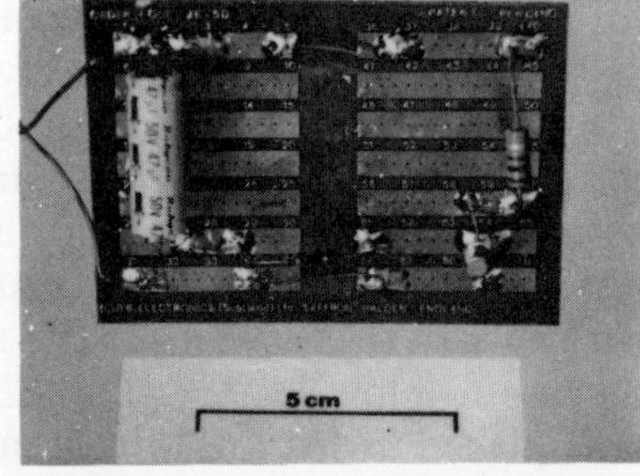

Blobboard

Making a Printed Circuit Board

We are going to examine the process of converting a printed circuit board (p.c.b.) diagram to the finished product. Although the process of making the master diagram is not too difficult, we shall not try it on this course, except for the nameplate on page 18.

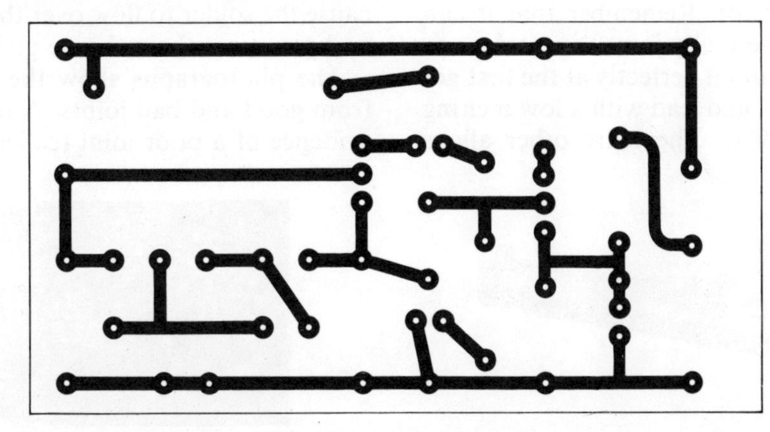

P.C.B. diagram

Preparation

1. Cut the board to size, using a hacksaw with a fine-toothed blade, copper side up and using downstrokes, so that the copper is not pulled away from the board. Smooth down any rough edges.
2. Remove the grease from the copper, using scouring powder and cotton wool, followed by rinsing and drying with a soft cloth.
3. From now on, keep your fingers off the copper as the grease from your fingers can stop the etch resist adhering.
4. The diagram is now transferred to the copper. Place a piece of carbon paper, carbon face downwards, onto the copper and Sellotape the diagram onto this. Now trace the outlines in the diagram onto the copper.
5. Paint the etch resist onto the copper, over the areas which are to be retained, and allowed to dry. The etch resist can be nail varnish, enamel paint or a Dalo pen.

Etching

1. Make up the etching solution. Iron (III) Chloride is usually used, with which great care must be taken, since it can burn skin, stain clothes and wood and attack iron and steel objects. ***You must observe the safety precautions that your teacher gives you and wear any protective gear that is supplied.*** There are etchants becoming available which are less dangerous to handle, and your teacher may have one of these.

   The proportions to use will be given on the packet but remember to add the crystals to the water and not vice-versa.
2. Place the solution in a shallow, non-metallic dish and lower the board carefully onto it, copper side down, where it will float on the surface. Leave it for approximately 20 minutes, although this time may be shortened by warming the solution.
3. Remove the board from the solution and check that all the unwanted copper has been removed and then rinse in running water.
4. Remove the etch resist either with propanone (for the Dalo and nail varnish) or white spirit (for the enamel).
5. Give the board a final wash in soapy water and a final rinse.

Drilling and Usage

Drilling can now be done with a very fine drill at high speed. Do not press hard or you could snap the drill or crack the board.

1. Solder the components in place but remember that too much heat could cause the copper to separate from the board.
2. The components should be inserted on the non-copper side and pushed in until they touch the board. The exception to this is the family of semiconductors as these can be damaged by heat so they should be left with long leads.
3. When a component has been soldered, the protruding leads should be snipped off.
4. The order for soldering is passive components (resistors, capacitors, etc.) first, then semiconductors and finally connecting wires.

When doing classwork, you may find it useful to put your initials on the board at the etch resist stage.

Soldering

Soldering is, of course, the fundamental process in the construction of circuits and, whilst it is easy to learn, lack of care and patience in soldering will cause many problems to the constructor. Remember that it can take far longer to find one faulty joint than it does to solder all the joints in a circuit perfectly at the first go.

Solder is an alloy of tin and lead with a low melting point, usually about 190°C. There are other alloys available for more specialised jobs. The purpose of soldering is to obtain good electrical conduction between the components in a joint. The method is to cause the solder to flow over the heated metal surfaces and penetrate them, hence giving a good join.

The photographs show the shapes to be expected from good and bad joints. A blob of solder is usually evidence of a poor joint (called a dry joint).

A good joint

A bad joint

Technique

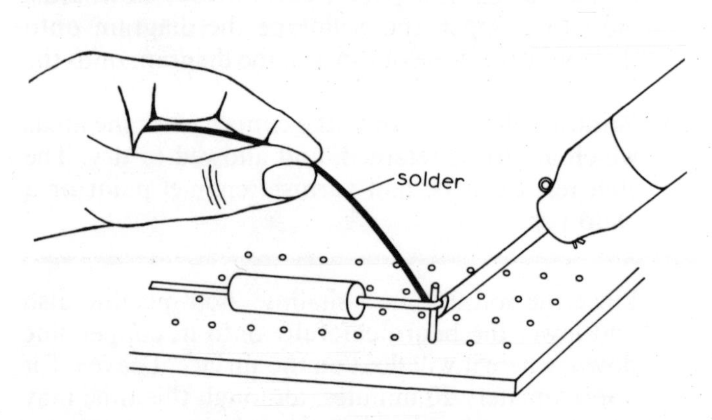

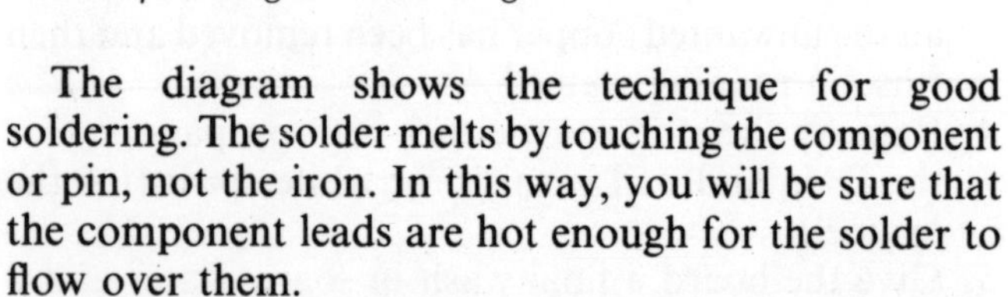

Technique for good soldering

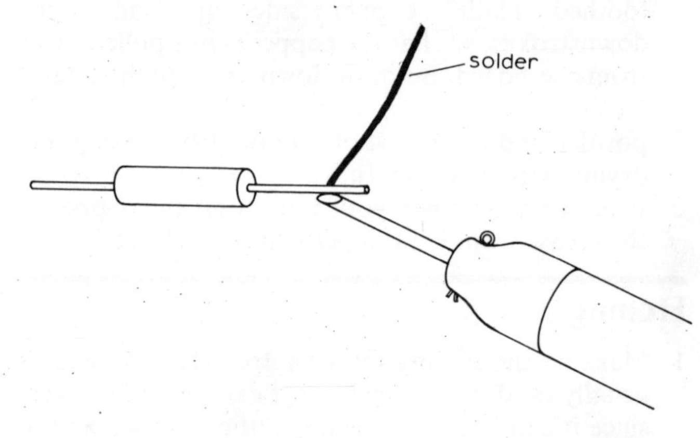

Tinning a lead

The diagram shows the technique for good soldering. The solder melts by touching the component or pin, not the iron. In this way, you will be sure that the component leads are hot enough for the solder to flow over them.

1. Ensure that the surfaces to be soldered are clean, dry and free of grease. The metals should be shiny.
2. Check that the bit of the iron is clean and shiny.
3. Most leads and the wire you will use are pre-tinned. This means they have a thin coating of solder already. Any wire not pre-tinned should have the iron applied and then the solder until the solder flows and gives a thin coating.
4. Ensure that all the joints are mechanically strong. The main function of solder is to provide good electrical conductivity at the joints, not to glue the components together!
5. Apply a *small* amount of solder to the bit.
6. Apply the iron to the joint and touch the solder to the joint, not the iron, so that it melts and flows. The molten solder will fill the gaps for you—you should not use the iron to shape the solder round a joint!
7. Allow the joint to cool before you disturb it.
8. Cut off the ends of the leads to give a neat finish.
9. Check joint visually and for mechanical strength.

Notes: At the temperatures used the metals would corrode on their surfaces. To prevent this, flux is used, but fortunately modern solders contain flux already. If, for any reason, a joint has to be kept hot then it must be protected from corrosion by the application of more flux. In effect, this means that more solder has to be used to prevent the corrosion.

Continued on p. 17

Heat Protection

Early semiconductor devices (transistors and diodes) were easily damaged by heat, and special precautions had to be taken during soldering. It is sufficient with the tougher modern devices to ensure that soldering is done reasonably quickly. A few seconds with the iron will produce a good joint and the possibility of damage should not arise.

Assembling a Circuit

Wherever possible, when using the different assembly systems, you should aim for every component to be fixed firmly in place before soldering commences. Check that the position of every component is correct and solder in the order resistors, capacitors, etc., then semi-conductors and finally any connecting wires. Take care that polar components are connected the right way round.

Veroboard and printed circuit boards require component leads to be fed through particular holes and matrix board requires the leads to be bent round particular pins. When using the first two, it is normal for the component leads to be pushed through until the body is flush with the non-copper side of the board, thus keeping the components firmly in place and the layout very neat. This requires accurate bending of the leads and this is achieved by holding the component as close as possible to the holes in question and bending the leads with narrow-nosed pliers. The diagrams below show four ways of mounting components, depending on the space available.

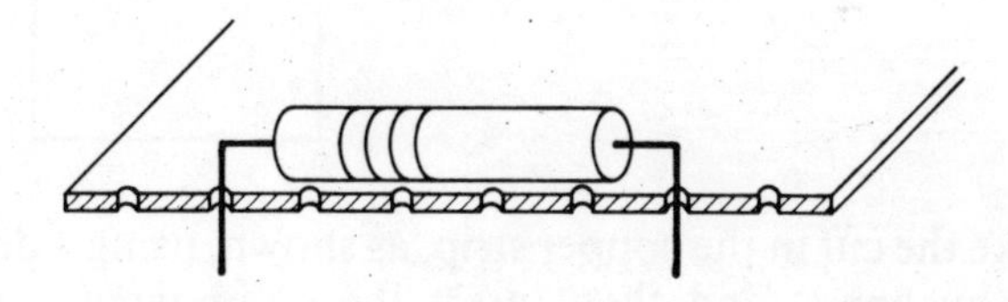

Accurate bending enables the component to be pushed flat to the board

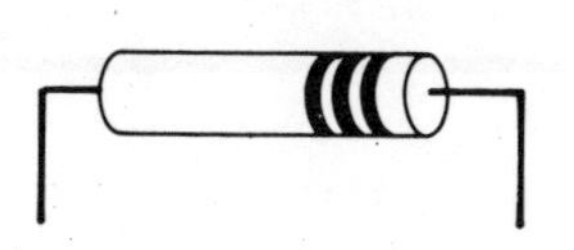

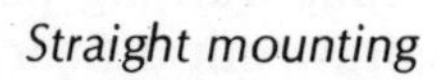

Straight mounting

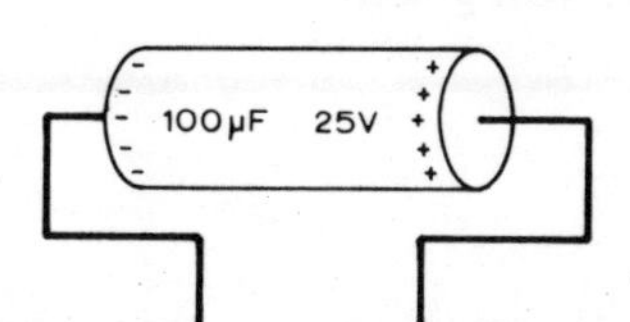

Bent leads

End-on mounting

Care

No component should have its leads bent, in a careless manner, where they enter the body since this can break the internal connections to the leads. Semiconductors and light-dependent resistors (page 8) are particularly delicate so make sure you always use narrow-nosed pliers for these.

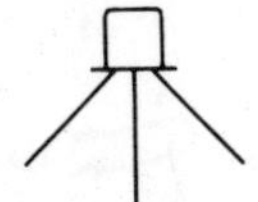

Damage may result from bending leads this way

Body of transistor is unaffected by careful bending like this, using pliers

Using Circuit Boards (Project: Warning Device)

This circuit can be put to several different uses with a little modification. The circuit emits a warning note through a loudspeaker when a drop of rain falls on the sensor pad (rain bleeper) or when the nameplate is touched by a finger (door bell). You could possibly adapt the circuit to warn of a rising water level.

Construction

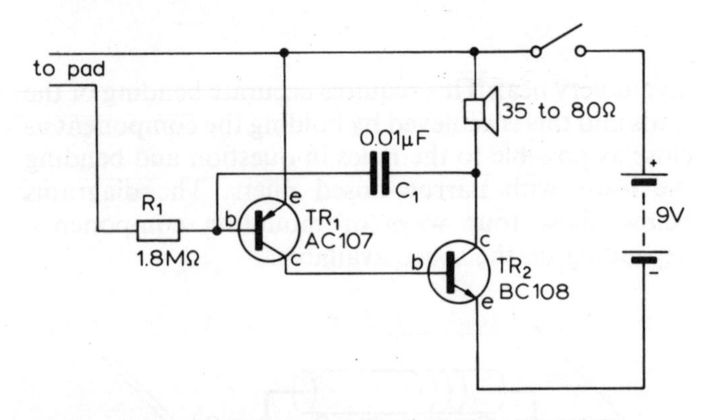

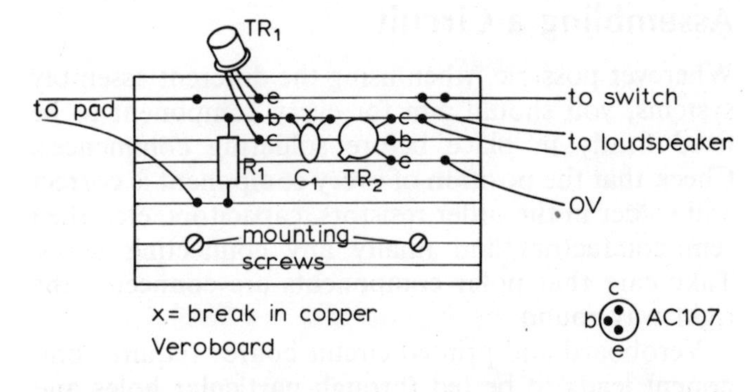

Make the cut in the copper strip, as shown, using a drill tip (by hand) and then insert the components and solder them in place. Remember to check with your identification sheets for any special instructions for each component. You need to solder three twin leads to the board for the pad, loudspeaker and battery. The battery has to go via a switch—make sure you get the polarity of the leads correct.

The pad is obviously mounted outdoors, copper side uppermost, but keep its lead as short as reasonably possible. The circuit board, loudspeaker, battery and switch are mounted in a suitable box, and this is kept where its sound can be heard.

The Circuit

The sound is produced when a current through the loudspeaker is switched on and off rapidly. This is the function of the rest of the circuit but this can only operate when a drop of rain 'joins together' the two wires to the pad and so completes the circuit.

You should try to design your own door bell pad so that when a finger is placed on the pad it is certain to bridge between two pieces of copper from the two different leads. The rain warning pad will be more effective if you can draw out a finer pattern. Check your design with your teacher before etching and read the instructions on p.c.b. construction.

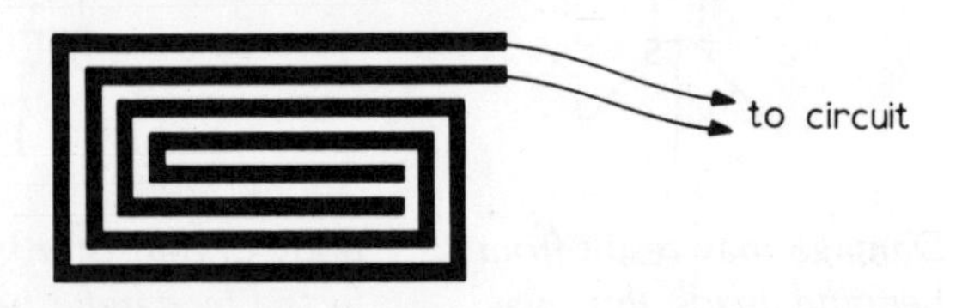

Rain warning pad

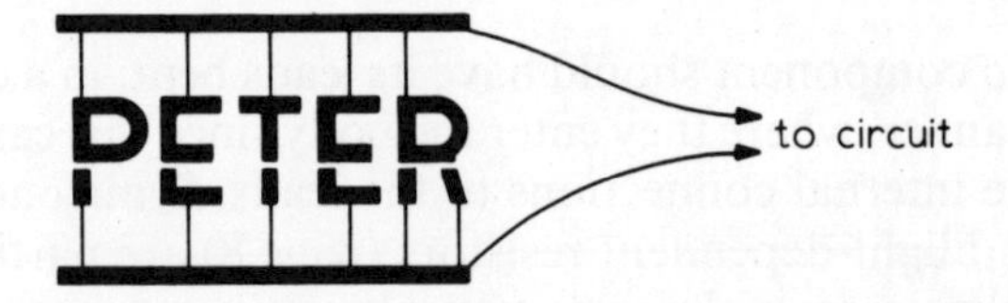

Door bell pad

Modifications

You can change the note from the device by altering the value of C_1. The easiest way to show this is merely to touch another capacitor to the two terminals of the existing one and you will notice the change in the note without even needing to do any soldering. If you wish to make the change permanent, then solder your new capacitor in place.

Numbers

In your electronics lessons, you will use numbers as small as one millionth and as large as several millions; these are discussed on page 25. You need to be familiar with certain prefixes and with fractions and decimals in general. There are at least four ways of writing one-tenth:

One-tenth	1/10	0·1	10^{-1}

If we now continue the series down to one millionth:

One-hundredth	1/100	0·01	10^{-2}
One-thousandth	1/1000	0·001	10^{-3}
One-tenthousandth	1/10 000	0·000 1	10^{-4}
One-hundredthousandth	1/100 000	0·000 01	10^{-5}
One-millionth	1/1,000 000	0·000 001	10^{-6}

We have abbreviations for some of these. For instance, instead of saying 'one thousandth of a metre' we can say 'one millimetre', where milli stands for 'thousandth of'.

There are four prefixes to learn:

For thousandth of	use milli (m)
millionth of	use micro (μ)
thousand-millionth of	use nano (n)
million-millionth of	use pico (p).

Notice that each of these is one thousand times smaller than the previous one.

This means:

there are one thousand milliamperes in one ampere
there are one thousand microamperes in one milliampere
there are one thousand nanoamperes in one microampere
there are one thousand picoamperes in one nanoampere.

To convert from one to the next one you have either to multiply or divide by a thousand:

400 mm (millimetres)	equals 0·4 m (metres)
3 m (metres)	equals 3000 mm (millimetres)
20 μA (microamperes)	equals 0·02 mA (milliamperes)
4000 ns (nanoseconds)	equals 4 μs (microseconds)

Notice how many of each unit are contained in 1 second:

1 second contains	1 000	milliseconds
1 second contains	1 000 000	microseconds
1 second contains	1 000 000 000	nanoseconds
1 second contains	1 000 000 000 000	picoseconds

Problems

1. What is the value of 0·1 as a fraction?
2. What is the value of 0·001 as a fraction?
3. What is the value of 1/100 as a decimal?
4. What is the value of 1/10 000 as a decimal?
5. Which prefix is used for 'millionth of'?
6. Which prefix is used for 'thousandth of'?
7. How many milliamperes (mA) are there in 1 ampere?
8. How many microamperes (μA) are there in 1 ampere?
9. How many times bigger is one milliampere than one microampere?
   Another electrical unit you will meet in this book is the *farad.*
10. How many picofarads (pF) are there in one microfarad (μF)?
11. How many picofarads (pF) are there in one farad (F)?
12. What is the value of 1 million picofarads (pF) in microfarads (μF)?
13. What is the length of time of 1000 nanoseconds (ns) in microseconds (μs)?
14. What is the length of time of 10 000 nanoseconds (ns) in microseconds (μs)?
15. Is 500 microseconds (μs) longer than 1 millisecond?
16. How many times bigger is 1 millimetre than 1 micrometre?
17. How many times bigger is 1 millimetre than 100 micrometres?
18. How many times bigger is 4000 milliamperes (mA) than 1 ampere?
19. Is 1 metre bigger than 3000 millimetres (mm)?
20. How many lengths of 10 millimetres make 1 metre?

Graph Work

One convenient way of showing information about electronic events is a graph. The readings below were taken of the voltage across the terminals of two different devices, when in use. The first is a dynamo starting to rotate from rest, and the second is a capacitor being used in a circuit which you will come across shortly.

Dynamo

Voltage (v)	Time (s)
0	0
0·8	1
1·7	2·1
2·25	2·8
3·28	4·1
4·0	5

Capacitor

Voltage (v)	Time (s)
0	0
1·3	0·5
2·3	1·1
2·8	1·5
3·5	2·4
3·75	3
3·9	3·5
4	5

Plot graphs of these two events, with the time as the horizontal axis and voltage as the vertical axis.

1. From your graph construct a table for each device like the ones above, showing the voltage at one second intervals.
2. Now construct another table for each, showing the increase in voltage during each second.
3. What can you say about the increase in the voltage every second across the dynamo?
4. What can you say about the increase in voltage every second across the capacitor?
5. Which device showed the greatest increase in voltage during the first second?
6. Which device showed the greatest increase in voltage during the last second?
7. The tables below show pairs of points, the time of one point being double the time of the other point. From your graph find the voltage at these points.

Dynamo

Time (s)	Voltage (v)
1·5	
3·0	
2·0	
4·0	

Capacitor

Time (s)	Voltage (v)
1·5	
3·0	
2·0	
4·0	

You should notice that a graphical display can achieve the following:

1. It can show you when a special relationship exists between measurements. For instance, the dynamo readings are directly proportional to each other. This means that if you double one measurement, the other one is doubled.
   This is a characteristic of straight line graphs which pass through the origin. (The origin is the point where the two axes intersect.)
2. It makes interpolation easy—that is, the finding of values between the ones already available.
3. It can give you an at-a-glance display of the behaviour of events. For instance, it is obvious that the voltage rises much more quickly during the first second than in the last second for the capacitor.

2 CIRCUITRY

Simple Circuits

Your study of electronics is based on your understanding of circuits, starting with this, the simplest.

Assemble the circuit shown below, but don't leave it switched on for too long or you will run down the batteries.

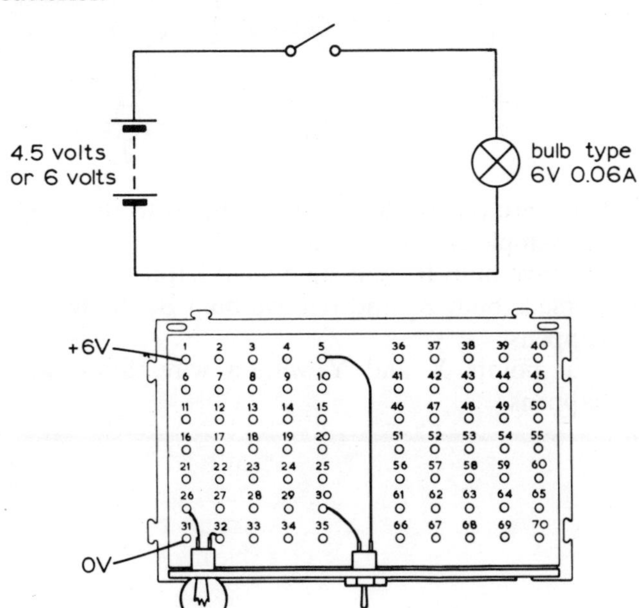

Here are some of the terms which apply to the circuit and which you will use for subsequent circuits.

Circuit

For the bulb to light, the electricity must be able to get from one terminal of the battery, through the bulb and back to the other battery terminal. This complete route makes a circuit.

Make/Break

When the switch is closed, the bulb comes on and you make the circuit. Open the switch and you break the circuit.

Current Path

This is the route followed by the electricity. In all your circuits, you should be able to trace with your finger the current path to help you understand the circuit.

The Battery

The function of the battery is to provide the electrical energy. The energy the battery has means it can force tiny particles called electrons round the circuit. You now need to know of the measurement called electromotive force (e.m.f.) which tells you how quickly the battery can provide energy.

Current

As soon as you make the circuit, the electrons can move along the wires. This constant flow of electrons is called the current.

Current is measured in amperes

Resistance

The electrons in the circuit flow through a transducer and here the electrical energy is changed into another form. (In your circuit the transducer was a light bulb which changed electrical energy into heat and light energy.)

Every transducer has a resistance. It is this resistance which fixes the size of the current—the rate at which electrons get round the circuit.

Resistance is measured in ohms

E.M.F.

Look what happens if you replace the battery with another with a higher e.m.f. (Again, do not leave the circuit connected for very long.) You can see that the battery is supplying more energy.

E.M.F. is measured in volts

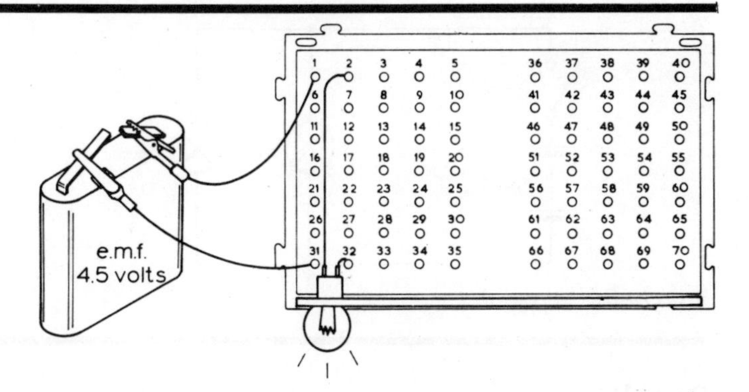

The battery with the low e.m.f. supplies energy at a slow rate

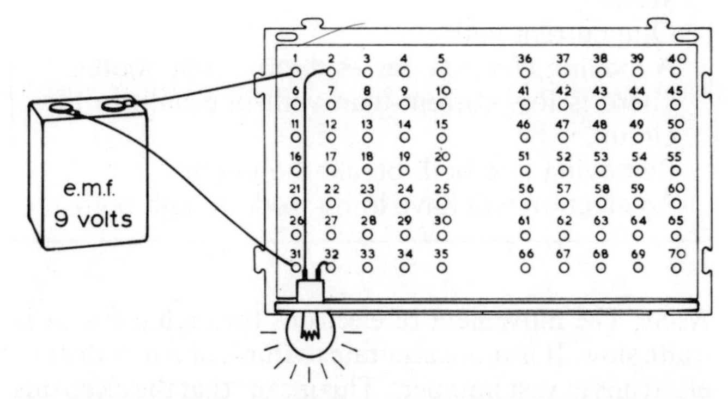

The battery with the higher e.m.f. supplies energy at a faster rate

Series and Parallel Circuits

Series

There are two basically different ways to assemble circuits when you have two or more components. These are called *series* and *parallel* methods. Assemble the series circuit shown and answer the following five questions.

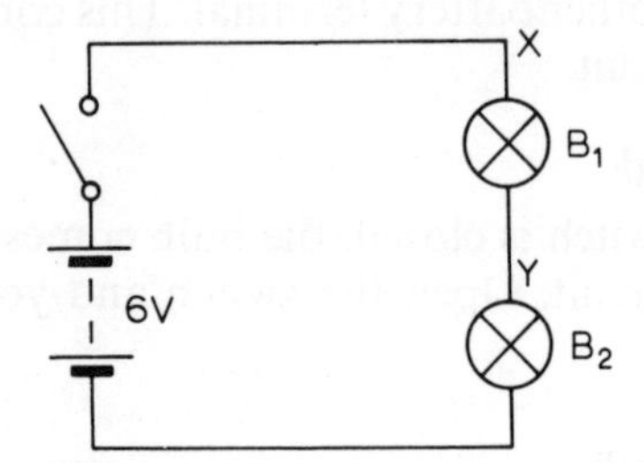

Two bulbs in series

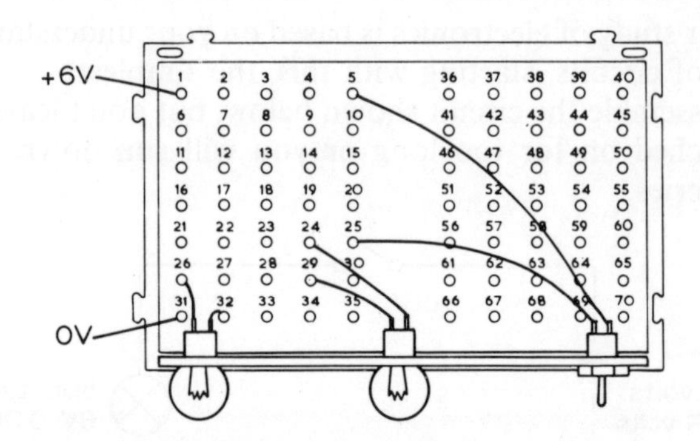

1. Using the numbers of the board, give the current path through the circuit.
2. Compare the brightness of the bulbs to the one in the sample circuit on page 23.
3. Remove bulb B_1 and state what happens.
4. Replace bulb B_1 and remove bulb B_2. State what happens.
5. Join points X and Y with a wire. State what happens.

Parallel

Assemble the parallel circuit and give your responses to the first four instructions shown above. Your teacher will discuss with you the effect produced by the fifth instruction.

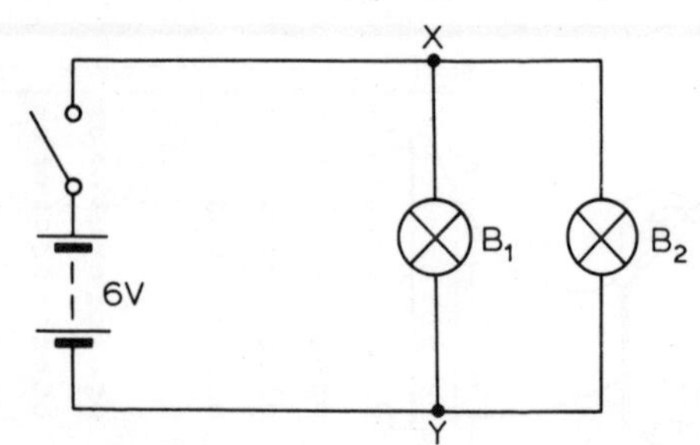

Two bulbs in parallel

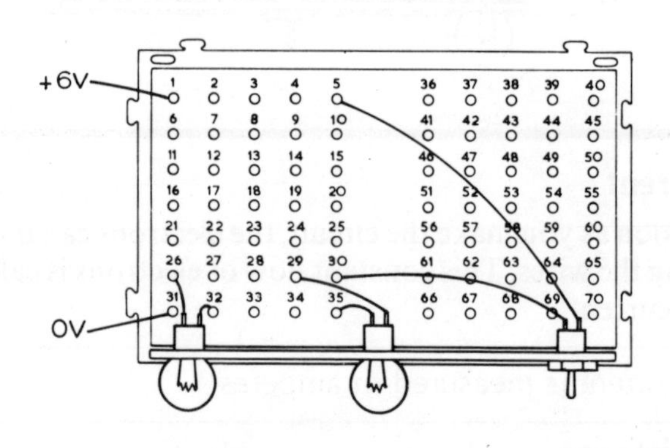

Results

Series
One current path.
A moving electron passes through both bulbs.
There is less current than with one bulb in the circuit.
Removing one bulb breaks the circuit.
An electron will have to do work in both bulbs.

Parallel
Two current paths.
Current divides and each electron can go through *either* one bulb *or* the other, but not both.
Each bulb receives the same current as if it were in the circuit on its own.
Removing one bulb does not affect the other.
Each electron does all its work in one bulb only.

Note: The movement of electrons through a circuit is quite slow. It is more accurate to think of it as a drift of electrons in vast numbers. This means that the electrons that arrive back at the battery are not necessarily those that left it. When a circuit is switched on, the electrons leaving the battery force other electrons which were free in the circuit to move also. So the actual current does not consist only of the electrons supplied by the battery.

Circuits—Problems

Answer these questions and check your answers by experiment.

1. What must be done to light the bulb, B_1?
2. What will now happen if you press switch 2?

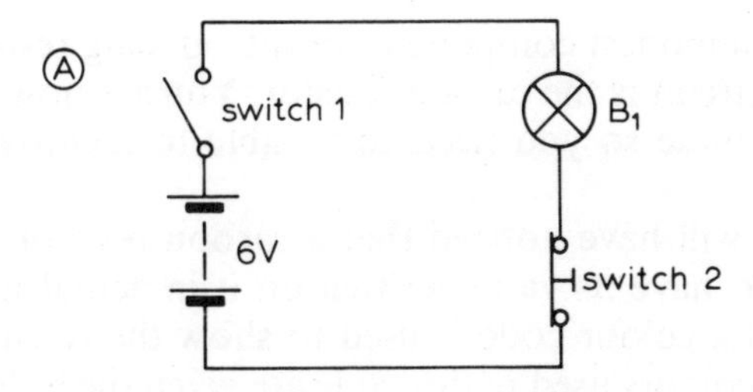

1. What must be done to light the two bulbs?
2. What can you say about the current flowing through bulbs B_1 and B_2?
3. What will happen if bulb B_2 is removed?

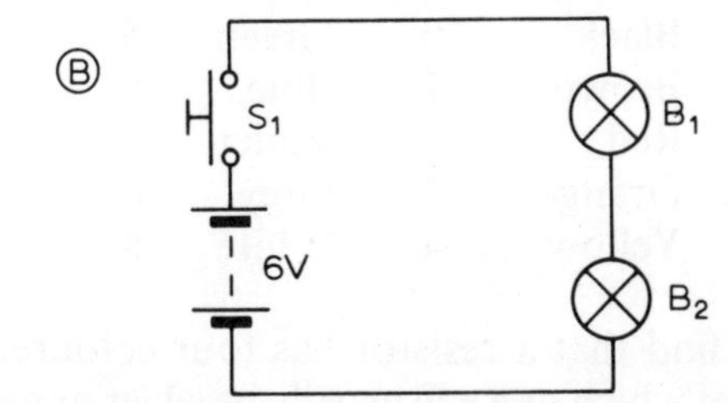

1. What must be done to light both bulbs?
2. What happens to the current at point B?
3. What can you say about the current flowing through bulb B_1 and that flowing through bulb B_2?
4. What happens when bulb B_2 is removed?

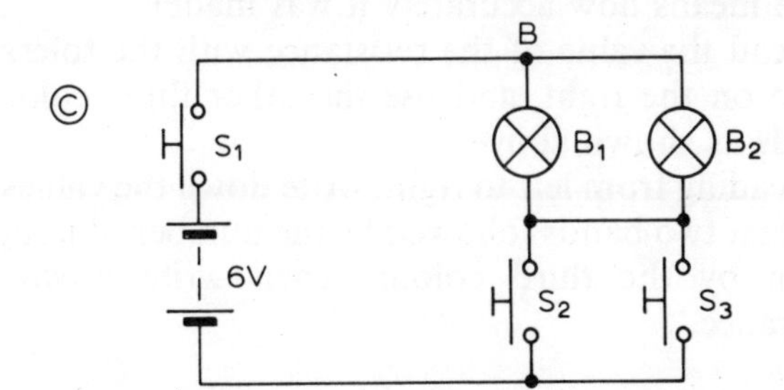

1. There are four different routes by which an electron could get round this circuit. Using the letters of the diagrams say what the four routes are.
2. What happens to the remaining bulbs if bulb B_4 is removed?
3. What happens if bulbs B_3 and B_4 are removed?
4. What happens if points G and L are connected by a wire?

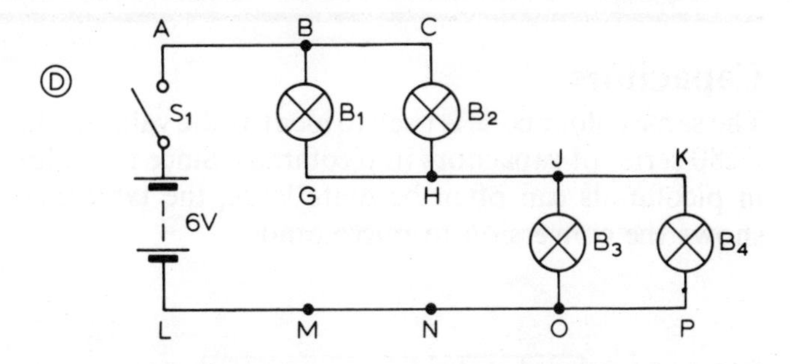

1. What must be done to light bulbs B_1 and B_3?
2. What would then happen if switch S_6 were pressed?
3. Draw out those parts of the circuit carrying current when bulbs B_1 and B_3 are lit.

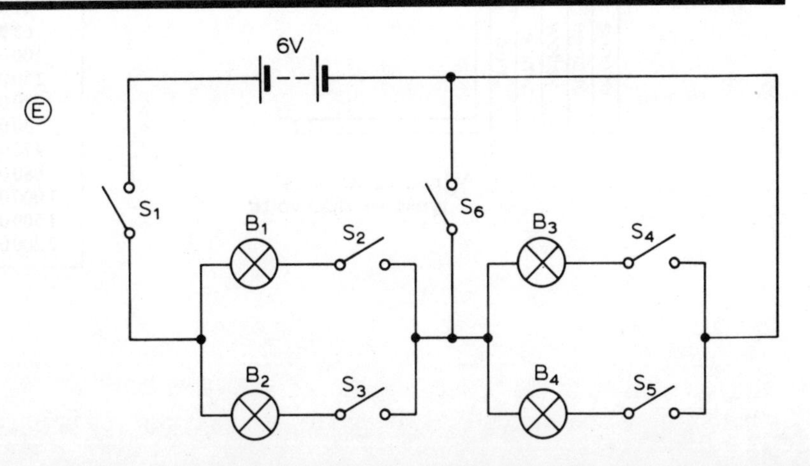

Colour Code of Resistors and Capacitors

Resistors

The commonest component for introducing resistance into a circuit is the carbon resistor. You are now going to use these so you need to be able to identify their value.

You will have noticed that a carbon resistor is too small to have its value written on it in actual figures. Instead a colour code is used to show the resistance.

The colours used in this code are given the following values:

Black	0	Green	5
Brown	1	Blue	6
Red	2	Violet	7
Orange	3	Grey	8
Yellow	4	White	9

You will find that a resistor has four coloured bands round it, of which one will usually be silver or gold. This silver or gold band tells you the tolerance of the resistor (that means how accurately it was made).

Read the value of the resistance with the tolerance band on the right, and use the other three coloured bands as shown above.

Reading from left to right, write down the values for the first two bands followed by the number of noughts given by the third colour. Then write down the tolerance.

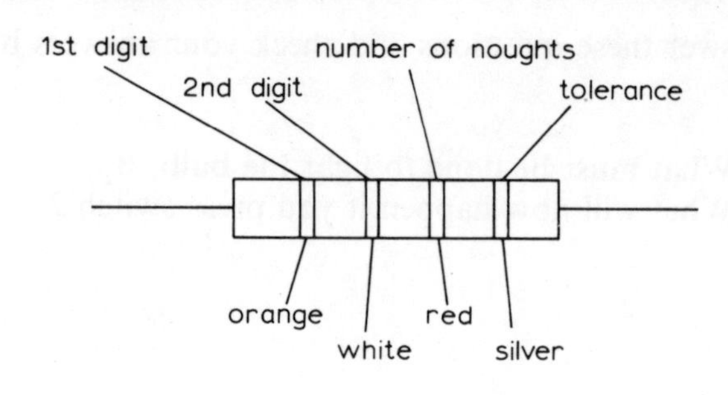

A 3900 ohm resistor

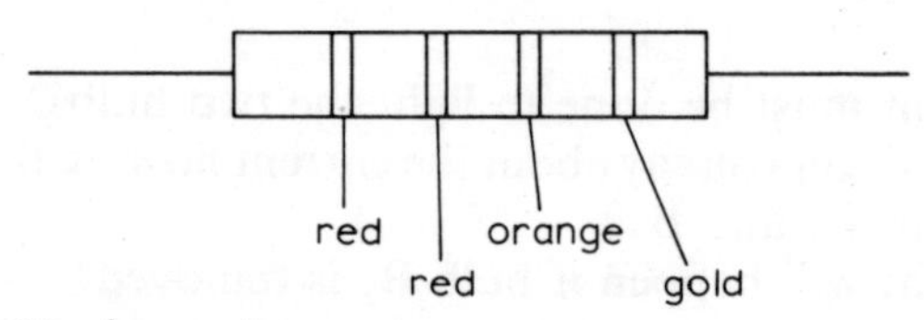

A 22 000 ohm resistor

Numbers between 1 and 10

If you think carefully about this system, you will see that the smallest number you can write is 10 (brown, black, black). For a number less than 10, only two number bands are used, instead of three.

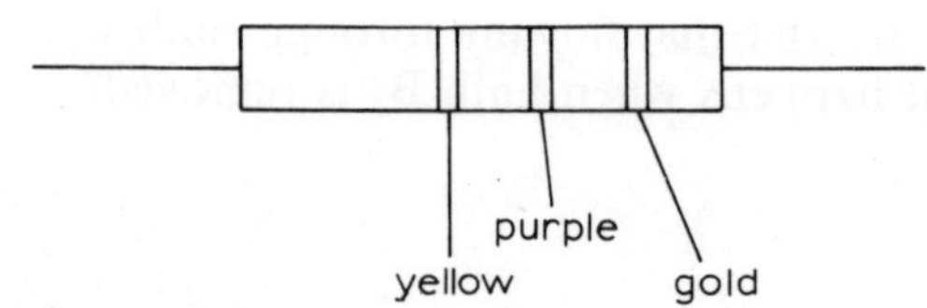

A 4.7 ohm resistor

Capacitors

The same colour code is used to identify the value of the C280 series of capacitors in picofarads. Since the value in picofarads can often be quite large, the table also shows the conversion to microfarads.

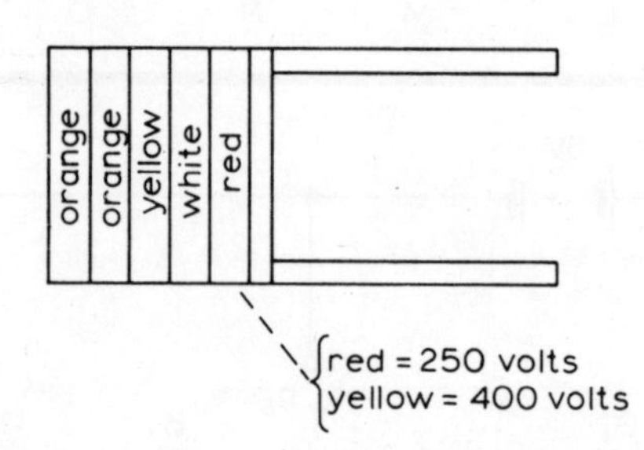

C280 Capacitors

Capacitance (pF)	Capacitance (μF)	Colours of bands 1	2	3	4
10 000	0·010	brown	black		
15 000	0·015	brown	green		
22 000	0·022	red	red		
33 000	0·033	orange	orange	orange	
47 000	0·047	yellow	violet		
68 000	0·068	blue	grey		black ±20% tolerance
100 000	0·10	brown	black		
150 000	0·15	brown	green		
220 000	0·22	red	red		
330 000	0·33	orange	orange	yellow	
470 000	0·47	yellow	violet		
680 000	0·68	blue	grey		white +10% tolerance
1 000 000	1·0	brown	black		
1 500 000	1·5	brown	green	green	
2 200 000	2·2	red	red		

Tolerance

It is an unfortunate fact of life that if you wish to buy a carbon resistor for 1p then it probably isn't going to be exactly the value you asked for. The fourth band on your resistor tells you the tolerance: that is, how far from the stated resistance the actual value may be.

Your resistors will almost certainly be marked with a silver, gold or red band. These have the following meanings:

Silver $\pm 10\%$ Gold $\pm 5\%$ Red $\pm 2\%$

Examples

A resistor is marked $1000\ \Omega \pm 10\%$
Highest possible value $= 1000\ \Omega + 10\%$ of 1000 $= 1000 + 100 = 1100\ \Omega$
Lowest possible value $= 1000\ \Omega - 10\%$ of 1000 $= 1000 - 100 = 900\ \Omega$
Thus a resistor marked $1000\ \Omega \pm 10\%$ could actually have a value anywhere between $900\ \Omega$ and $1100\ \Omega$.

A resistor is marked $220\ \Omega \pm 10\%$.
Highest possible value $= 242\ \Omega$
Lowest possible value $= 198\ \Omega$
Thus a resistor marked $220\ \Omega \pm 10\%$ could have a value anywhere between $198\ \Omega$ and $242\ \Omega$

Preferred Values

In the series of 10% tolerance resistors, only certain values are commonly available. As you have seen, the value of a resistor is written as two figures followed by a certain number of noughts.

The numbers available for the first two figures are

10, 12, 15, 18, 22, 27, 33, 39, 47, 56, 68, 82.

This is called the E12 series. You know that a resistor marked $10\ \Omega \pm 10\%$ could have a value anywhere between 9 and $11\ \Omega$. If you work out the 'range' of each resistor, you will see that the series includes all possible values. These twelve numbers are called *preferred values*.

If you need a resistor whose value is more accurately known then you can either test the values of a number of 10% resistors with a resistance meter until you find the value you want or you can buy a 5% or 2% tolerance type or even a 1% (brown band) type.

Large Numbers

We use the prefixes kilo and mega to denote thousand and million. Thus:

$5000\ \Omega = 5$ kilohms ($5\ k\Omega$)
$2\,000\,000\ \Omega = 2$ megohms ($2\ M\Omega$)
$3900\ \Omega = 3{\cdot}9$ kilohms ($3{\cdot}9\ k\Omega$)
$22\,000\ \Omega = 22$ kilohms (22 k)
$4\,700\,000\ \Omega = 4{\cdot}7$ megohms ($4{\cdot}7\ M\Omega$)
$3\,900\,000\ \Omega = 3{\cdot}9$ megohms ($3{\cdot}9\ M\Omega$)

Colour Code—Problems

Resistors

1. Give the values indicated by the following colours on a resistor.
   (*a*) Red, red, yellow, silver
   (*b*) Green, blue, brown, gold
   (*c*) Orange, white, red, silver
   (*d*) Brown, grey, red, silver
   (*e*) Brown, green, green, gold
   (*f*) Grey, red, black, gold

2. For the following resistors, work out the range indicated (the highest and lowest possible values)
   (*a*) $1000\ \Omega \pm 5\%$
   (*b*) $10\,000\ \Omega \pm 10\%$
   (*c*) $150\ \Omega \pm 10\%$
   (*d*) $1200\ \Omega \pm 10\%$
   (*e*) $33\,000\ \Omega \pm 10\%$

3. State the colour code that should be used for the following resistors:
   (*a*) $1000\ \Omega \pm 5\%$
   (*b*) $220\ \Omega \pm 5\%$
   (*c*) $33\ \Omega \pm 5\%$
   (*d*) $470\ \Omega \pm 10\%$
   (*e*) $27\,000\ \Omega \pm 10\%$
   (*f*) $680\,000\ \Omega \pm 10\%$

4. Write out the following resistances in ohms:
   (*a*) 1 megohms
   (*b*) 2·2 megohms
   (*c*) 22 megohms
   (*d*) 47 kilohms
   (*e*) 4·7 kilohms
   (*f*) 10 kilohms

Capacitors

Some capacitors, notably the Mullard C 280 series, use the same colour code to give capacitance in picofarads. (Although you have not studied capacitors yet you may still have to be able to choose the right capacitor for the job!)

5. Write down the value of the following capacitors in picofarads or microfarads.
   (*a*) Red, red, orange, white
   (*b*) Brown, black, yellow, white
   (*c*) Brown, black, orange, black
   (*d*) Green, blue, yellow, black
   (*e*) Blue, grey, red, white
   (*f*) Green, blue, red, black

6. Change the following values in picofarads to microfarads. (Remember, you will have to divide by one million.)
   (*a*) 500 000 pF
   (*b*) 100 000 pF
   (*c*) 10 000 pF
   (*d*) 1 000 000 pF
   (*e*) 2 200 000 pF
   (*f*) 4 700 000 pF

The Ammeter and the Voltmeter

The ammeter and voltmeter are going to tell you about the current and potential difference in a circuit. These are two extremely important readings and it is vital that you are able to use these instruments in any circuit you come across.

The Ammeter

The ammeter measures the current, in the units amperes, sometimes abbreviated to amps or A.

This instrument tells you how many electrons are flowing through it (and the bulb) every second.

We say that it is measuring the rate of flow of the electrons.

You may wish to compare this with our use of the word current for water flow, since it has a very similar meaning. When a river has a large current it means a lot of water passes in a short time. A large electric current means a lot of electrons pass in a short time.

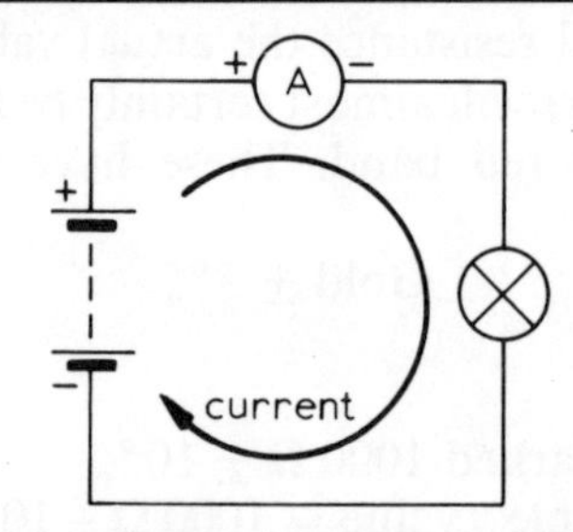

The Voltmeter

The voltmeter tells you the potential difference measured in volts, abbreviated to V.

The current flows through the bulb because there is a potential difference across its terminals, A and B. This potential difference is the force pushing the electrons through the bulb.

A voltmeter measures this potential difference by finding the difference in the energy of the electrons at A and B.

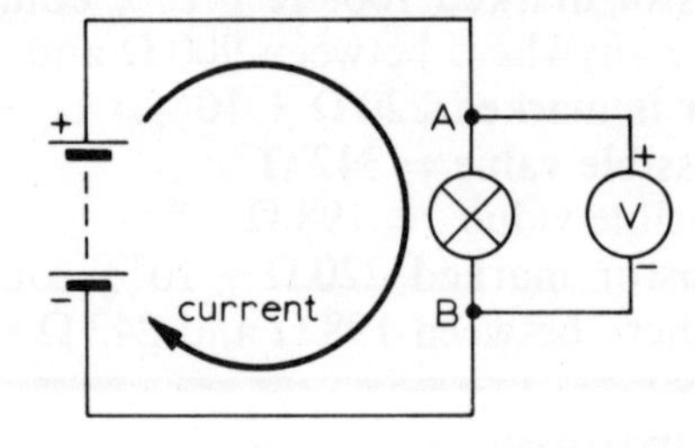

This potential difference is also called the voltage drop across the bulb or just the voltage across the bulb.

Using the Meters

Notice the following points about how these meters are used in circuits.

The Ammeter

An ammeter is connected in series so that all the current has to flow through the meter.

This means you will have to break the circuit when you wish to insert an ammeter.

The meter has positive and negative terminals which must lead towards the terminal of the battery that carries the same sign.

The Voltmeter

A voltmeter is connected in parallel in a circuit. It will then tell you the potential difference (voltage drop) across the two points to which it is connected.

The circuit need not be disturbed when the voltmeter is connected.

The meter has positive and negative terminals which must lead towards the terminal of the battery that carries the same sign.

Safety Rules

1. Never connect any meter into a circuit that has a supply voltage of more than 24 V unless your teacher is with you.
2. Work out, roughly, the size of current and voltage in the circuit and use meters which have a full scale deflection (f.s.d.) larger than the expected value. (The f.s.d. is the highest reading a meter will give.)
3. If you do not know what values to expect, use your meters which have the largest f.s.d. and work down to your sensitive meters when you are sure you will not overload them.

Series Circuits

Voltage

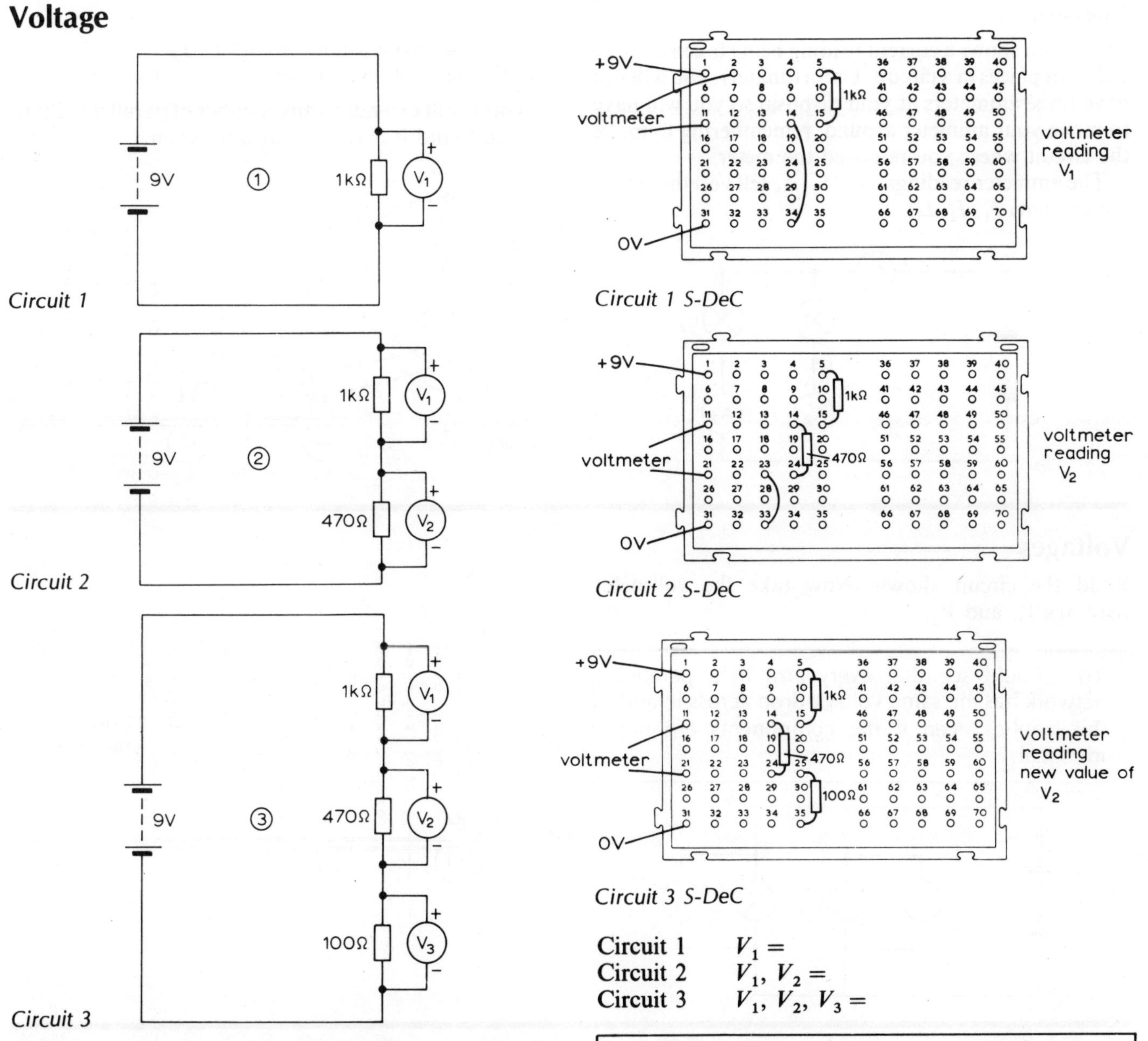

Circuit 1

Circuit 1 S-DeC

Circuit 2

Circuit 2 S-DeC

Circuit 3

Circuit 3 S-DeC

Circuit 1 $V_1 =$
Circuit 2 $V_1, V_2 =$
Circuit 3 $V_1, V_2, V_3 =$

Use your voltmeter to find the voltages marked in the above circuits. You can of course take the readings without disturbing the circuit being tested. You only have to clip the voltmeter across each resistor in turn.

Your results should show that voltages across components in series always add up to the supply voltage.

Current

Break circuit 3, from above, and insert the milli-ammeter as shown. You can, of course, only do this in one place at a time if you only have one meter. Take readings from A_1, A_2, A_3.

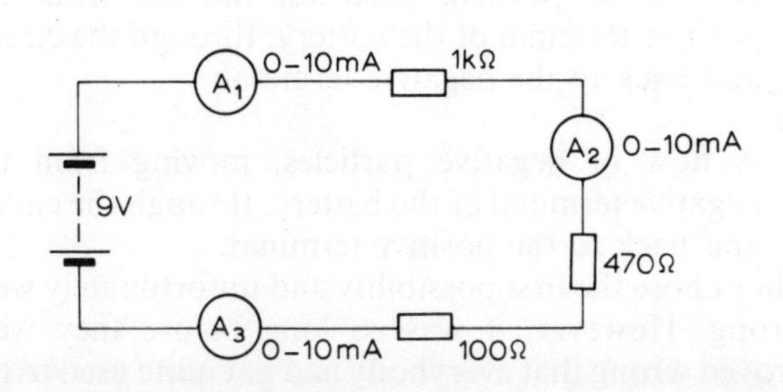

Your results should show that the current in a series circuit is the same at all points in the circuit.

Parallel Circuits

Current

The circuit shows a current reading being taken in three different places in a circuit. Unfortunately, you will not have three ammeters at your disposal so you will have to move your ammeter around, remembering to make the circuit where you removed the meter.

The ammeter readings, A_1, A_2, A_3 tell you the size of the currents I_1, I_2, I_3.

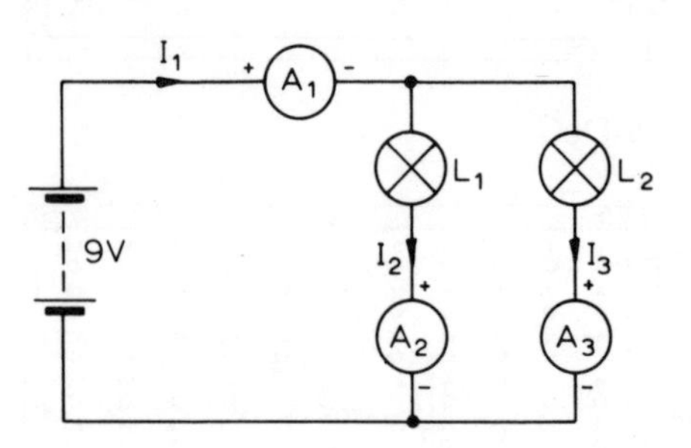

Write down your values for I_1, I_2, I_3.
In the circuit shown above $I_1 = I_2 + I_3$

This result extends to any number of parallel paths in a circuit, and is a very important result.

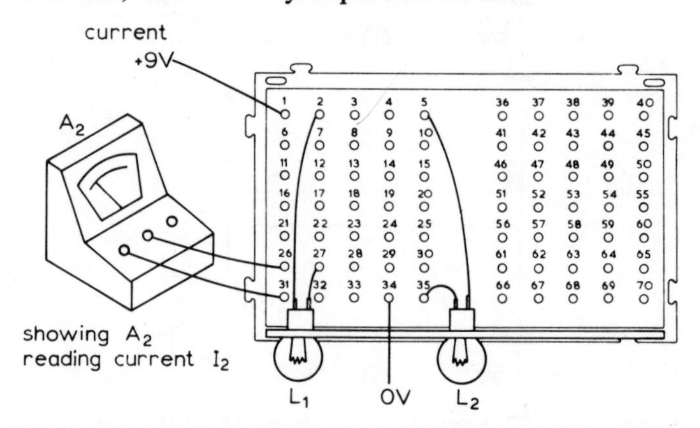

Voltages

Build the circuit shown. Now take the voltmeter readings V_1 and V_2.

> You should see that every arm in a parallel network has the same voltage drop across it, and this applies even if the components are not identical.

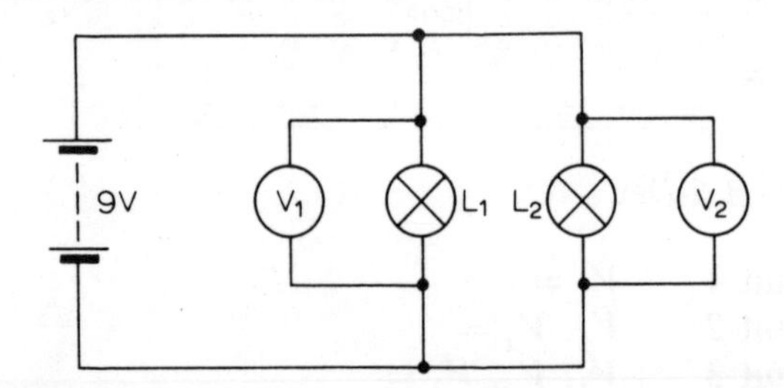

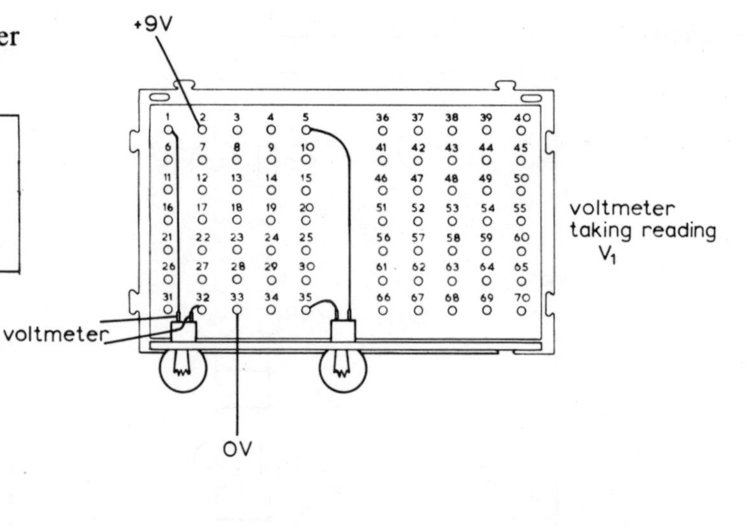

Direction of Current

In the nineteenth century, scientists tried to decide on the direction that an electric current takes in a circuit. They decided that an electric current could be either

1. A flow of positive particles, moving from the positive terminal of the battery, through the circuit and back to the negative terminal;

or

2. A flow of negative particles, moving from the negative terminal of the battery, through the circuit and back to the positive terminal.

They chose the first possibility and unfortunately were wrong. However, it was so long before they were proved wrong that everybody had got quite used to the system. So, to this day, you will find most circuit diagrams marked with the current flowing from positive to negative.

This is called the *conventional current.*

Circuits marked with the current flowing from negative to positive are showing the flow of electrons.

This is called the *electron current.*

Most of the time was shall use conventional current. It may seem strange that we should continue to use a system which we have shown to be incorrect. However, it is really only a matter of use of language and does not affect the performance of any circuits or our appreciation of them.

The Multimeter

The multimeter is an extremely useful 3-in-1 meter which will measure current, voltage and resistance. In addition, it may have, say, five ranges for each of these quantities so it is a very flexible instrument.

Construction

The heart of the multimeter is a very sensitive ammeter which can measure down to a few microamps. The switching on the panel connects this meter to different arrangements of resistors inside the meter to enable it to perform all its tasks. When it is used as a resistance meter an internal battery is brought into use, so a periodic check is needed on the battery's condition to ensure continued accurate readings.

Use

It is not possible to give an exact sequence of steps since meters do vary considerably but remember the following points:

1. Select the function you require.
2. Switch to the scale with the largest full scale deflection for first reading.
3. Connect the meter with the correct polarity when it is being used as an ammeter or voltmeter.
4. Find which scale on the dial you are supposed to read.
5. Switch to a more sensitive scale, if needed.

Reading the Scale

One problem with a meter which has a total of, say, 15 different ranges is the lack of space on the dial for marking in all the values. You will usually have to do some simple conversion with the reading you obtain from the meter to get the correct value for the range you selected.

With the meter switch set as shown in the photograph, a dial reading of 30 V represents an actual value of 120 V. Thus the actual value, in this case, is four times the dial reading, e.g., if the dial reads 10 V, then the actual value is 40 V.

Whenever you use a multimeter you must find out this conversion factor for the readings you get. Your teacher may set up an exercise for you to practise using the multimeter, so that you are proficient when you come to use it in the circuits you build.

Practice Circuits

Take the readings asked for in the following circuits. You should be proficient in the use of meters by the end of this exercise and understand the basic rules of circuits. The numbers refer to the hole numbers on the S-DeC.

$I =$
Voltage drop across resistor, $V =$

$I =$
V across R_1 =
V across R_2 =

$I =$
V across 1 kΩ resistor =
V across 10 kΩ resistor =

$I =$
V across R_1 =
V across R_2 =

$I =$
V across 1 kΩ resistor =
V across 10 kΩ resistor =

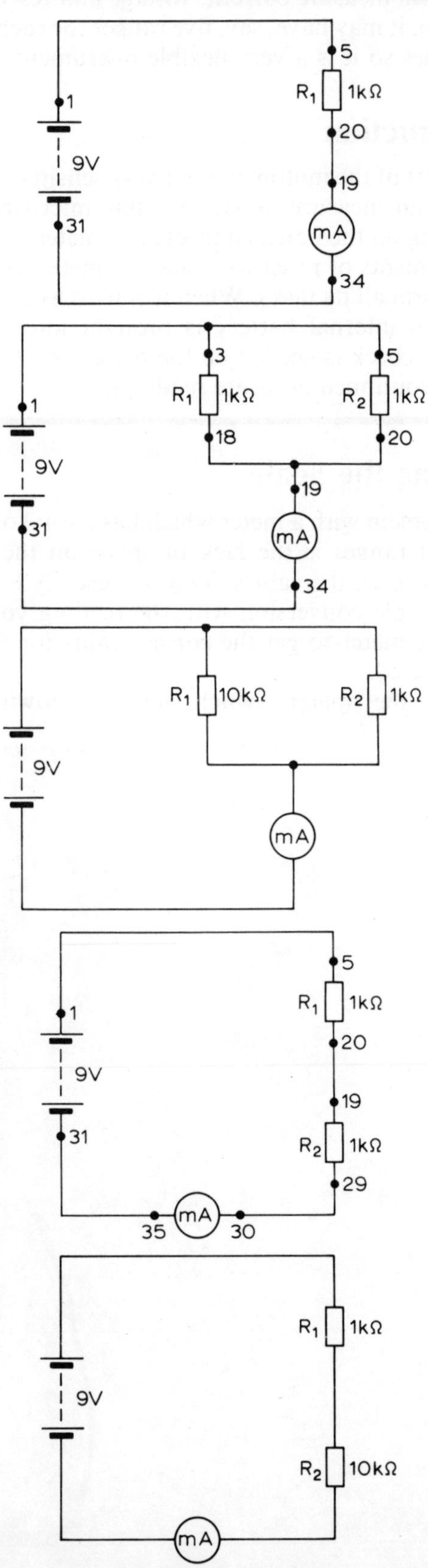

Practice Circuits

In the following questions, supply the values asked for.

1\.

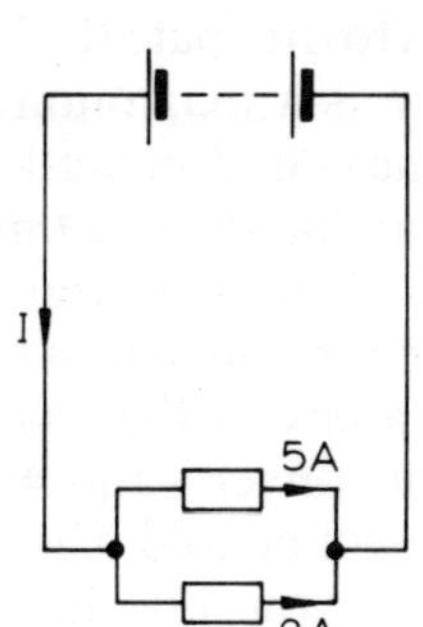

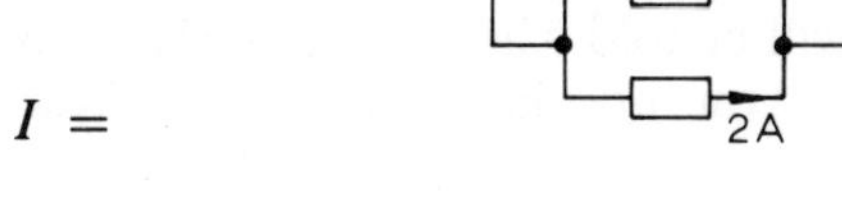

I =

2\.

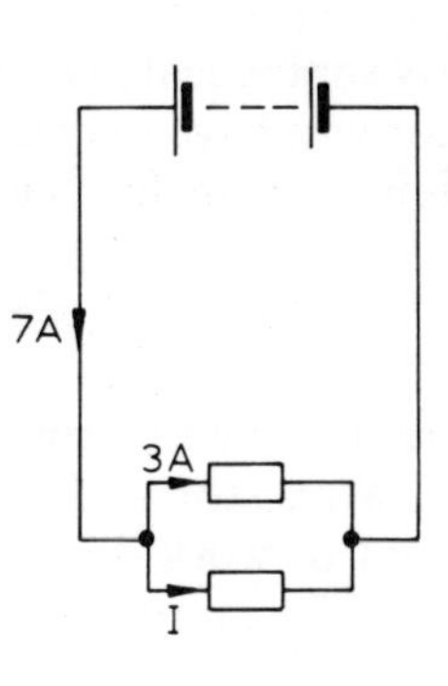

I =

3\.

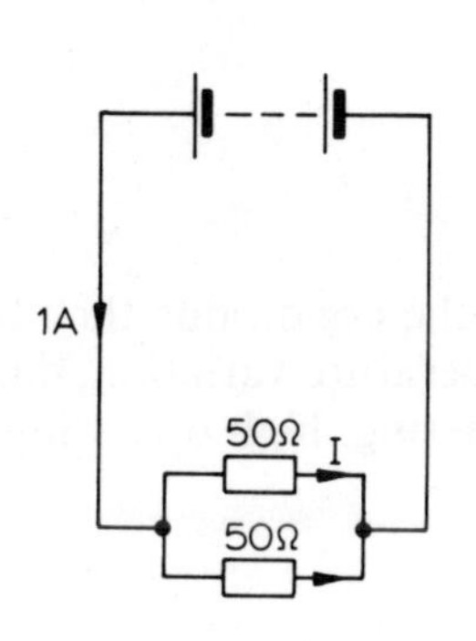

I =

4\.

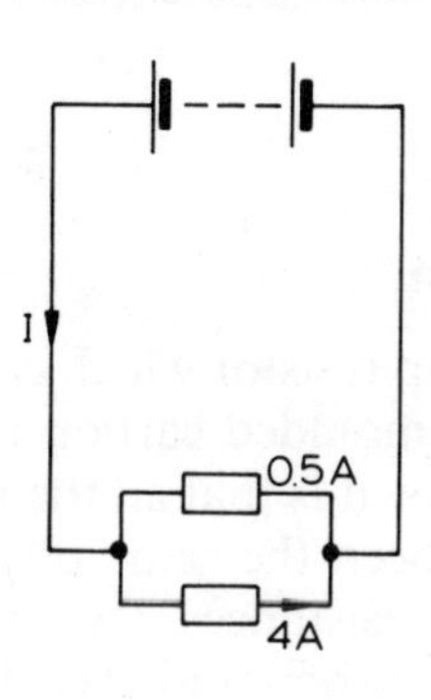

I =

5\.

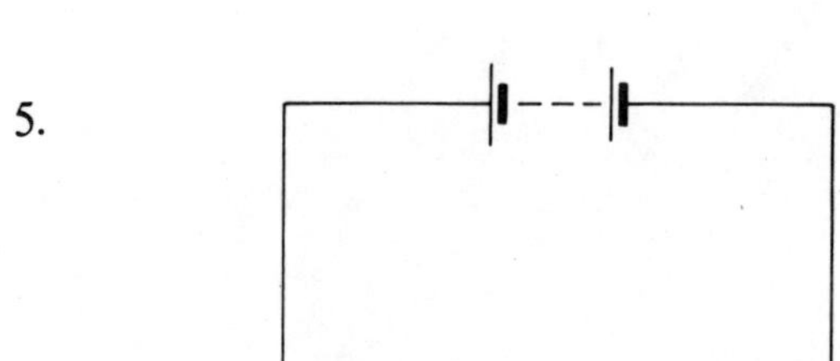
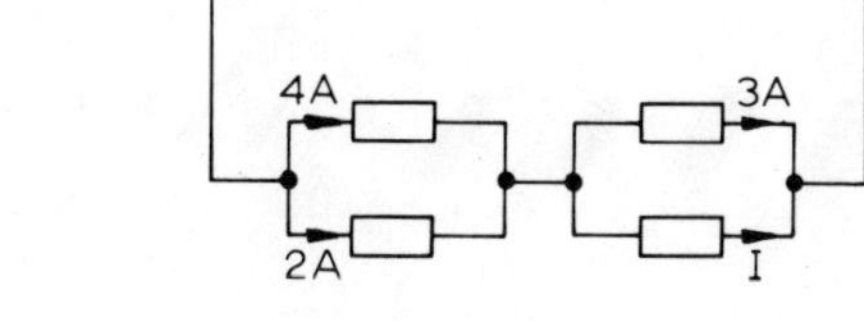

I =

6\.

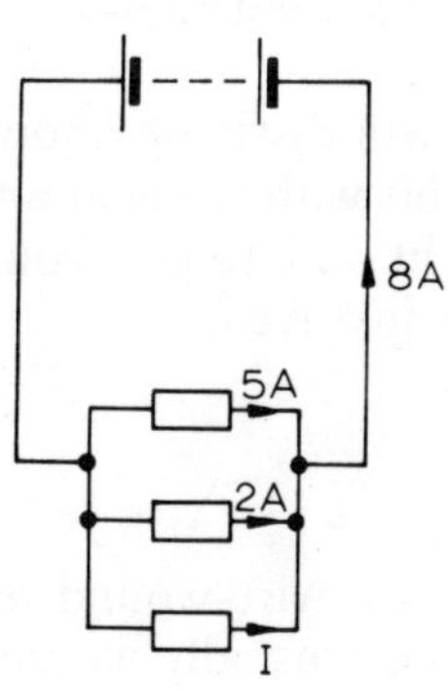
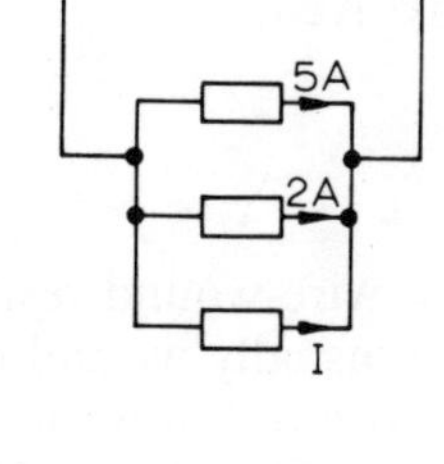

I =

7\.

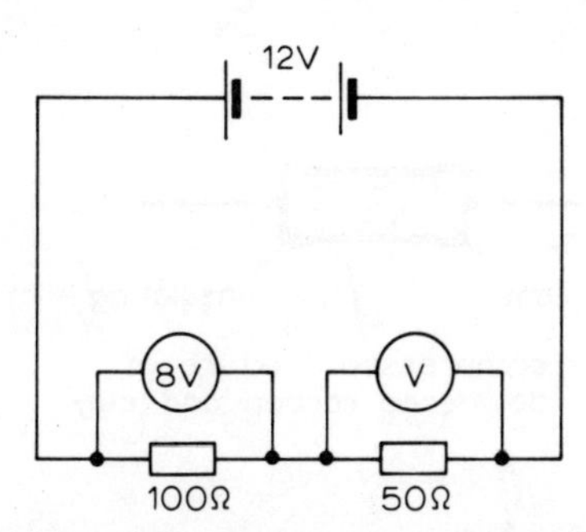

V =

8\.

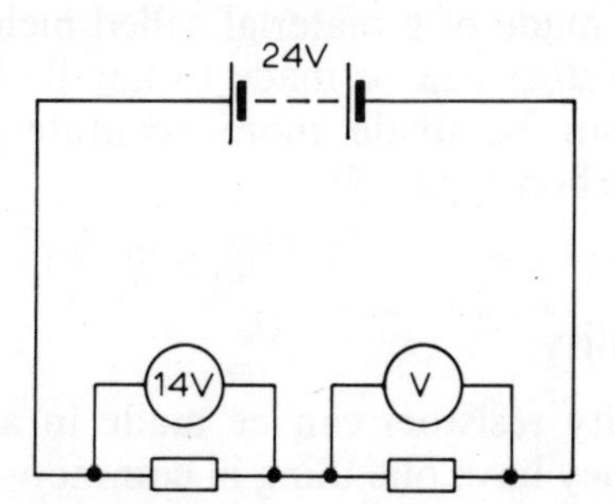

V =

Resistors

There is now a large range of resistor types available. You will not need to use most of them unless you start to make some fairly sophisticated equipment. There is, however, one factor you have to consider immediately in selecting the right resistor for the job.

Power

You are now aware that the electrons have to do work in getting through a resistor and that this results in heating. The amount of work done every second in a resistor is the power. The more power, the more heat a resistor produces.

The resistor then has to lose all the heat it receives every second or it will overheat and burn up. We say that the resistor has to dissipate the heat it receives. For every resistor, there is a maximum power that it can dissipate. This is measured in watts (W).

A 5 W resistor can handle five times as much power as a 1 W resistor. You will learn shortly how to calculate the power consumed by a resistor. Every time somebody designs a circuit they have to calculate the power every resistor converts and select a suitable one. In this course, you will be told what type of resistor to use but it will do you no harm to check these calculations.

Types

Moulded Carbon

The common cheap resistor which can be used in most circuit is called a moulded carbon resistor. You may have wondered how it is that all the resistors you have used so far have been the same physical size and yet have had different resistances. Two substances are used in these resistors, carbon and clay, finely ground and mixed together. If you wish to make a low-value resistance you use a mixture containing a lot of carbon and not much clay and for a high-value resistor you use plenty of clay.

Typically, they will dissipate about $\frac{1}{2}$ W. You may wish to compare this with the heat and light given out by, say, a 60 W light bulb to give you some idea of the quantity of power involved.

Wire-Wound

As the name suggests, wire-wound resistors are made from a length of wire, usually wound on a former of ceramic and covered with a thin protective coating of cement. Obviously the wire will not be copper wire, since this has too low a resistance for most uses but it is likely to be made of a material called nichrome.

These resistors can be made to handle high power and also can be made more accurately than the moulded carbon type.

High Stability

High stability resistors can be made in a variety of ways, but they have one thing in common—their high stability. Essentially this means that whatever happens to them (within reason, of course) they keep the same value. Some of the conditions they have to be proof against are temperature variation, dampness, storage, heat during soldering, high operating voltages and so on.

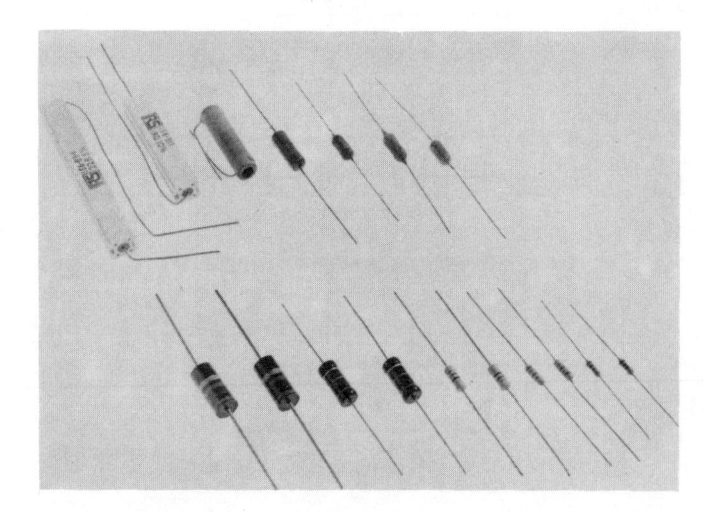

Wire-wound and moulded carbon resistors

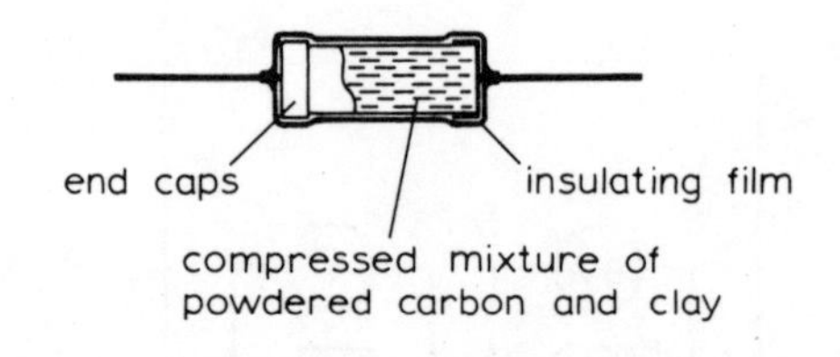

Cross section of a moulded carbon resistor

Variable Resistors

The diagrams show two common variable resistors which are essentially the same in operation. The photograph shows a large heavy-duty type which your school probably possesses. The commonest type of variable resistor consists of a carbon track and a wiper, which can easily be seen in the skeleton pre-set type. This type is adjusted by means of a screwdriver which rotates the wiper. This is usually done in the last stages of construction of the device being made. The enclosed type is used where adjustment is needed whilst the device is being used—for instance, the volume control of a radio.

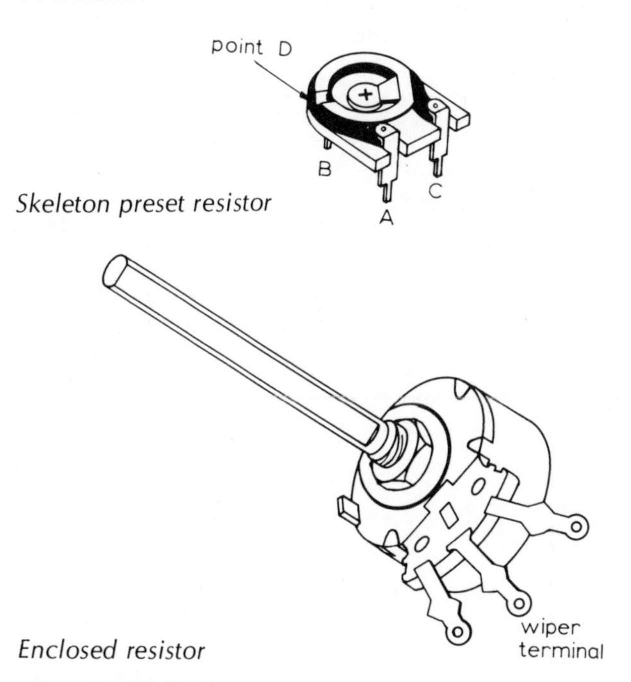

Skeleton preset resistor

Enclosed resistor

Operation

In the skeleton preset resistor if only terminals A and C are used then the current path is through the complete length of the carbon track and so it is operating as a fixed resistor. However, if the wiper terminal (usually the centre one) is used then it operates as a *variable* resistor.

For instance, if terminals A and B are used the current has to pass through the carbon track as far as point D, then it passes through the wiper arm and out through terminal B. If the wiper arm is now rotated *clockwise*, then the current has to pass through a greater length of carbon resistor and so the resistance *increases*.

Heavy duty variable resistor

Specification

The value quoted for a variable resistor is its maximum. Its minimum is normally zero. Thus a 10 kΩ variable resistor will give you a resistance anywhere between 0 and 10 kΩ.

As with a fixed resistor, the variable resistor must be able to dissipate the power being put through it.

Typically, the moulded carbon track type will take a maximum of 1 W for a large one (say 25 mm diameter) whilst a skeleton pre-set will take say $\frac{1}{4}$ W. For power loads of more than 1 W, variable resistors with a track which is wire-wound are usually used.

Use

Either of these two experiments will show you the effect of a variable resistor in controlling the current in a circuit. Operate the wiper and note the effect on the bulb brightness.

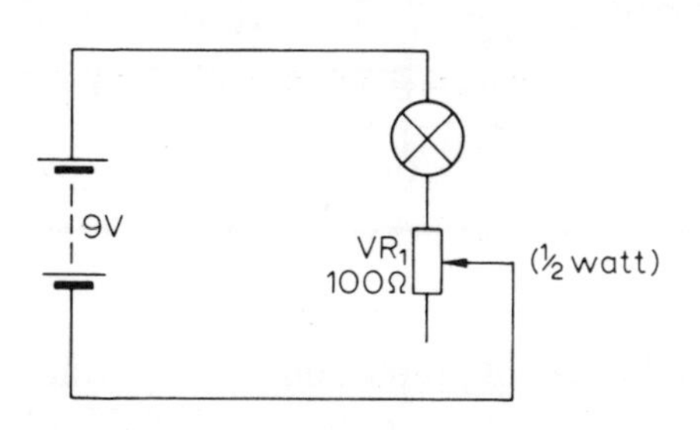

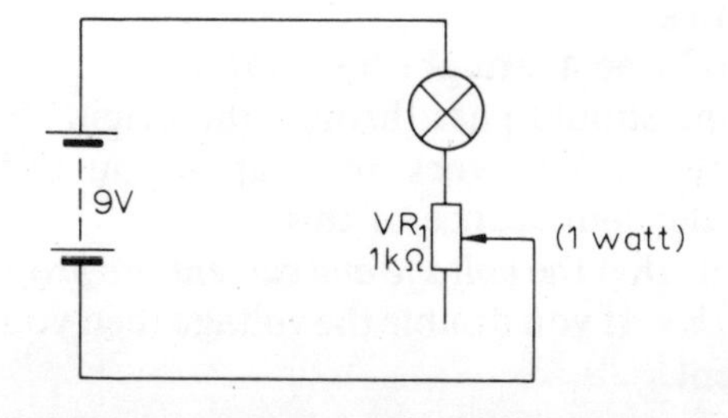

Ohm's Law

You have now been using three electrical quantities, current, voltage and resistance for some weeks, or perhaps longer, and these three are in fact related to each other. This circuit enables you to discover the relationship between them. Your teacher may have had to alter some of the values in the circuit to suit the milliammeters you have.

Verifying Ohm's Law

Instructions

In this experiment you are to take pairs of readings of the current through, and voltage across, the fixed resistor. After each pair of readings, use the variable resistor to alter the value of the current in the circuit and read your meters again.

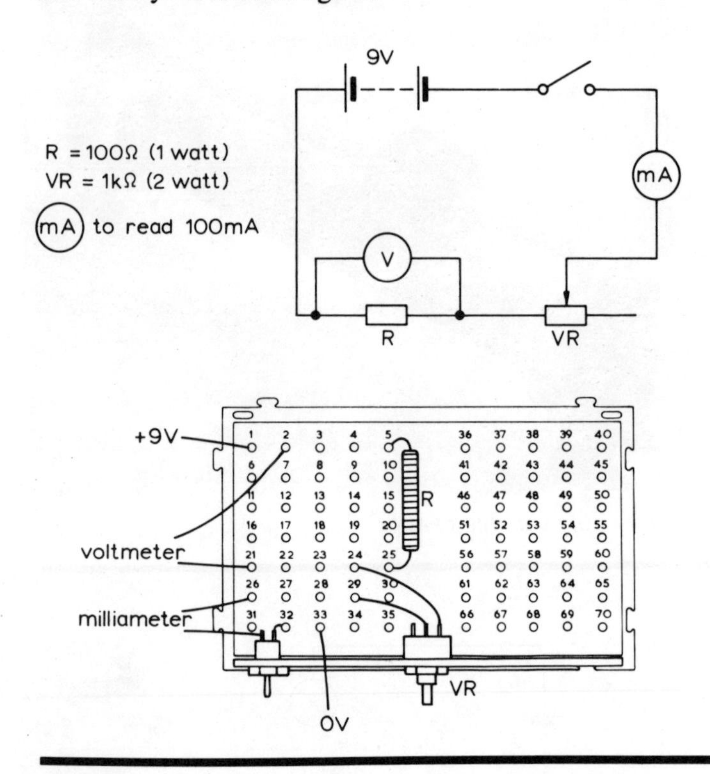

Take six pairs of readings, spaced out between the maximum and minimum values, put them in a table and plot a graph of voltage against current.

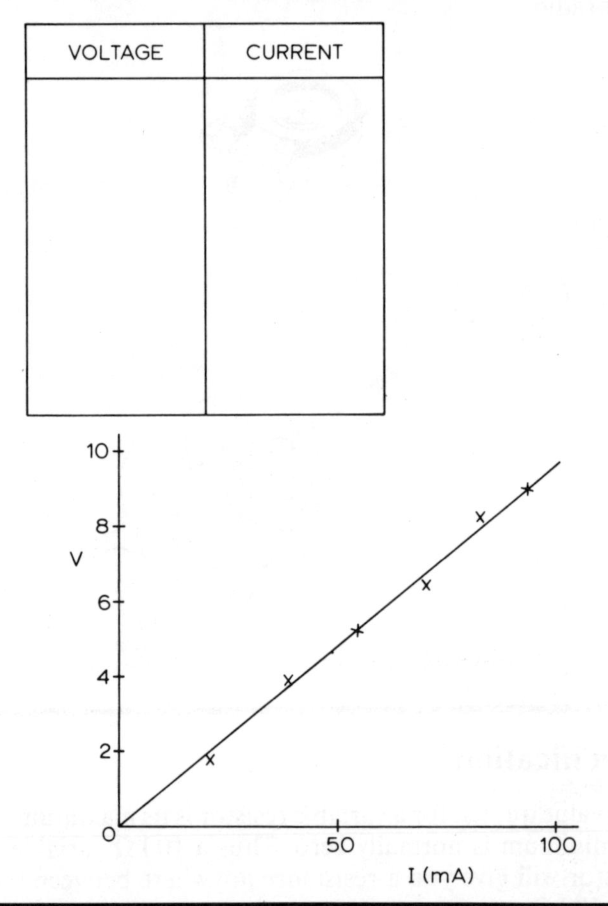

VOLTAGE	CURRENT

Results

You should obtain a graph similar to the one shown. It will not go through the same points (unless your resistor and battery were identical to the one used for the author's experiment) but it should have two important features:

1. **It should be a straight line graph.**
2. **The line should pass through the origin.**

From your earlier work on graphs you should remember the significance of this.

It means that the *voltage and current are proportional to each other*. If you double the voltage then you double the current.

Check this on your graph by writing down the current readings for a voltage of 4 V and of 8 V. Your second current reading should be double the first. Is this so?

This experiment (or similar one) was performed by George Simon Ohm in the nineteenth century and he summarised it thus:

> **Ohm's Law**
> If the temperature and physical conditions of a conductor do not change, then the voltage across it and the current through it are proportional.

Continued on p. 35

Ohm's Law can also be expressed as

$$\frac{V}{I} = \text{constant}$$

It is this constant which we call resistance. As you know, the unit of resistance is the ohm.

The size of the ohm was chosen such that, if a voltage of 1 volt caused a current of 1 ampere in a resistor, then the value of that resistor was 1 ohm.

It follows from this that

$$\boxed{\frac{V}{I} = R}$$

when the units are volts, amperes and ohms.

This equation can also be written as:

$$\boxed{V = I \times R} \quad \text{and} \quad \boxed{I = \frac{V}{R}}$$

The easy way to remember these three equations is to use the 'magic triangle'.

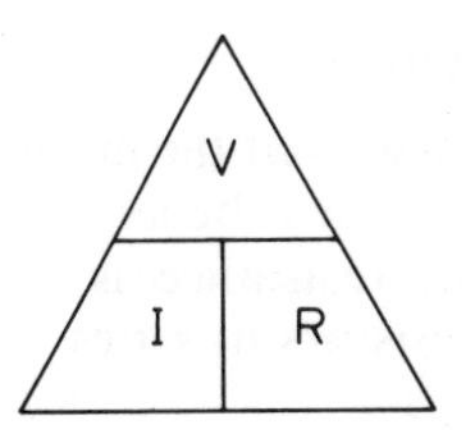

Cover up the quantity you wish to find and it is equal to the other two, as they are shown in the triangle, for example, to find I, cover up I and read off the triangle $I = \frac{V}{R}$

Calculations

These three equations can be used to tell you what current, voltage or resistance to expect in a circuit.

Examples

1. What current would flow if a 9 V battery were connected across a 180 Ω resistor?

   $I = \frac{V}{R} \qquad I = \frac{9}{180} = \frac{1}{20} A$ or 50 mA

2. What voltage is needed to pass a current of 0·2 A through a 56 Ω resistor?

   $V = I \times R \qquad V = 0{\cdot}2 \times 56 = 11{\cdot}2$ V.

3. A 9 V battery is found to pass a current of 20 mA through a resistor. What is value of the resistor?

   $R = \frac{V}{I} \qquad R = \frac{9}{0{\cdot}02} = 450\,\Omega$

   (Note that the current of 20 mA must be converted to 0·02 *amperes* to be used in the equation.)

So you can see that, if you know two of the values in a circuit, you can work out the third one using one of the three equations above.

Resistors in Series and Parallel

You will often come across circuits with resistor networks containing more than the two resistors we have used so far. It is useful to know the rules for these larger combinations.

Series Resistance

You know, of course, that the more resistors you have in series in a circuit then the less current flows. This is because the total resistance is a result of adding together all the resistors in series.

This means that the four resistors shown cut down current in exactly the same way as one 139 Ω resistor. Use an ohmmeter to check this calculation.

$$R_{total} = R_1 + R_2 + R_3 + R_4 = 22 + 22 + 39 + 56 = 139\ \Omega$$

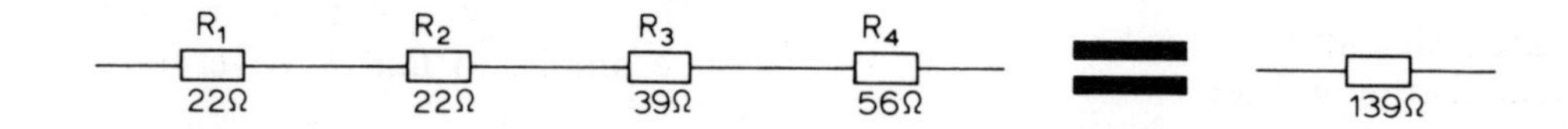

Voltages in Series

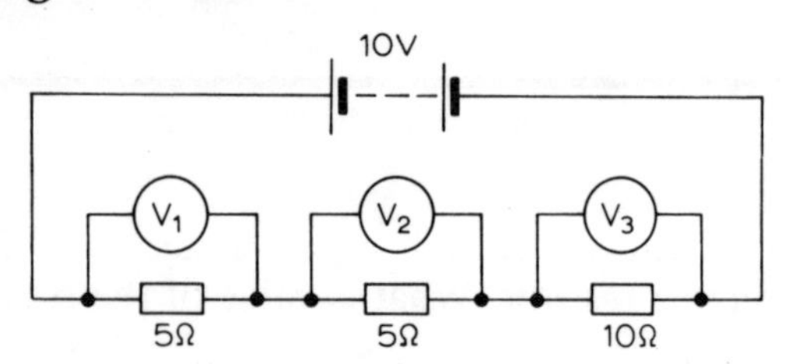

$$R_{total} = 5 + 5 + 10 = 20\ \Omega$$

In the same way as before, these three resistors have the same effect in the circuit as one 20 Ω resistor.

Now we shall find the current flowing.

Using the equations:

$$I = \frac{V}{R} \qquad I = \frac{10}{20} = \frac{1}{2} A.$$

Now let us find the voltage across each resistor:

for R_1 $\quad V_1 = I \times R_1 \quad V_1 = \frac{1}{2} \times 5 = 2\frac{1}{2}\ \text{V}$

for R_2 $\quad V_2 = I \times R_2 \quad V_2 = \frac{1}{2} \times 5 = 2\frac{1}{2}\ \text{V}$

for R_3 $\quad V_3 = I \times R_3 \quad V_3 = \frac{1}{2} \times 10 = 5\ \text{V}$

In fact, this effect is in direct proportion: the 10 Ω resistor has twice the voltage drop across it that the 5 Ω resistor has. Two important conclusions are:

1. The total resistance of a number of resistors in series is found by adding together all the individual resistances.
2. For a number of resistors in series, the total voltage available to them is 'divided' between them according to the size of each resistor.

Parallel Resistance

As you know, when you put more resistors *in parallel* in a circuit the total current increases. This is because you increase the number of current paths every time you put a resistor in parallel with others. There is an equation which enables you to work out which single resistor would replace a number of resistors in parallel.

$$\frac{1}{R\text{Total}} = \frac{1}{R_1} + \frac{1}{R_2} + \frac{1}{R_3} + \ldots$$

This equation can be used for any number of resistors.

Example 1

Let us say that $R_1 = 10\ \Omega$ $R_2 = 10\ \Omega$ $R_3 = 20\ \Omega$

$$\frac{1}{R_{total}} = \frac{1}{10} + \frac{1}{10} + \frac{1}{20}$$

$$= \frac{2+2+1}{20} = \frac{5}{20}$$

$$\frac{1}{R_{total}} = \frac{5}{20} \qquad R_{total} = \frac{20}{5} = 4\ \Omega$$

So these three resistors pass as much current (between them) as one 4 Ω resistor.

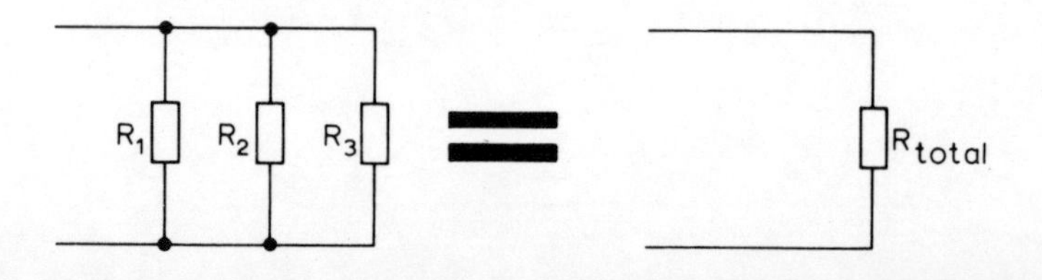

Ohm's Law—Problems

1. Find the current flowing in the following circuits, given the battery voltage and total resistance as shown. Use the equation $I = V/R$.
   (*a*) $V = 9$ volts $R = 18\ \Omega$
   (*b*) $V = 10$ volts $R = 5\ \Omega$
   (*c*) $V = 6$ volts $R = 120\ \Omega$
   (*d*) $V = 10$ volts $R = 1\ \text{k}\Omega$
   (*e*) $V = 10$ volts $R = 500\ \Omega$

2. In the following circuits you are told the current flowing and the battery voltage. Calculate the total resistance in the circuit. Use the equation $R = V/I$. Find the nearest E12 value to the calculated value.
   (*a*) $I = 3$ A $V = 12$ volts
   (*b*) $I = 1$ A $V = 6$ volts
   (*c*) $I = 0{\cdot}1$ A $V = 10$ volts
   (*d*) $I = \frac{1}{3}$ A $V = 9$ volts
   (*e*) $I = 50$ mA $V = 10$ volts

3. Find the battery voltage needed to cause the current shown to flow in the given resistance. First write down which equation you would use.
   (*a*) $I = 2$ A $R = 18\ \Omega$
   (*b*) $I = \frac{1}{2}$ A $R = 22\ \Omega$
   (*c*) $I = \frac{1}{4}$ A $R = 12\ \Omega$
   (*d*) $I = 10$ mA $R = 560\ \Omega$
   (*e*) $I = 1$ mA $R = 3{\cdot}9\ \text{k}\Omega$

4. Voltage $= 10$ V. Find the ammeter reading.

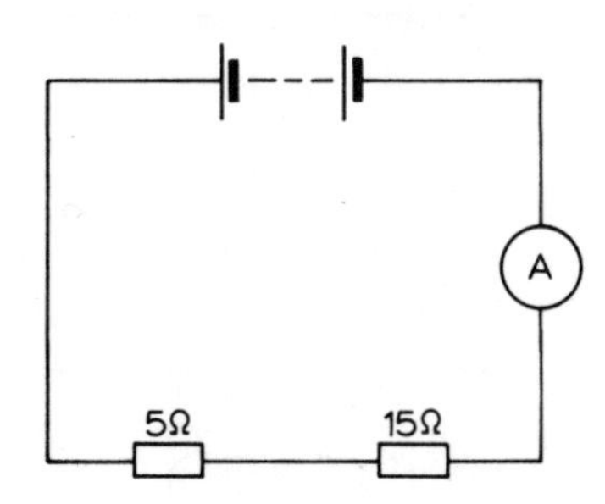

5. $I = 0{\cdot}2$ A. Find the battery voltage.

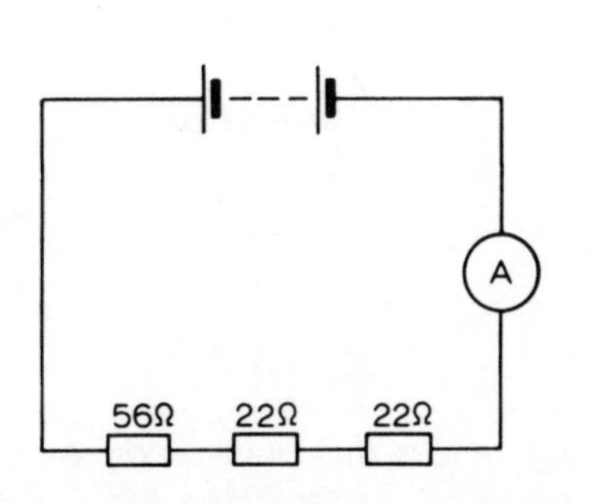

6. $I = 5$ mA.
   Find the value of R_1.
   Find the value of V_1.
   Find the value of V_2.

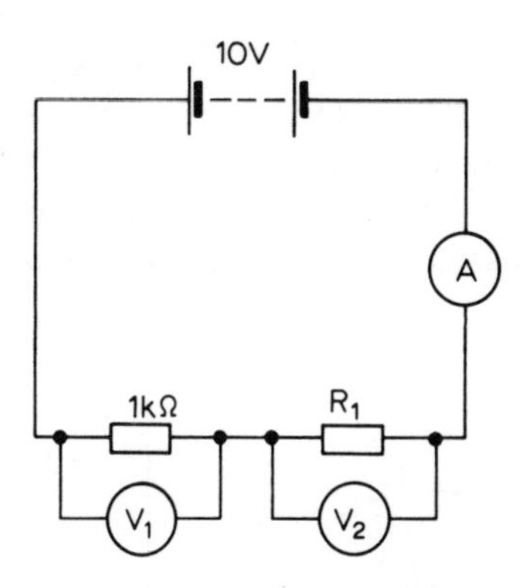

7. In the following circuits, two resistors in parallel, called R_1 and R_2, are connected to a battery with a voltage of value V. Calculate the total current which the battery supplies in each circuit.
   (*a*) $R_1 = 2\Omega$ $R_2 = 2\ \Omega$ $V = 12$ volts
   (*b*) $R_1 = 3\ \Omega$ $R_2 = 6\ \Omega$ $V = 12$ volts
   (*c*) $R_1 = 1\ \text{k}\Omega$ $R_2 = 1\ \text{k}\Omega$ $V = 24$ volts
   (*d*) $R_1 = 1\ \text{k}\Omega$ $R_2 = 1{\cdot}5\ \text{k}\Omega$ $V = 10$ volts
   (*e*) $R_1 = 18\ \Omega$ $R_2 = 27\ \Omega$ $V = 12$ volts

8. The following circuits each contain two resistors in parallel, R_1 and R_2, and a battery. Draw a circuit diagram of each and calculate the current which flows through each resistor and mark it on the diagram. (Calculate the current through R_1 by ignoring R_2 and vice-versa).
   Now mark the total current flowing in the circuit on the circuit diagram.
   (*a*) $R_1 = 4\ \Omega$ $R_2 = 8\ \Omega$ $V = 12$ volts
   (*b*) $R_1 = 1\ \text{k}\Omega$ $R_2 = 1{\cdot}5\ \text{k}\Omega$ $V = 6$ volts
   (*c*) $R_1 = 220\ \Omega$ $R_2 = 330\ \Omega$ $V = 9$ volts
   (*d*) $R_1 = 100\ \Omega$ $R_2 = 1\ \text{k}\Omega$ $V = 10$ volts
   (*e*) $R_1 = 150\ \Omega$ $R_2 = 220\ \Omega$ $V = 9$ volts

Quantity of Electricity and Power

Quantity of Electricity

You have now used an ammeter to measure current many times. The ammeter measures how many electrons pass through every second. We say it is measuring the rate at which the electrons pass. For instance, if your ammeter is reading 1 A, it means that approximately 6 000 000 000 000 000 000 electrons have passed through the meter in one second.

Obviously, if you wished to know how many electrons had passed in two seconds you would multiply that rather large number by two, for three seconds multiply by three and so on.

So you need the ammeter and the stopwatch to measure the total quantity of electricity used. The ammeter tells you the rate at which the electrons are going round and the stopwatch tells you for how long the current has passed.

However, we cannot measure the quantity of electricity by the number of electrons that have passed because the numbers are just too large. Fortunately, whenever we wish to know how much electricity has passed there is a simpler unit.

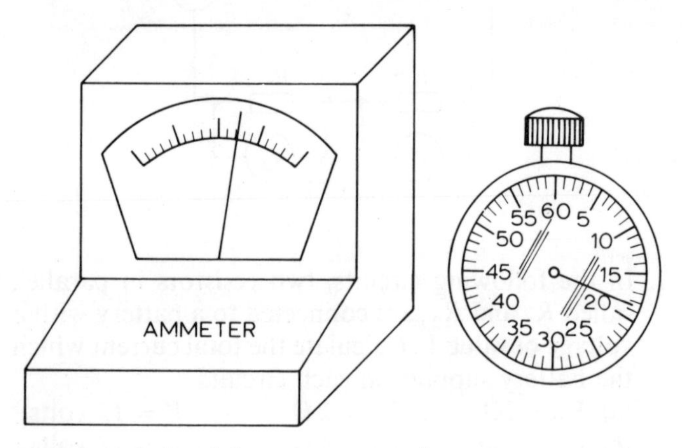

Quantity of electricity = current × time

> When one ampere has flowed for one second, we call the quantity of electricity one coulomb (C)

If two amperes flow for five seconds, then the quantity of electricity is 10 coulombs. The general rule is

$$Q = I \times t$$

Q is quantity in coulombs
I is current in amperes
t is time in seconds.

Power

Now you may think that your electricity meter at home is measuring coulombs. It would seem reasonable to be charged on the quantity of electricity used. However, it is not just quantity that counts!

The electrons pass through the two light bulbs shown at the same rate but the electrons through the mains light bulb are carrying and releasing more energy. This is because they are being driven by a higher voltage source. As a result, the mains bulb gives out more light. To calculate how much power is produced by any device, use the following equation.

Power =	Voltage ×	Current
(watts)	(volts)	(amps)

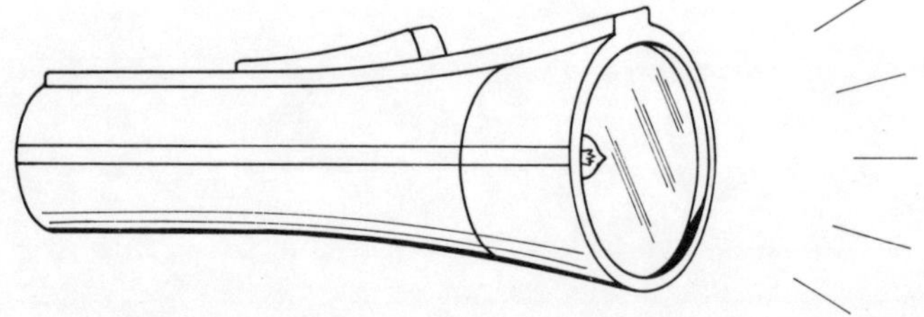

Low voltage results in a low power output

A high voltage across this bulb produces a higher power output

Power

Remember that the power converted by any component is calculated by:

Power = voltage × current

If we use our usual units in the above equation, volts and amperes, then the power is given in *watts.* It is important to understand the significance of the term power:

Power measures the energy being converted by a device, every second

You can test this by examining the effect of using different resistors, each working at different powers. When a resistor converts power it produces heat, but remember that this is not the reason why we use them. In electronic circuits, this production of heat is a positive drawback. Now look at the connection between power and heating effect.

Heat from a Resistor

Instructions

Use resistors rated at 5 W. This means they can operate at this power without damage. Use each in turn for 5 min in the apparatus shown and record the current and temperature rise. Start with cold water each time and the same quantity of water and use a 10 V supply. Calculate the power converted by each resistor.

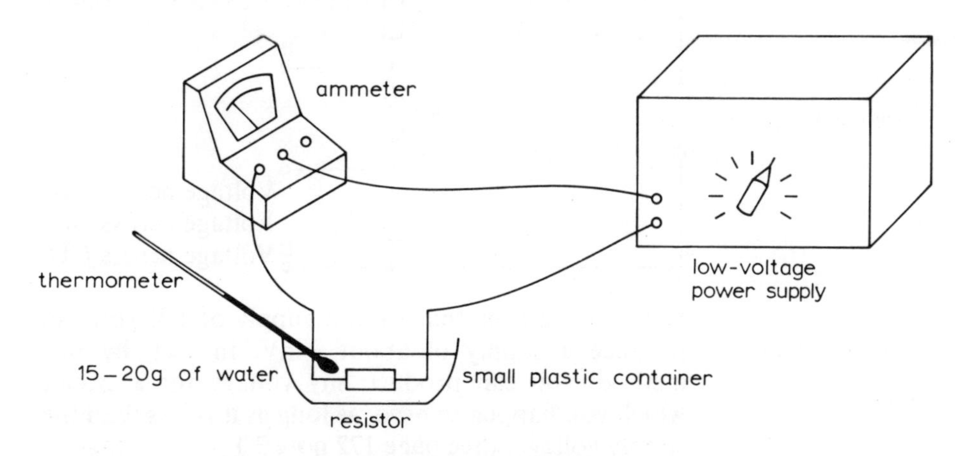

Resistance (Ω)	Voltage (V)	Current	Power ($V \times I$)	Temperature rise
22	10			
33	10			
47	10			

Results

It should be clear that the heating effect produced depends on the power converted.

More power converted—more heat released

The Importance of Power rating

It is important that every component in a circuit is capable of handling the power it converts. Repeat the above experiment with a 22 Ω resistor, rated at $\frac{1}{8}$ W. You will readily see that the resistor overheats because it cannot *dissipate* the heat it produces. Hold the resistor in pliers and do not immerse it in the water. Take care not to burn yourself. It should be clear that mis-use of a resistor in this way has to be avoided.

The Potential Divider

You have seen in the circuits used so far that the supply voltage is divided between all the series components. This result is put to very good use in electronic circuits. In the circuit shown,

V_1 and V_2 add up to the supply voltage.
V_2 is five times larger than V_1
V_1 is one-sixth of the supply voltage
V_2 is five-sixths of the supply voltage.

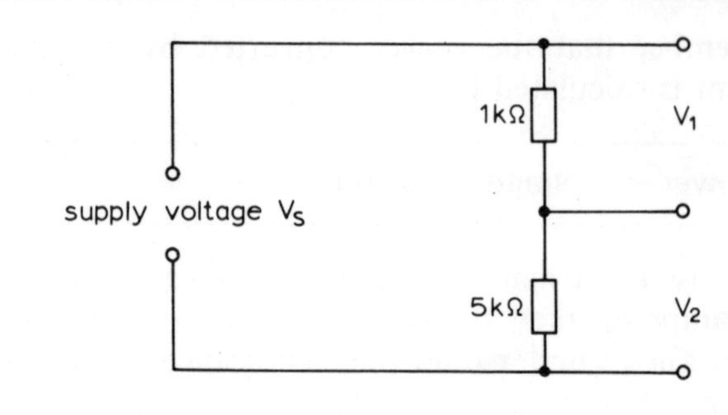

From your work on resistors you should have been able to work out the above figures for yourself, given the values of the two resistors, 1 kΩ and 5 kΩ. This arrangement is found very frequently in transistor circuits and is called a *potential divider*. Obviously, it divides the total potential (voltage) into parts.

Measuring the Voltages

Instructions

Construct the two potential dividers shown below and take the measurements asked for.

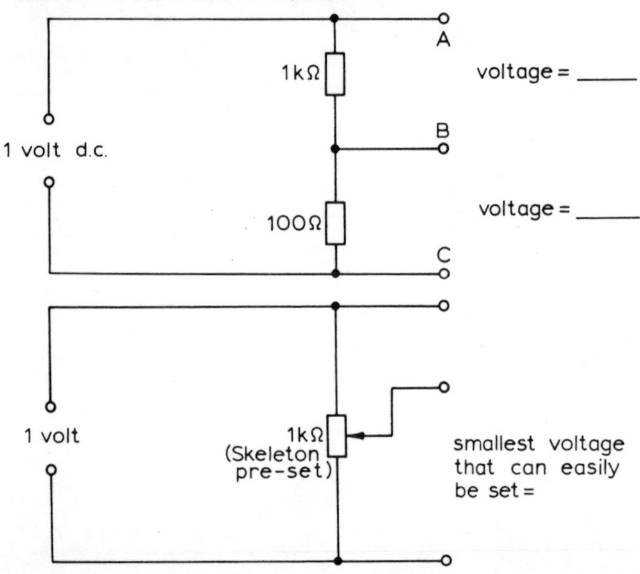

Although the variable divider is very useful, you can see that there is a limit to how finely you can adjust the variable resistor to produce a very small voltage. The multi-tap divider below is a much better way of giving a very small output. Take the readings asked for.

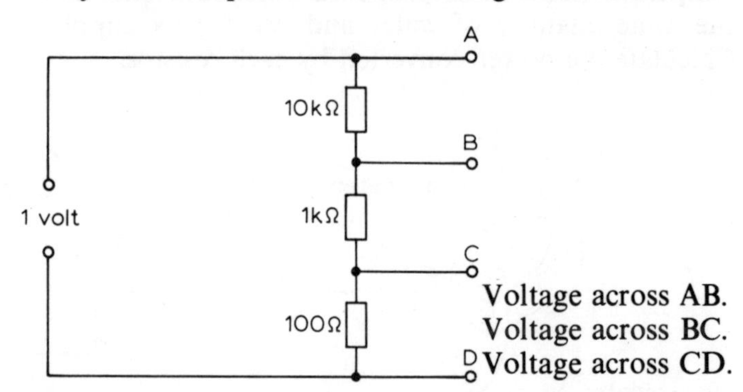

Here you can see that from a supply of 1 V you can produce a supply of about 9 mV. In fact, by this method, you can produce any voltage for a circuit which you happen to need, as long as it is less than the supply voltage. (See page 172 note 3.)

Taking Current from a Voltage Divider

In the above arrangements, you did not actually use the voltage produced. Let us imagine that you connect your 'tapped' voltage to a circuit which has an overall resistance of 100 kΩ. This is called a *load*. We can simulate this circuit with one 100 kΩ resistor, called a *load resistor*, connected across your voltage, as shown.

Instructions

Take the voltage reading as before. Now use smaller resistors of 10 kΩ, 1 kΩ and 100 Ω as loads. This will increase the load current. Repeat the voltage measurements across AB. Page 172 describes the type of voltmeter needed.

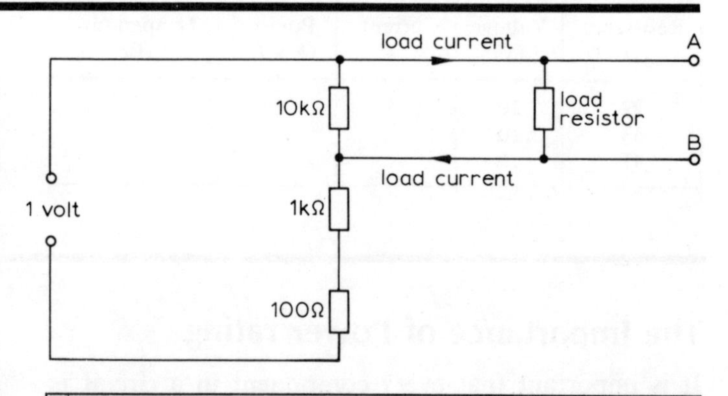

The voltage from a potential divider falls when current is drawn.

Signal Generator

The signal generator is a useful piece of equipment which produces (generates) an electrical signal which will give a musical note with a loudspeaker. The note is approximately 2 octaves above middle C on the piano. Signals that lie in the range which we can hear are said to be *audible*.

Construction

Assemble the circuit as shown on Veroboard. The front panel of your case needs a switch and four output terminals—one black (D) and three red terminals (A, B, C), labelled high, medium and low. You may also wish not to include a battery in the case but have a pair of input terminals on the front panel for a battery connection.

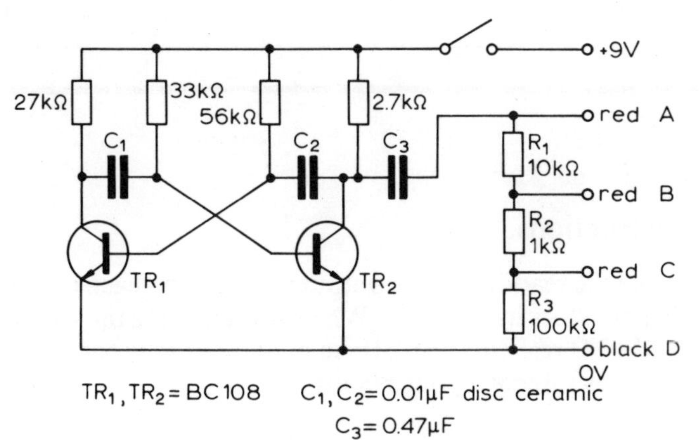

TR_1, TR_2 = BC108 C_1, C_2 = 0.01μF disc ceramic
C_3 = 0.47μF

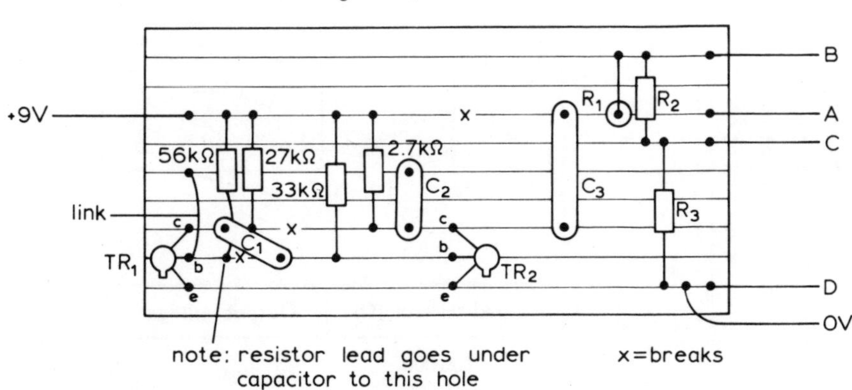

The Circuit

The circuit consists of a *multivibrator* (see page 119) which produces the note and a *potential divider* which gives the output at three different voltages.

Testing

Connect the battery, switch on and connect an earphone (2000 Ω) to the highest output (A and D). You should hear the note of your signal generator. Now test the medium output (B and D) and the lowest (C and D). You should hear your note reduced in volume in each case.

Now look more closely at your output. Use an a.c. voltmeter and find the output voltages of the three different levels. These should, of course, be in the same ratio as the resistors R_1, R_2 and R_3, namely, related one to the next by a factor of ten. Record the voltages and check that they come close to your expected values.

Output	Output voltage
High (AD) Medium (BD) Low (CD)	

Drawing Current

As with all voltage supplies, we must check how constant the output is when we use it. Use an 80 Ω loudspeaker, which will take a large current, and then use a 2000 Ω earphone, which will take much less current. Record what happens to your voltage in each case.

Output voltage with 80 Ω loudspeaker =
Output voltage with 2000 Ω earphone =

Clearly, this unit is not designed to give a large current.

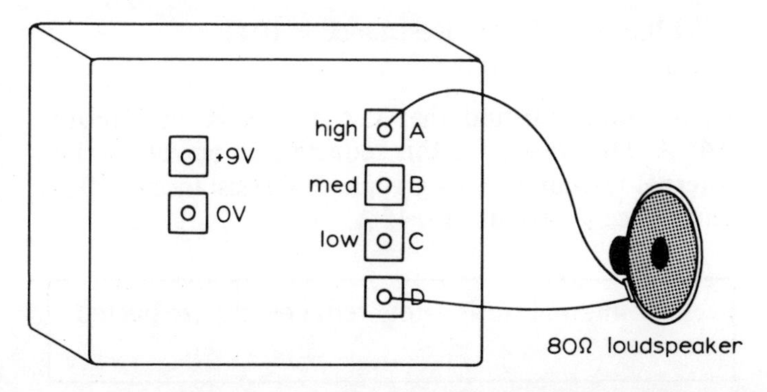

Internal Resistance

It is time to look at an over-simplification that we have made in our circuits up to now. There is one factor that complicates all circuits.

> Every electricity supply has its own (internal) resistance

To take a simple example, a PP6 battery behaves like the components shown inside the square. The battery produces about 9 V but it has its own internal resistance (this is about 10 Ω for a PP6). Any current has to flow through this internal resistance as well as through any resistance you connect to the battery. (Notice the use of the term *load* for any resistance that is connected to a supply and takes current from it.)

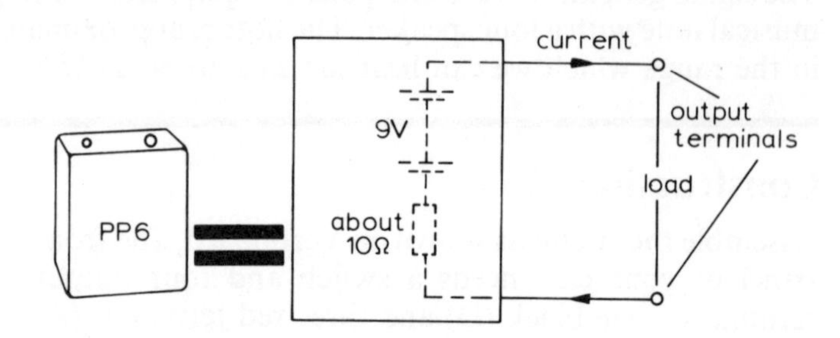

> The internal resistance always acts in series with any load

Effects of Internal Resistance

Instructions

Connect this circuit for a few seconds only, and read the current.

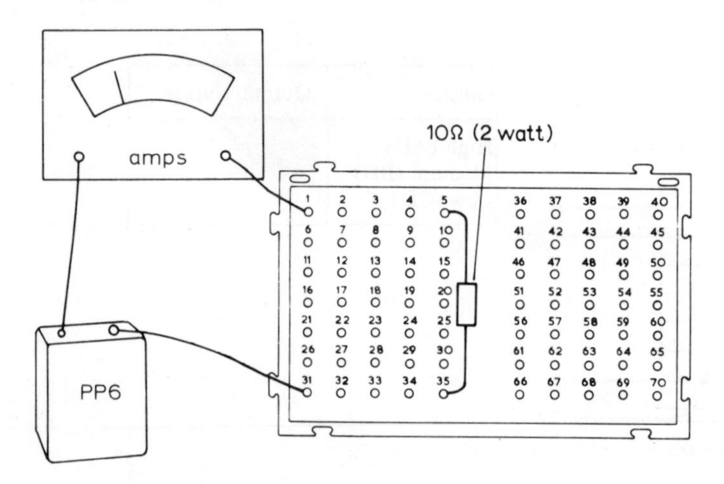

In theory, up to now, you would have calculated the current in the following way:

Voltage = 9 V. Resistance = 10 Ω
Current = 9 ÷ 10 = 0·9 A.

In practice, you find that the current is only about 0·45 A. The reason for this reduction in current is the internal resistance. This gives a total resistance of 20 Ω and hence a current of 0·45 A.

> The internal resistance reduces the expected current.

Instructions

Take the reading for the output voltage with each of the five resistors in the table. (When you are using the 100 Ω and 10 Ω resistors, leave them connected only for a few seconds.) Record your results.

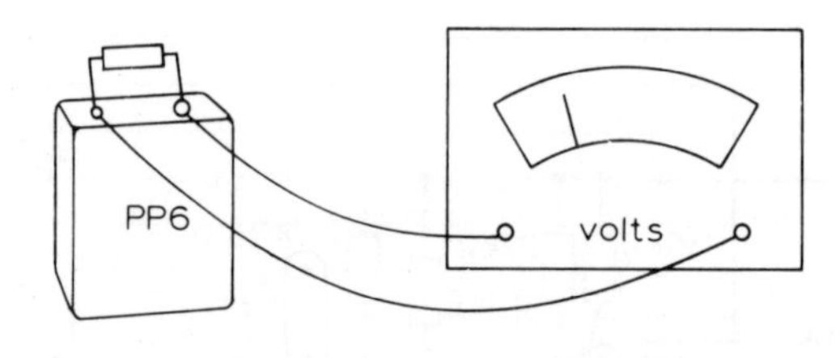

Load resistor (Ω)	Output voltage
100 000	
10 000	
1000	
100	
10	

> As the current drawn from a supply increases, its output voltage decreases

The reason for this behaviour is that the available voltage has to be 'shared' between the load resistor and the internal resistor. As the load resistor decreases in size, the internal resistor takes a bigger share of the available voltage.

Resistors—Summary

Energy Change

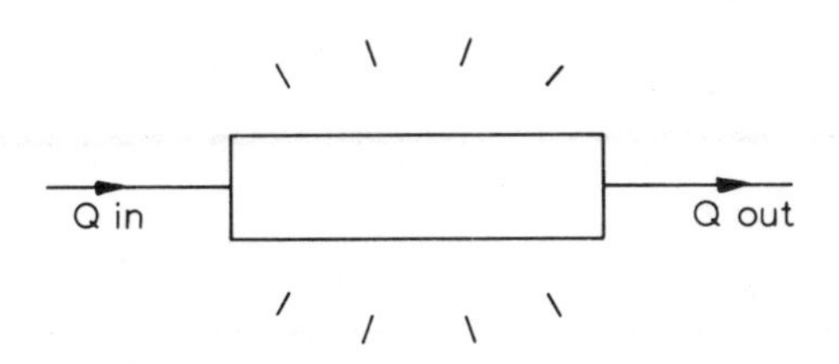

The resistor changes electrical energy into heat energy. No electricity is lost—only energy is removed.

Compare it with a hydro-electric dam.

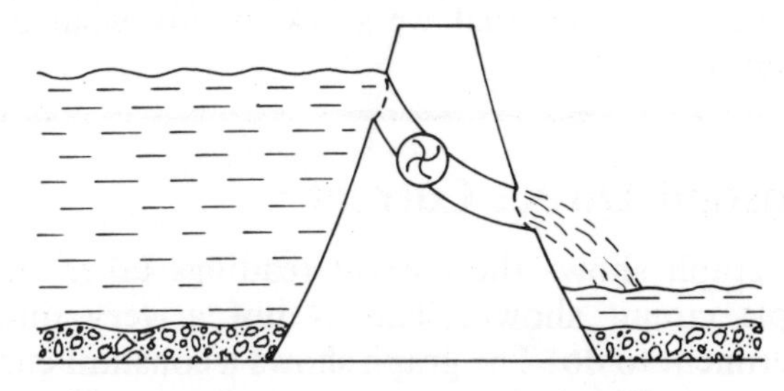

No water is lost—only energy is removed.

Voltage Divider

The voltage across each resistor is proportional to that resistor's value.

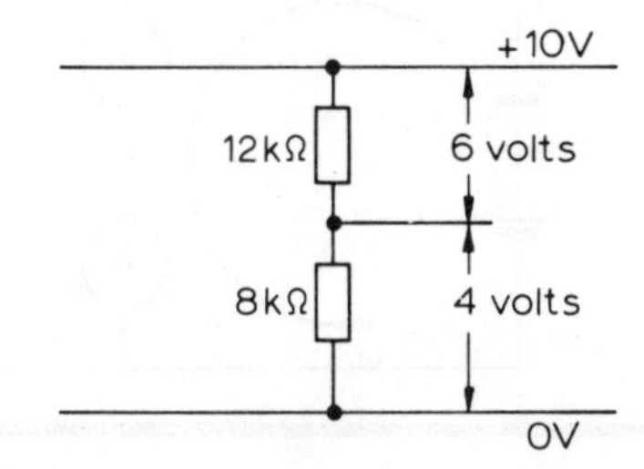

Ohm's Law

The resistor obeys Ohm's Law. The voltage drop is proportional to the current.

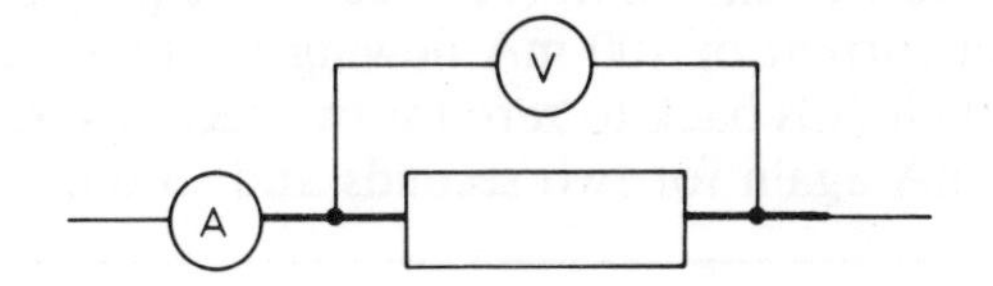

Cutting Down Current

The 1 kΩ resistor limits the current to less than 10 mA, which is safe for the light emitting diode (l.e.d.).

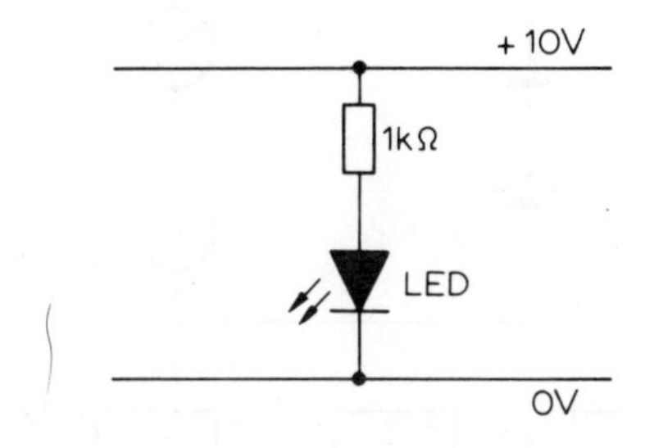

Conversion

The resistor converts current change into voltage change. If I_c varies between 0·5 mA and 0·7 mA, then the output voltage varies between 5 V and 7 V

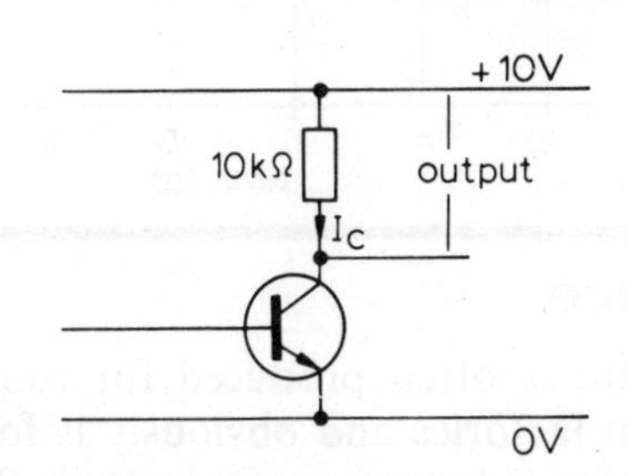

3 ALTERNATING CURRENT

Introduction

This chapter is devoted to studying alternating current but first let us see that we know exactly what direct current is.

Constant Direct Current

The graph shows the current readings taken in the simple circuit shown. This is not a very difficult experiment to do! The graph shows a constant current of 100 mA. This is a direct current at its simplest and is what you have used in your circuits so far.

Direct current flows in one direction only round a circuit

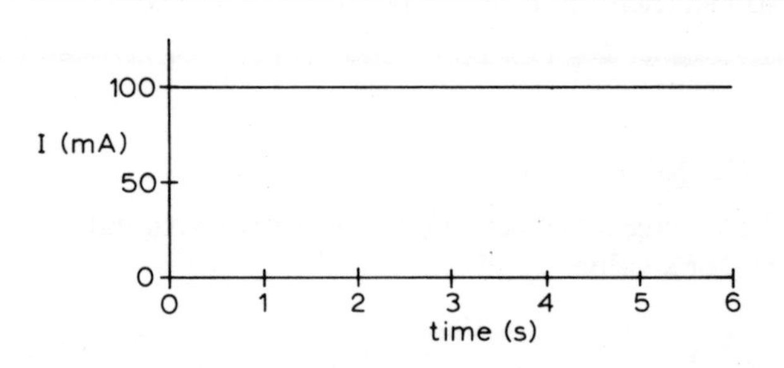

Variable Direct Current

With a switch in the circuit, the operator can produce current flow as shown in the circuit. The graph below shows a current of 100 mA flowing for two seconds, which then falls back to zero for two seconds, rises up to 100 mA again for two seconds and so on.

This is variable direct current

Below are displays of two fairly common direct current supplies. Here you see more gradual changes in the sizes of the current. Again these produce current which flows in one direction only and hence are direct current. On page 46 you will see how we represent a current which changes its direction.

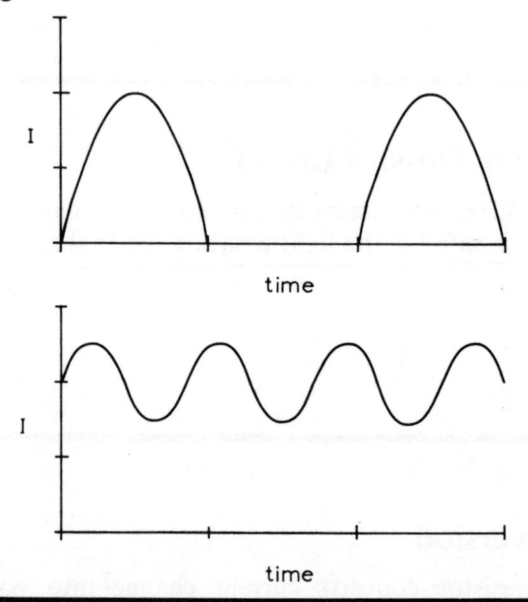

Occurrence

Direct current is often produced for running heavy machinery in factories and obviously is found in any circuits containing batteries—for instance the electrical instruments in a motor car. It is also produced by some of the transducers you met earlier. For instance, the selenium cell in a light meter produces direct current, the size of which depends on the brightness of the light. Measuring the current on the ammeter in the light meter then gives a guide to the strength of the light.

Displays of Direct Current Supplies

The graphs below are all displays of either direct current or direct voltage supplies. Describe what the graphs show you about the behaviour of these supplies. The first one has been done for you.

1.

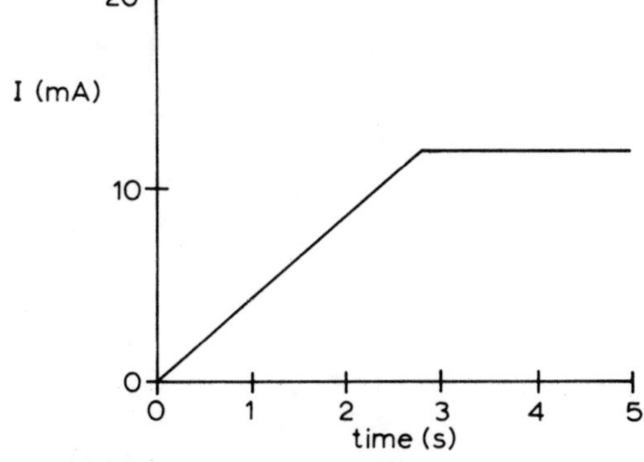

The current starts at 0 and rises uniformly to 12 mA after 2·6 s and then stays constant.

2.

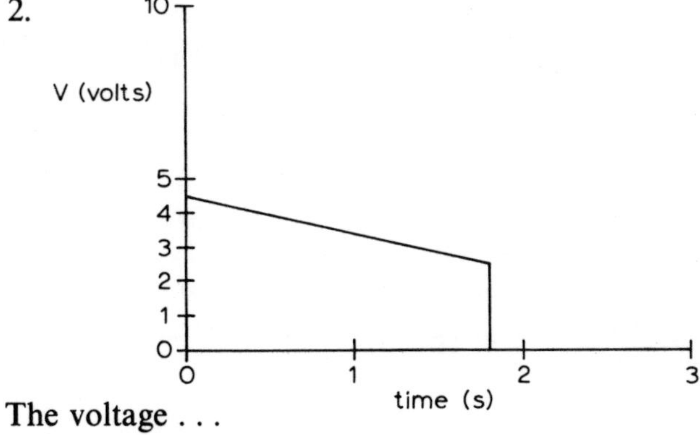

The voltage . . .

3.

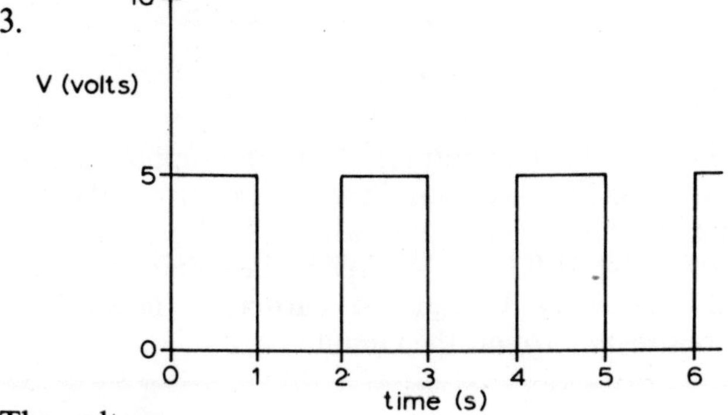

The voltage . . .

4.

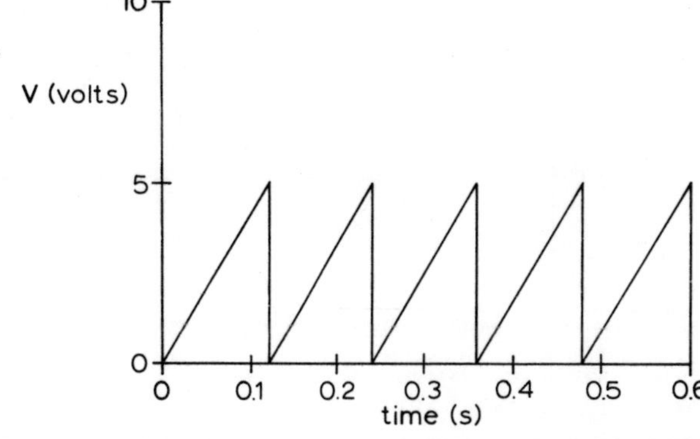

The voltage . . .

5.

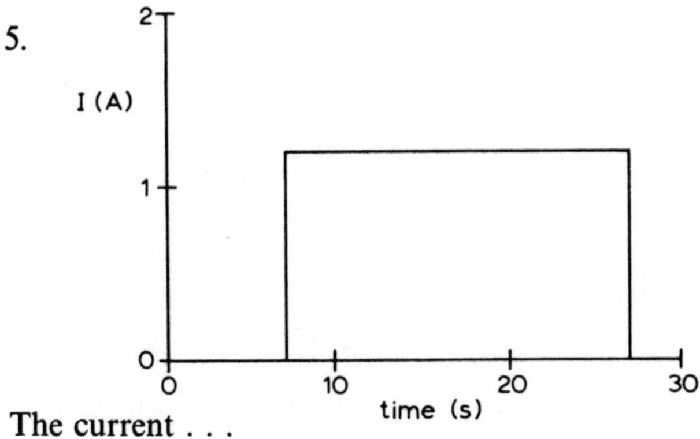

The current . . .

6.

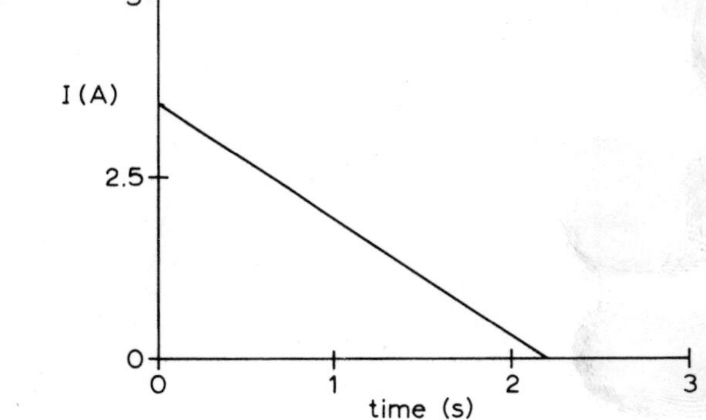

The current . . .

7.

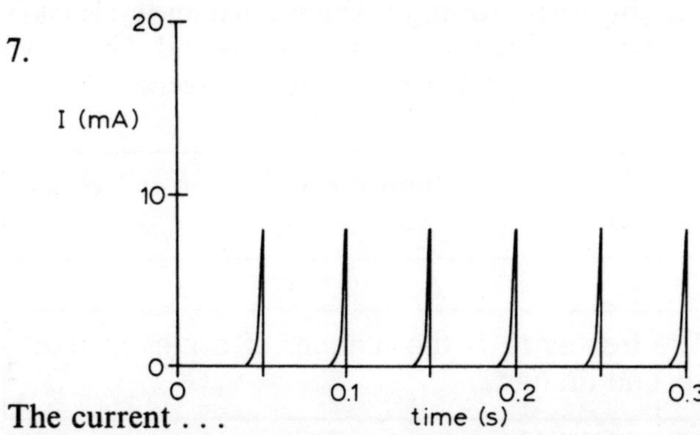

The current . . .

8.

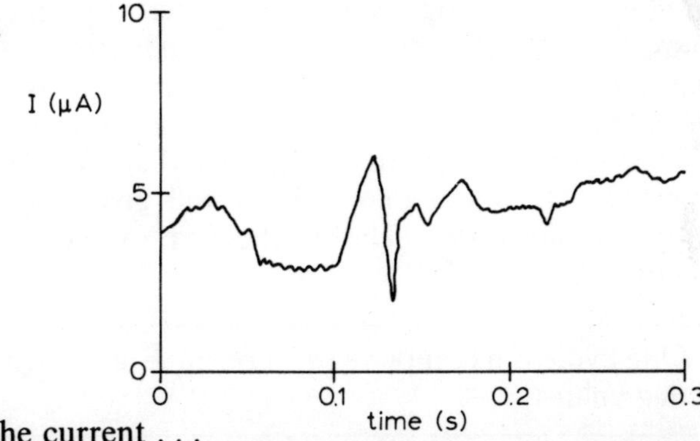

The current . . .

Properties of Alternating Current Supplies

It is now time to examine the properties of electricity supplies which are said to be alternating current (a.c.) supplies.

> An alternating current supply is one that changes direction at regular intervals

The diagrams opposite show the nature of a.c. supplies, in the simplest of circuits. It is the regular change in direction of the current which distinguishes alternating current from direct current. It means that we now have to be able to describe extra factors in a circuit—in what way and how often the current changes direction.

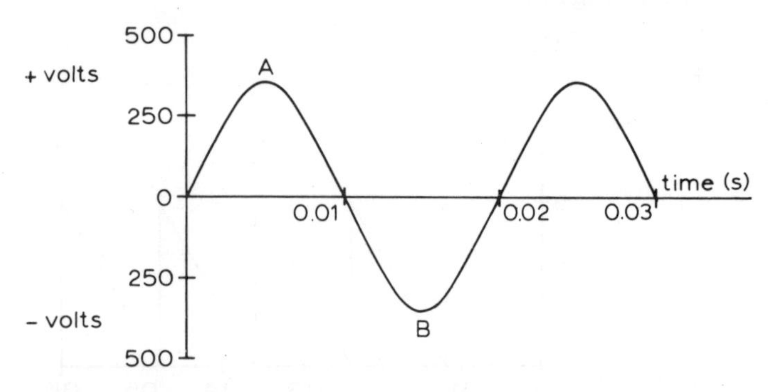

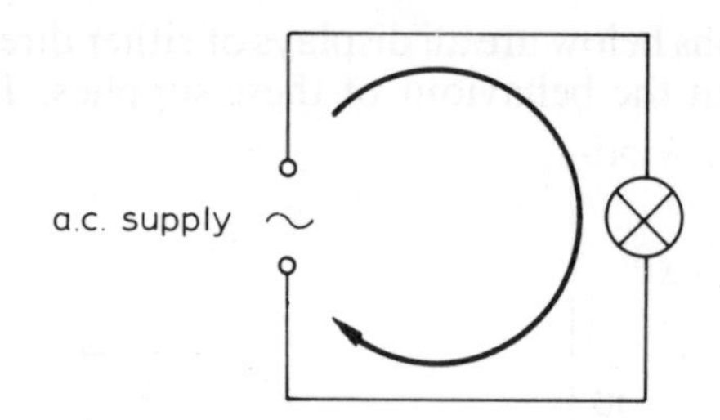

One half cycle

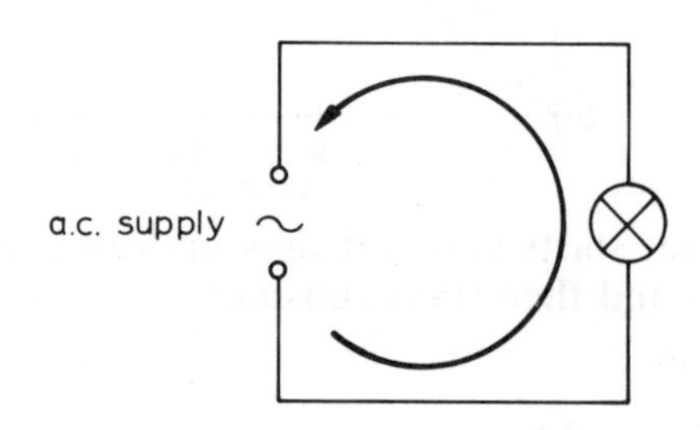

The other half cycle

Look at how we display the mains supply—we use positive and negative numbers. The positive and negative signs indicate current flowing in *opposite directions.* The voltage at B is just as large as the voltage at A. The only difference is that the current is flowing the opposite way round the circuit.

Important Properties—Cycle, Period, Frequency

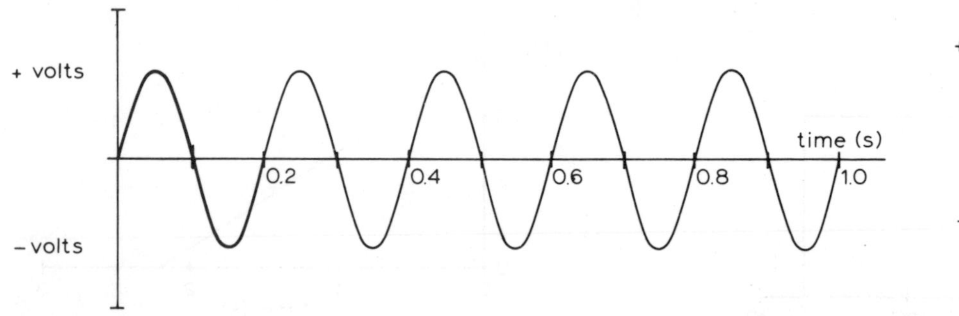

An a.c. waveform

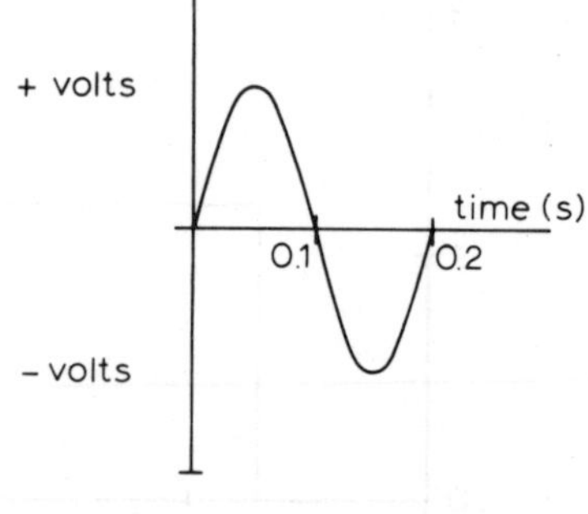

One cycle of the a.c. waveform

An alternating voltage will rise from zero to a maximum, fall back to zero, rise to a maximum in the opposite direction then fall back to zero. This is *one cycle.* This is shown by the thick line in the above graph and shown on its own on the right-hand graph. If you repeatedly draw out one cycle, it will give you the complete waveform. (This is the test for correctly identifying one cycle.)

> One cycle is a complete set of positive and negative values

In the above waveform, you can see that one cycle takes 0·2 s and there are five cycles in one second. These two numbers are called the period and frequency.

> The period is the time taken for one cycle (in seconds)

> The frequency is the number of cycles in one second (in hertz)

Frequency of Alternating Current Supplies

The two graphs show two a.c. waveforms. This particular waveform (like those on the previous page) is called a *sine wave*. The a.c. sine waveform of our mains electricity is, in fact, the shape as produced by the generators at the power station and is not altered in shape in any way for the home. If you examine the two graphs you will see that they have the same time scale, so clearly the second supply is 'alternating' faster than the first.

Graph 1

Period (time for one cycle) $P = 0{\cdot}2$ s

Frequency (number of cycles in one second) $f = 5$ Hz

Graph 2

Period $P = 0{\cdot}1$ s

Frequency $f = 10$ s

You should be able to take these readings from the graphs yourself by now and you will get more practice in the coming pages. You may have noticed that there is a connection between period and frequency—look at the two sets of numbers above.

(0·2 and 5) (0·1 and 10)

Perhaps you can see that each number is the reciprocal of its partner (that is, each number is equal to the other number 'upside-down').

For example $0{\cdot}2 = \frac{1}{5}$ and $5 = \frac{1}{0{\cdot}2}$

$$P = \frac{1}{f}$$

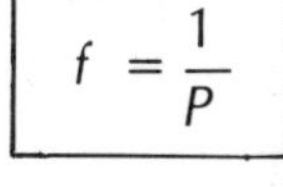

Graph 1

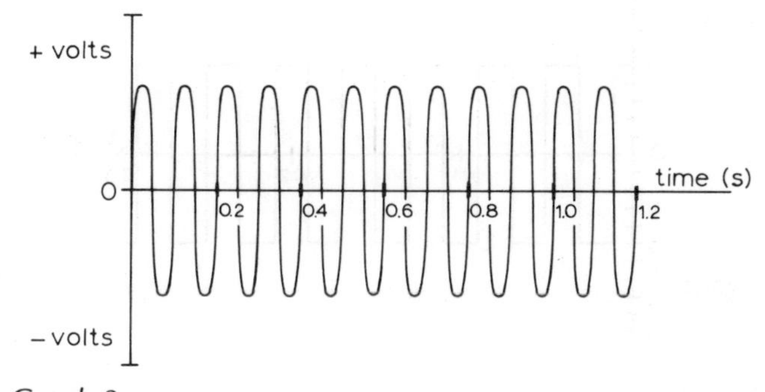

Graph 2

Radio Frequencies

An understanding of frequency is vital for the work later in this book on radios. The sound information your radio receives is transmitted on a carrier wave. This is a sine wave like the ones you have studied but of a much higher frequency. Every radio station transmits its programmes on one fixed frequency which only that station is allowed to use. When you tune your radio to a particular station the tuning section of your radio picks out that frequency and ignores all the others.

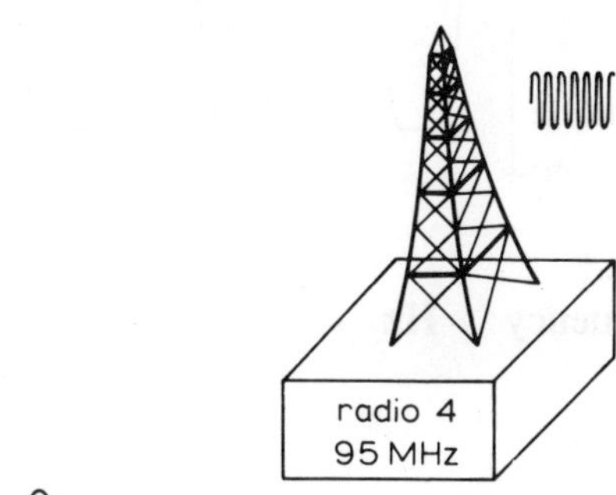

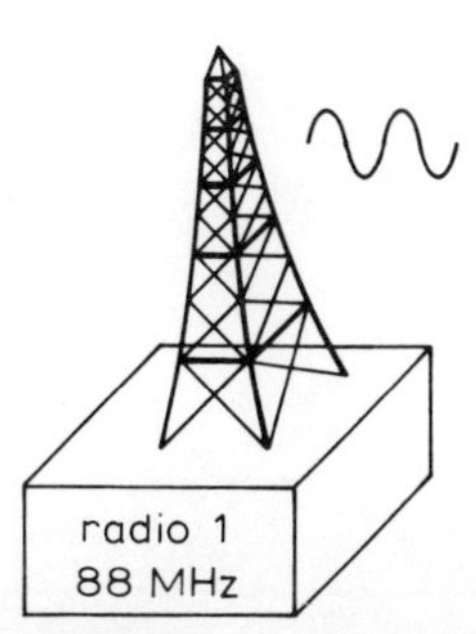

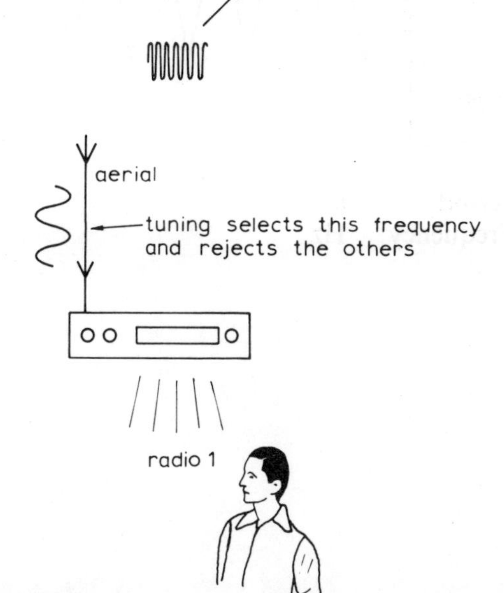

Alternating Current—Problems

For each of the following a.c. graphs, state the period and frequency.

1.

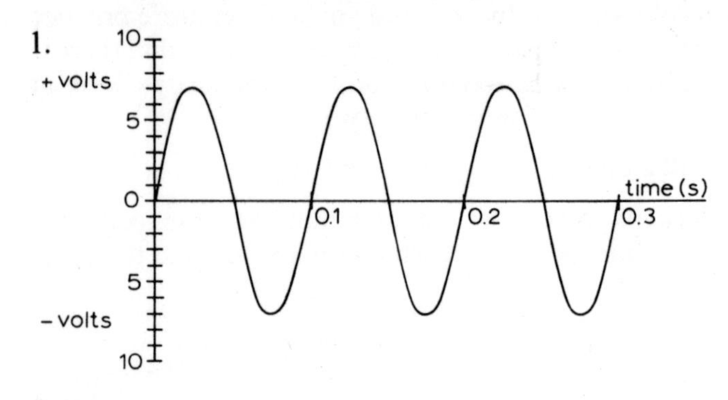

Period s
Frequency Hz

2.

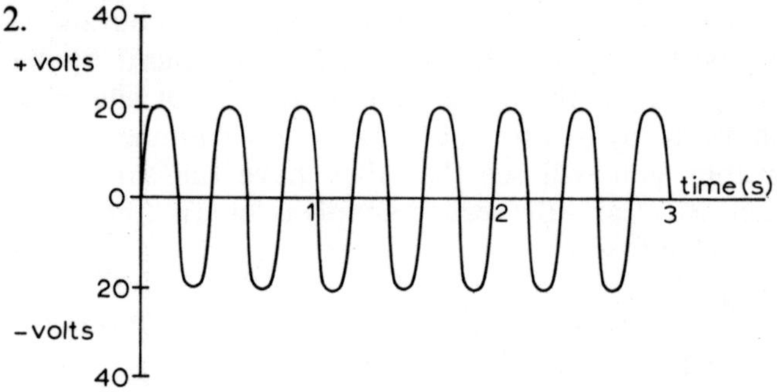

Period s
Frequency Hz

3.

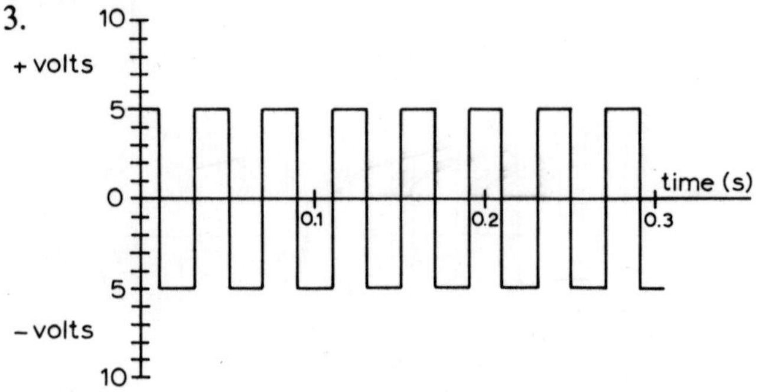

Period s
Frequency Hz

4.

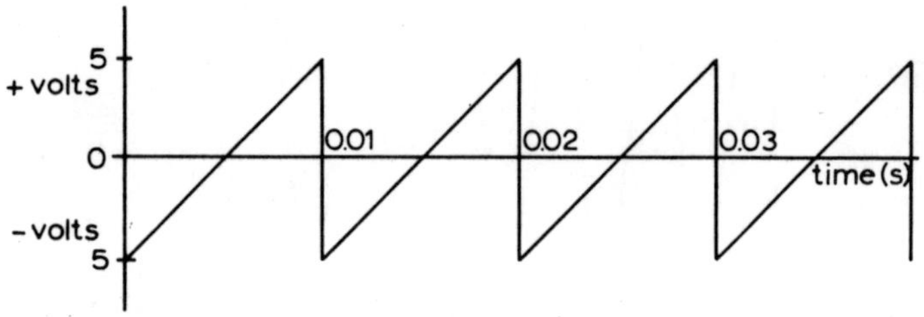

Period s
Frequency Hz

5.

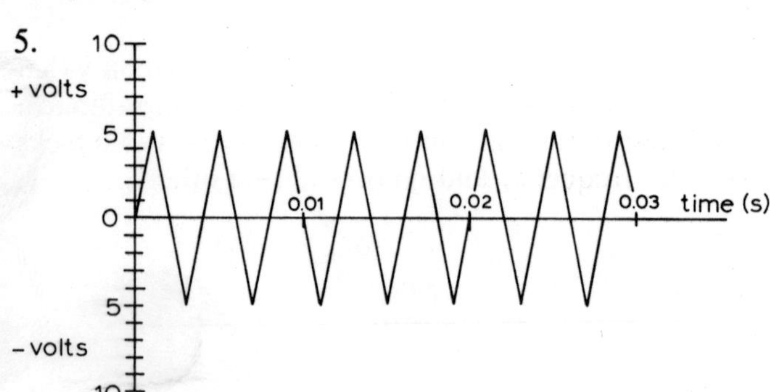

Period s
Frequency Hz

6.

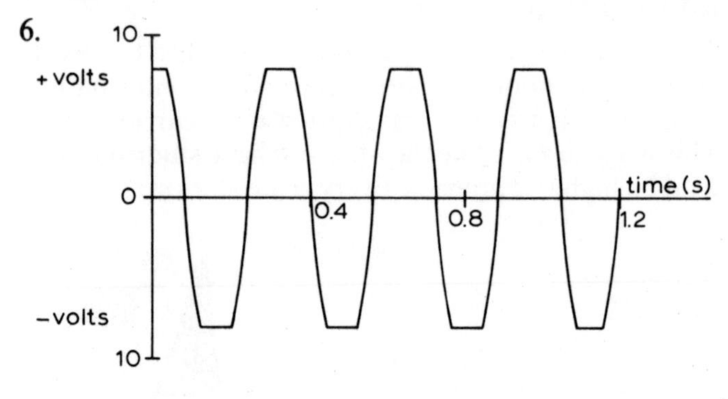

Period s
Frequency Hz

Root Mean Square Values

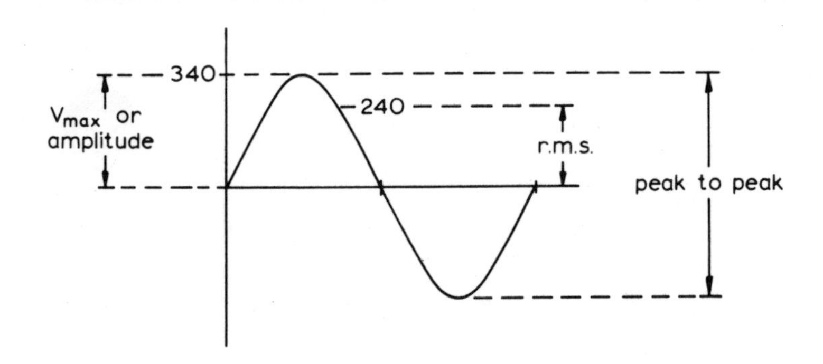

Graph 1

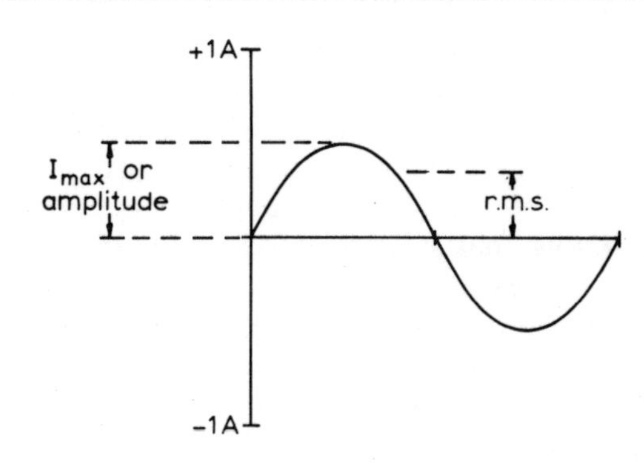

Graph 2

Graph 1 shows one cycle of our mains supply voltage and graph 2 shows the current that flows when the supply is connected to a light bulb of resistance 680 Ω. Some of the terms used on the graphs will have an obvious meaning: the maximum current and voltage are often abbreviated to I_{max} and V_{max}. The peak-to-peak voltage is measured from the maximum positive value to the maximum negative value and will be twice V_{max}.

Amplitude is the term commonly used in talking about waveforms in general, not just electrical ones, and means the maximum displacement of a wave from its zero position—the same as V_{max} and I_{max} in our graphs.

You know how to calculate the current flowing in a d.c. circuit given the voltage and the resistance. However, it is another matter with an a.c. supply, since the voltage is continually changing. What you would like to know is the effective value of the supply voltage—a single value for your calculations to replace those countless changing values. Fortunately, there is such a value.

> The effective value of an a.c. waveform is its root mean square (r.m.s.) value.

Note:

> A.C. meters always give r.m.s. values.

> A.C. supplies are quoted as r.m.s. values.

For sine waves

> R.M.S. value = 0·7 × maximum value.

> Maximum value = 1·4 × r.m.s. value.

(The numbers 0·7 and 1·4 are only approximate values for $1/\sqrt{2}$ and $\sqrt{2}$, but they will do for our purposes.) What does this all mean? Simply, the r.m.s. value of a supply in a circuit would have the same effect as the supply itself has. The r.m.s. value is the single value which you can use in your calculations to work out the effect of an a.c. supply in a circuit.

Alternating Current— Problems

All the supplies on this page are of sine wave type.

1. What is the peak value of the following a.c. supplies?
   (*a*) 240 V (*b*) 12 V (*c*) 10 V
   (*d*) 110 V (*e*) 50 V
2. The following voltages were found stamped on the casing of a set of components. They are the maximum peak voltages they will tolerate without damage. What value of r.m.s. voltage could safely be used with them?
   (*a*) 350 V (*b*) 24 V (*c*) 1000 V
   (*d*) 30 V (*e*) 6 V
3. What will an a.c. ammeter read when the following a.c. supplies (r.m.s.) are connected to a circuit of total resistance 10 Ω?
   (*a*) 12 V (*b*) 24 V (*c*) 10 V
   (*d*) 16 V (*e*) 30 V
4. An old transformer has its outputs quoted as peak-to-peak values. Find the maximum voltage then the r.m.s. value for each output.
   (*a*) 6 V
   (*b*) 9 V
   (*c*) 12 V

The Nature of Sound

In this course you will spend quite a lot of time building equipment to handle sound so it is useful to know something about the nature of sound.

Sound has to have a substance through which to travel and this is, of course, normally air, although you may know that whales are thought to 'talk' to each other underwater and ships detect submarines by bouncing sound waves off them and listening for the echoes. You may even have used the old car mechanic's trick of locating a fault in a car engine by using a piece of wood as a stethoscope.

The substance through which the sound travels, be it air, water, wood, is called the *medium*. Sound will not travel without a medium.

Sound is movement

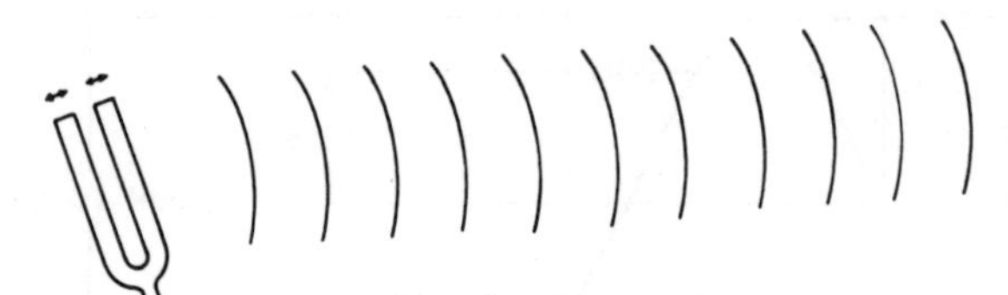
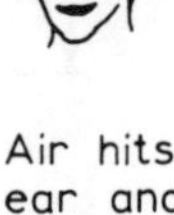

To make a sound, something must move. To carry a sound, a medium must vibrate. To detect a sound, something must receive the vibration and move in time with it.

Alternating Current from Sound

Electronic systems can be connected to the world of sound by two transducers: the microphone and the loudspeaker. These convert sound into alternating current and back again. The important thing is that these transducers make exact copies of what they receive (or, at least, they try to). If the vibration of a tuning fork has a pattern like a sine wave and a frequency of 256 Hz, then the air it hits will vibrate in a similar way with $f = 256$ Hz and any microphone this reaches will produce alternating current in sine wave-form with $f = 256$ Hz. (It is hard to picture a tuning fork 'vibrating with a sine wave pattern' but your teacher may be able to help you.)

The diagram shows a typical electronics set-up. The electronic circuitry could well be a public address amplifier. This takes in alternating current from the microphone, makes a magnified copy and sends it to a loudspeaker so that the listener hears the speech more loudly.

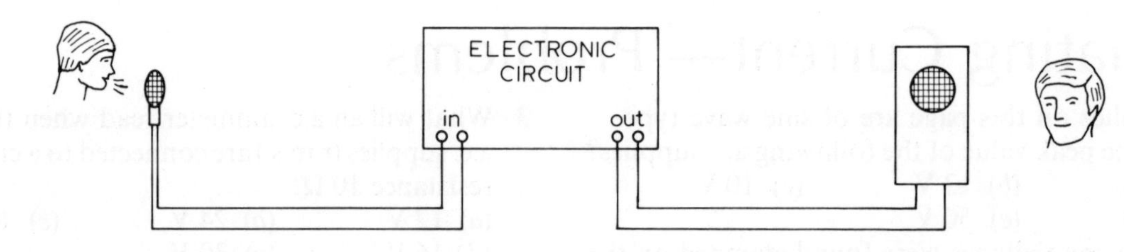

Looking at Sound

Using two instruments, a microphone and an oscilloscope, we can look at how any sound makes the air vibrate. You know that a microphone changes the air vibrations into an exact copy in alternating current so let's look at the oscilloscope.

An oscilloscope takes in electricity and draws a voltage-time graph of it.

The oscilloscope continuously measures the voltage of the incoming signal and displays the readings as a graph with the voltage measured vertically and time measured horizontally, just like the graphs on the last few pages. A spot of light takes the place of your pencil point. There are two problems with this:

1. The spot soon reaches the other side of the screen (you run out of 'graph paper') so it has to start again at the beginning.
2. The spot may move too quickly for you to see so you have to rely on two things if you are to see a graph:
   (*a*) the spot leaves a faint trace behind it;
   (*b*) your brain tends to take a sort of photograph of the path.

Now this could mean that you just see a jumble of graphs. The oscilloscope (with a little help from the operator) overcomes this problem in a very ingenious way: if the incoming signal is always the same shape, then the 'graph' it draws on each 'sweep' is put in exactly the same place as the one before. The result is that you see what you think is a stationary graph whereas what you are actually seeing is a spot covering the same ground, over and over again, at high speed. If your teacher alters the speed at which the spot travels you will easily see what is happening.

Looking at Waveforms

Now draw the 'graphs' for the following sound sources, as your teacher demonstrates them.

Examine the waveforms of any other sound sources you have, such as musical instruments and the rain warning device from Chapter 1.

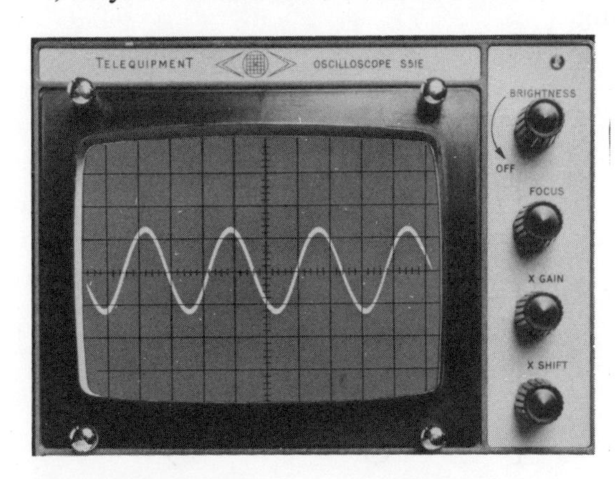

Tuning fork—low note

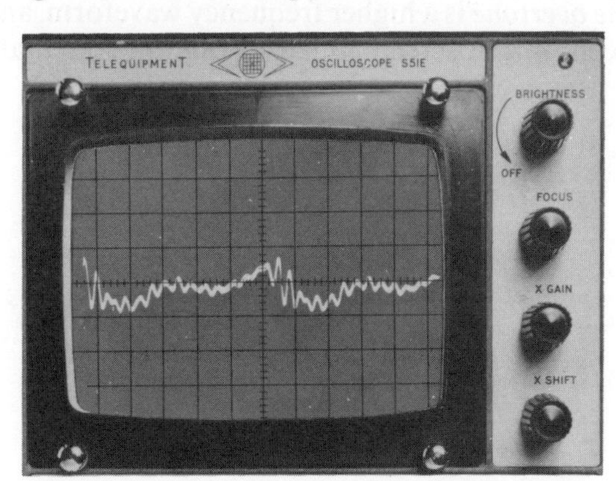

Piano note

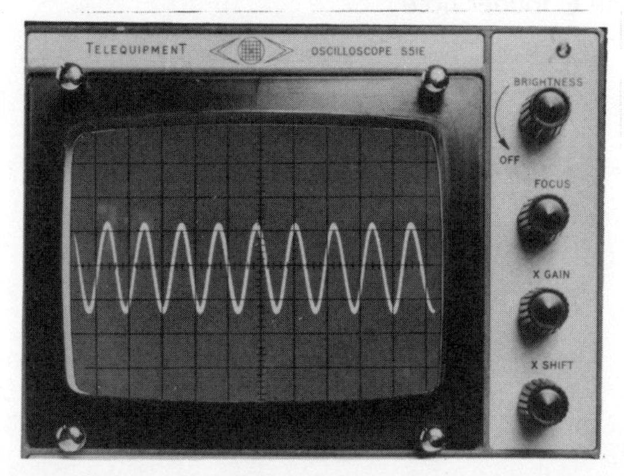

Tuning fork—higher note

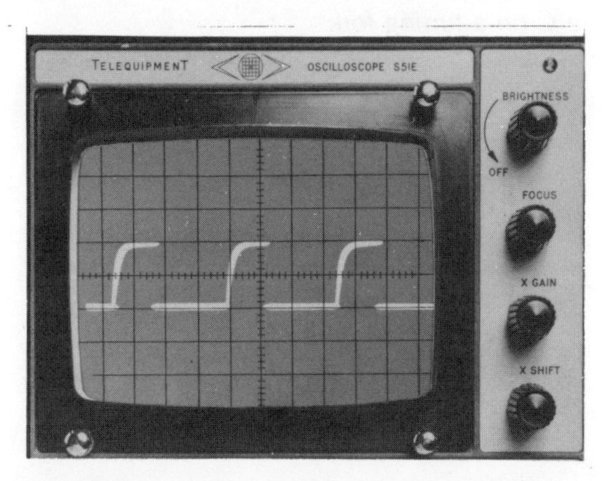

Signal generator

Musical Instruments

You have examined a selection of sounds and your oscilloscope has shown you at least four things:

1. A single note has a single waveform with a constant frequency.
2. The pitch of a note is fixed by the frequency of vibration:
   high frequencies are high notes
   low frequencies are low notes.
3. The loudness of a note depends on the amplitude of the vibration,
4. The shape of the waveform fixes the quality of the note.

Timbre or Quality

You have seen that when two instruments (or the voice) play exactly the same note, you can still tell the difference between the two sounds. This is because each instrument has its own quality or *timbre.*

Instruments do not produce pure sine-wave notes like the tuning fork, but *mixtures* of notes. Every note from a musical instrument contains two kinds of waveform.

The fundamental is the main waveform and this fixes the pitch. You can measure its frequency on the oscilloscope.

The overtone is a higher frequency waveform, and can be seen on the screen added to the fundamental.

For example, middle C on the piano has a main waveform of 262 Hz and a selection of higher notes added to this, but the effect is still that of a note of 262 Hz.

When you play middle C on, say, a trumpet you produce the same fundamental but a different selection of overtones and at different amplitudes. In fact, the 'brillance' of a trumpet is a result of the larger proportion of higher frequencies which it contains. These higher frequencies are also called *harmonics.*
In the displays shown the oscilloscope is on the same setting.

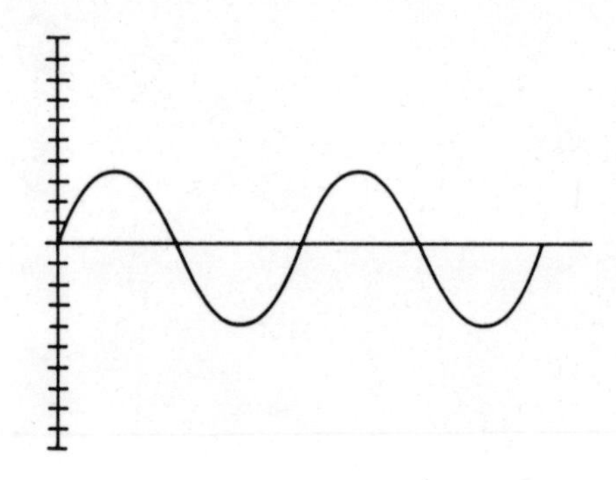

Middle C on a tuning fork

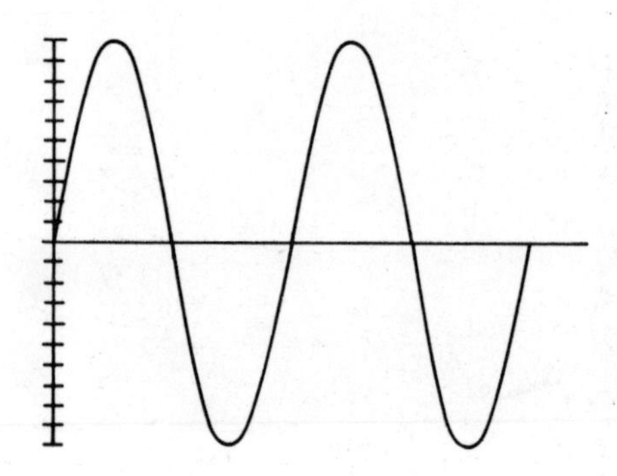

Middle C louder note

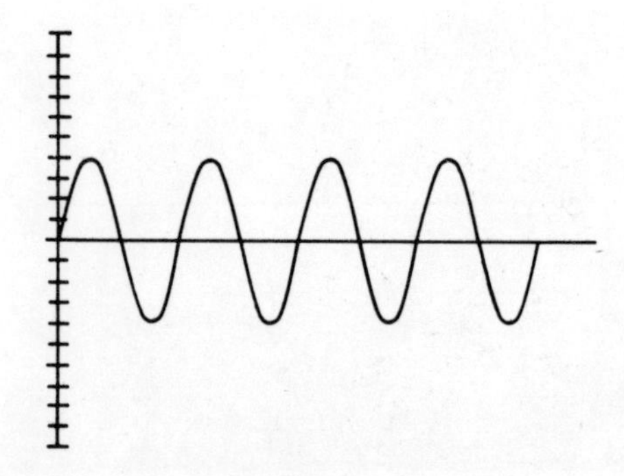

C above middle C on a tuning fork

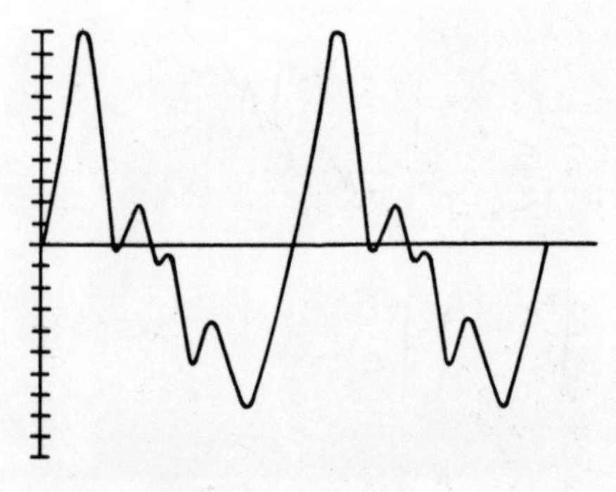

Middle C on a clarinet

Alternating Current—Summary

Changing Value

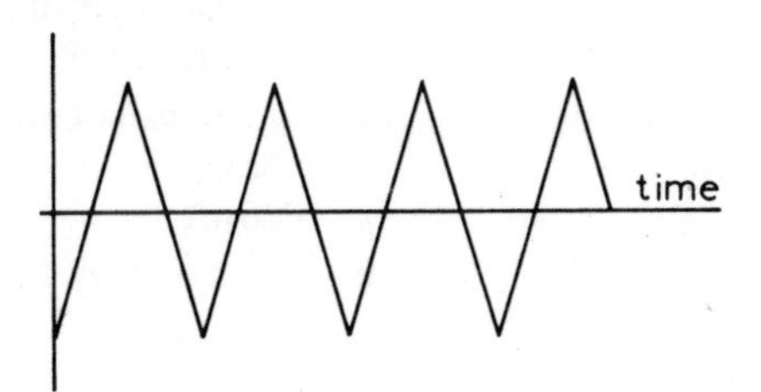

Alternating current has a continually changing value and changes in direction at regular intervals.

Sine Wave

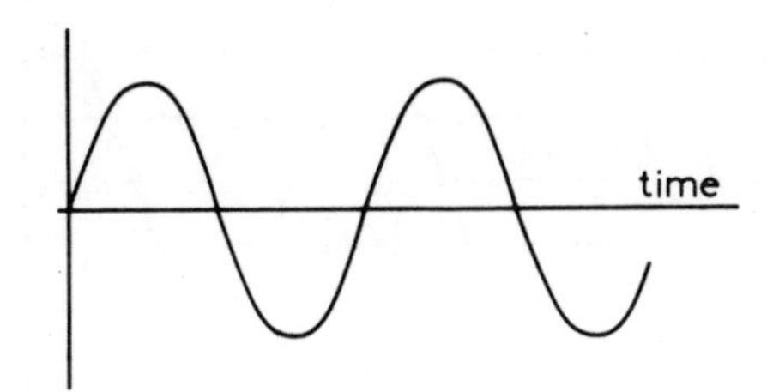

A waveform commonly found is the sine wave.

One Cycle

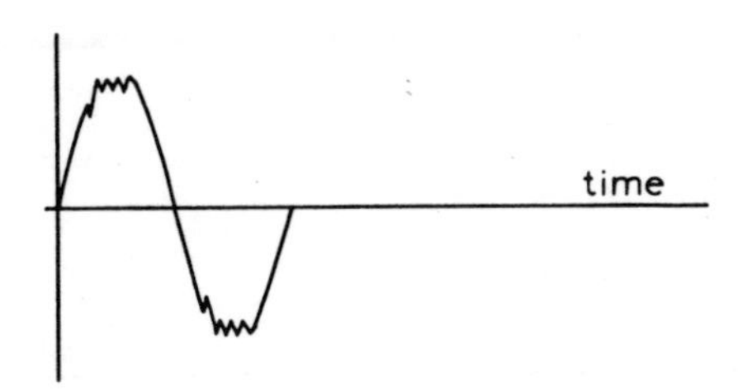

One cycle is a complete set of positive and negative values.

Waveform

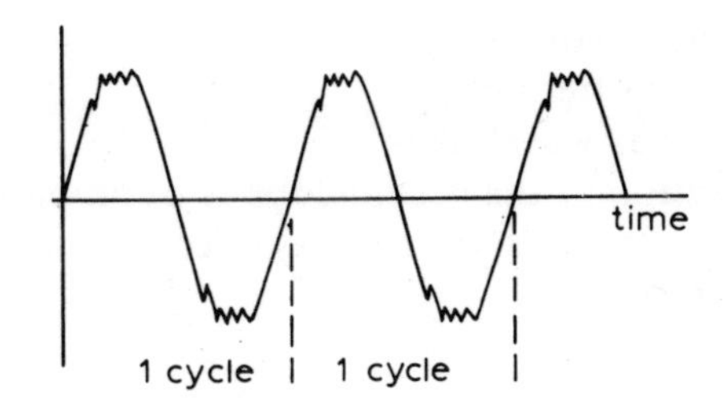

Cycles repeated show the complete waveform.

Root Mean Square Value

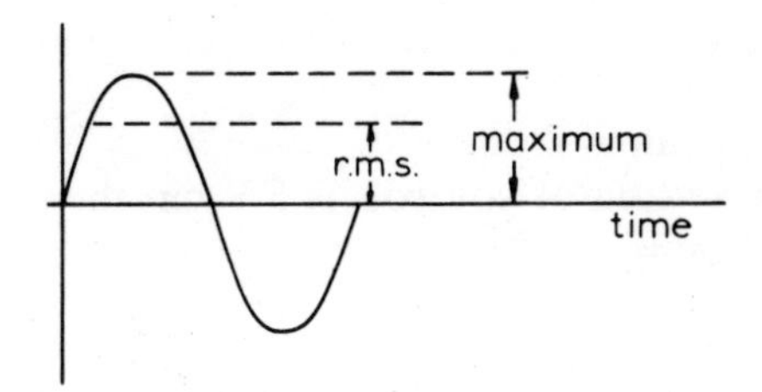

The r.m.s. (effective) value is the constant value that would produce the same power. For a sine wave r.m.s. value = 0·7 × maximum value.

Frequency

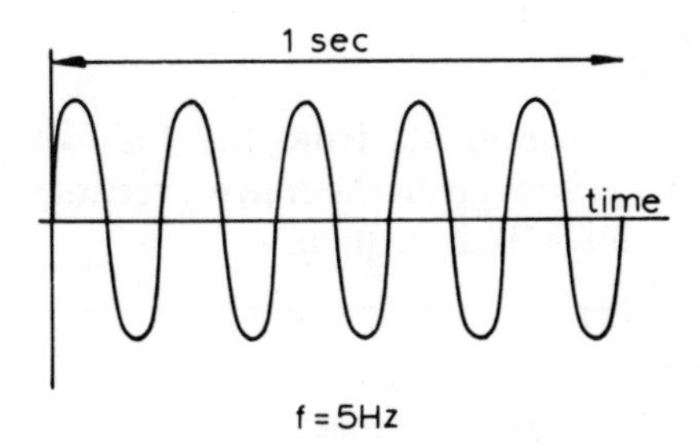

Frequency (f) is the number of cycles in one second.

The Oscilloscope

The oscilloscope is a very important instrument with three main uses:
1. Measuring voltage
2. Measuring frequency
3. Looking at waveforms.

It may seem that the oscilloscope is a complex instrument to use but most of the array of control knobs have a simple job to do.

Ask your teacher to set up the trace on an oscilloscope and connect a torch cell to the input terminals. Move the a.c./d.c. switch to a.c. and find and adjust the following controls until you see the effect of each.

1. Brightness.
2. Focus.
3. X-shift.
4. Y-shift
5. Time-base (or time/cm).
6. Y-sensitivity (or volts/cm).
7. Velocity.

Measuring Voltage and Frequency

The method of measuring these two quantities is shown in the examples below.

Example 1

Height of peak = 2 cm
If voltage setting of controls is 5 V/cm then
Peak voltage = 10 V.

Example 2

Length of one cycle = 5 cm
If time base setting is 1ms/cm then time for one cycle is 5 ms (0·005 seconds)

$$\text{Frequency} = \frac{1}{\text{Time for one cycle}} = \frac{1}{0{\cdot}005}$$

Hence frequency = 200 Hz.

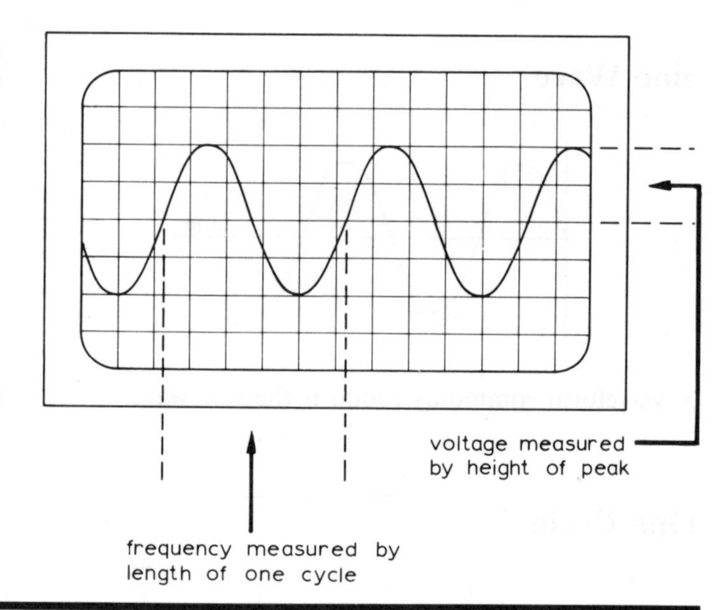

Instructions

Take the measurements from the following oscilloscope displays and, given the control settings, work out the peak voltage and frequency.

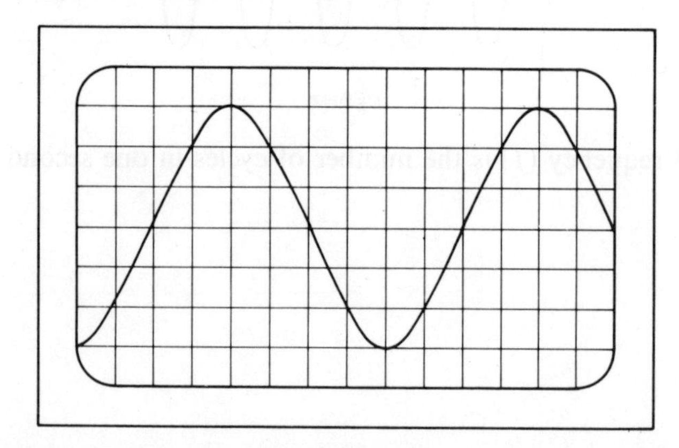

Voltage setting = 2 V/cm
Time-base setting = 5 ms/cm

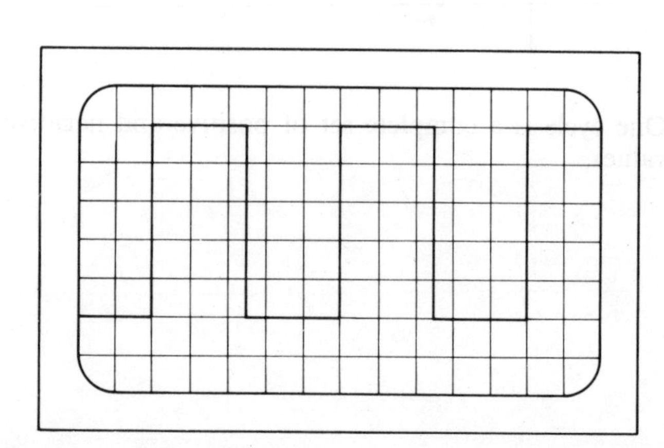

Voltage setting = 1 V/cm
Time-base setting = 2 ms/cm

Using the Oscilloscope

To acquaint yourself with the oscilloscope, you are going to use it to take measurements on some supplies. Try to set up the oscilloscope so that one cycle occupies as much of the screen as possible, in the case of the a.c. supplies.

Voltage Measurement

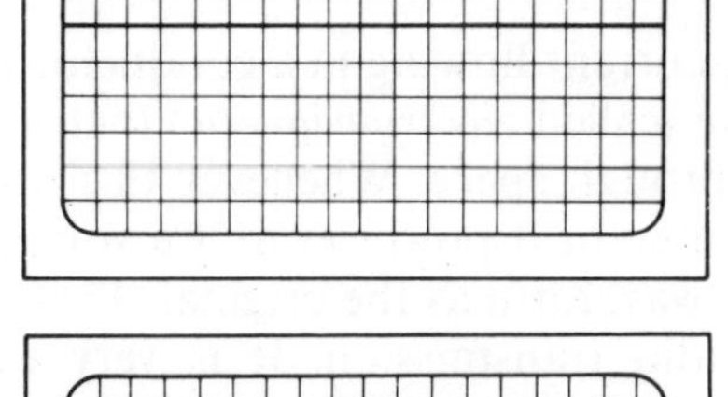

Torch Cell

Voltage setting —— V/cm
Height of trace —— cm
Voltage —— V

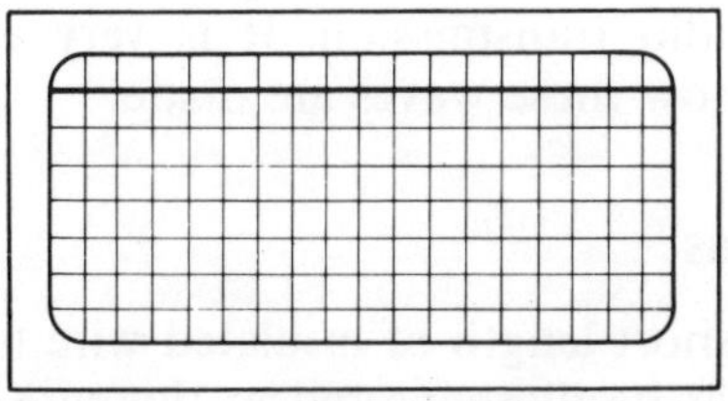

PP6 Battery

Voltage setting —— V/cm
Height of trace —— cm
Voltage —— V

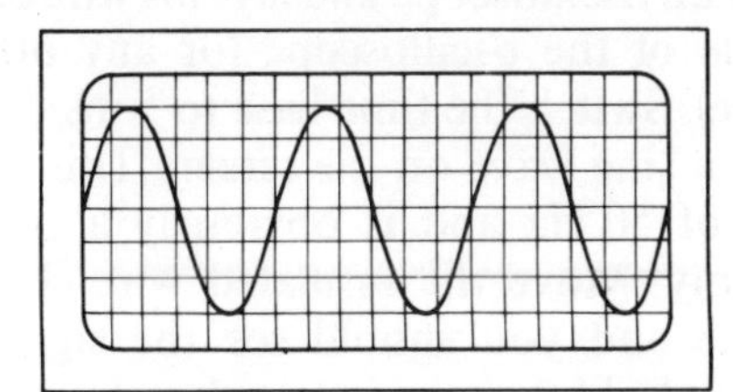

Sine Wave Signal Generator

Set the signal generator to its maximum output.
Voltage setting (oscilloscope) —— V/cm
Height of peak —— cm
Peak voltage —— V
Does your oscilloscope reading correspond with the stated output of the generator?

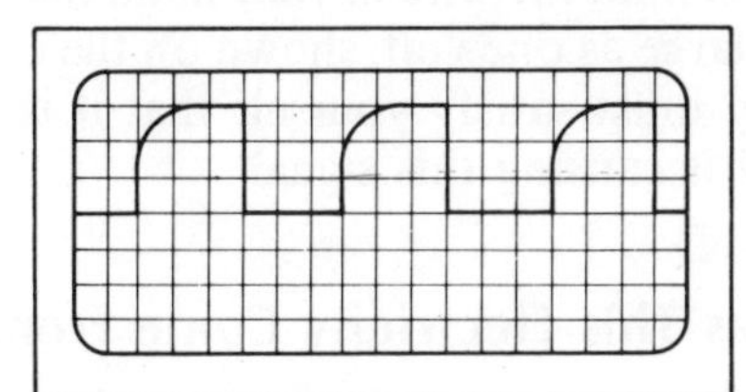

Project Signal Generator (See page 41)

Voltage setting —— V/cm
Height of peak —— cm
Peak voltage —— V

Frequency Measurement

Check that your time base is set to calibration. Set the generator on 1 kHz, or, better still, ask your partner to set the frequency and then cover up the dial.

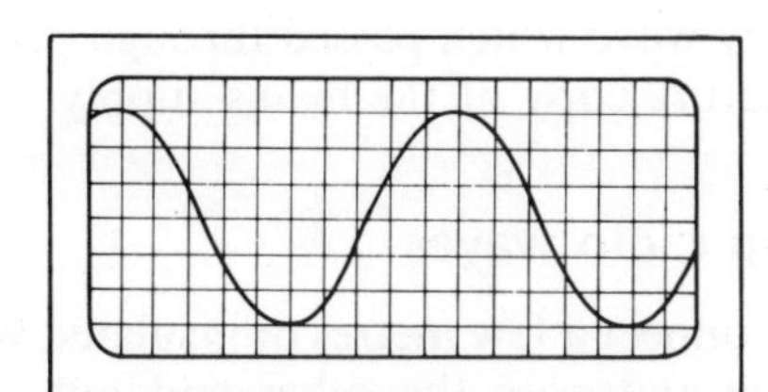

Signal Generator

Time setting —— ms/cm
Length of one cycle —— cm
Time for one cycle —— ms
Frequency —— Hz

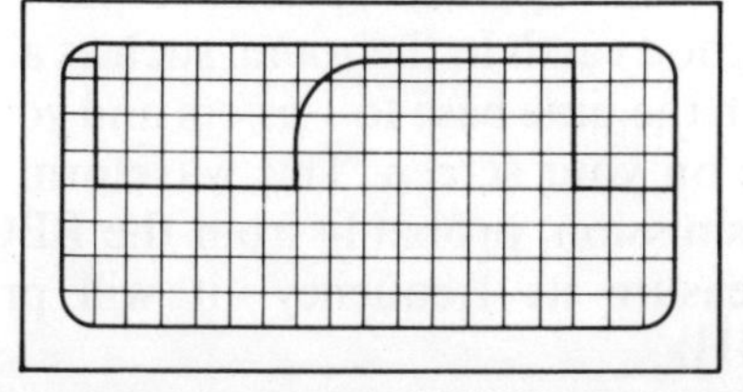

Project Signal Generator

Time setting —— ms/cm
Length of one cycle —— cm
Time for one cycle —— ms
Frequency —— Hz

Radio Transmission

This is a good opportunity to introduce radio transmission, because it depends on the use of sine wave signals, like those you have met recently. There are two important stages in radio transmission: the production of radio waves and modulation, and we shall look at the first of these now.

Electromagnetic Waves

Alternating current flowing in a conductor gives out a type of wave (called *electromagnetic*) that is capable of travelling through space. Whenever this wave crosses another conductor it produces in it a voltage with the same shape waveform as the original. These waves are used for radio transmission. It is very easy to demonstrate how these waves are made.

Instructions

Connect a short length of insulated wire to the input terminal of an oscilloscope and lay this wire close to the mains cable of the oscilloscope (or any other mains cable in use). Switch the time base to 5 ms/cm and you should see a sine wave on the screen. This wave has a frequency of 50 Hz and is obviously a copy of our mains supply. Move the insulated wire closer to the mains cable and you should see the signal become larger. If you hold the wire in your hand the signal may become as large as one volt, shown on the oscilloscope screen. You must satisfy yourself that it is the mains supply that is causing this signal.

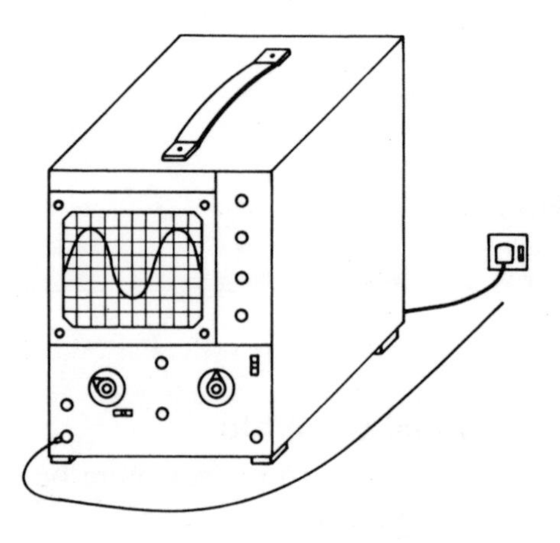

Waves given out by mains supply

Where Has This Electricity Come From?

Remember that the wire into the oscilloscope was connected to nothing else. Anything produced in the wire had to travel through the air to get there. The cable carrying the mains electricity was in fact behaving like a transmitter. The alternating current produces electromagnetic waves which move outwards through the air and the insulated wire you used acted like an aerial. Every wave which passed through produced an exact miniature copy of the mains supply.

Picking up radio waves

This time connect a few metres of insulated wire to the oscilloscope and pass the other end out through a window. Use another wire to connect the oscilloscope earth to a good earth in the room, such as a water pipe. Now switch the time base to 1 μs/cm and you will again see a wave on your screen. This waveform is in fact a radio transmission, probably from the BBC, and you should measure its frequency—it will probably be about 1 MHz.

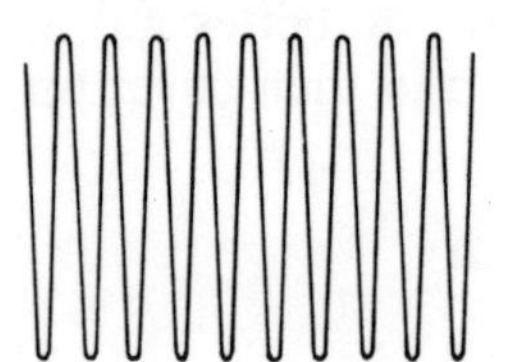

Radio waves received

Continued on p. 57

Do the experiments on the last page mean that we now hold the key to radio transmission? Unfortunately, this is not yet the whole story. Imagine that we wish to transmit the signal given by the signal generator, built earlier. In theory, we only need a cable carrying the signal and it will transmit the waves we need, sending them out into space, as demonstrated. There are still two drawbacks to this system:

Waves at the frequencies at which we can hear fade very rapidly in the air.

An aerial will pick up all the waves in the air and not be able to separate them.

Modulation

These problems are overcome by the use of a *carrier wave*. This is a very high frequency wave which will travel great distances. Every radio station is given a frequency for its own use and no other station may use it. For example, Radio 1 has the frequencies of 1 053 kHz and 1 089 kHz. To transmit a radio programme, the BBC then *modulates* this wave. The diagrams show how they would transmit the simple note from our signal generator.

What Happens Next?

When this waveform arrives at a radio receiver, the radio just has to pick out one frequency, say 1 053 kHz, from any others present. This means the radio can choose Radio 1 and ignore all the others. Most important though, this one frequency can be shaped (modulated) to carry any other signals or frequencies. Any wave pattern at all, even that produced by a whole symphony orchestra can be copied using this shaped carrier wave.

Advantages of Modulation

Radio waves can be sent great distances using high frequency waves.

The carrier wave can be modulated to give any other waveforms.

A radio receiver, by selecting just one frequency, can pick up everything transmitted by one station.

(a)

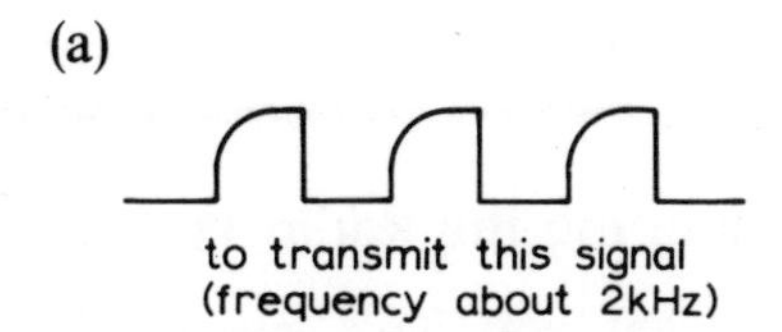

to transmit this signal
(frequency about 2kHz)

(b)

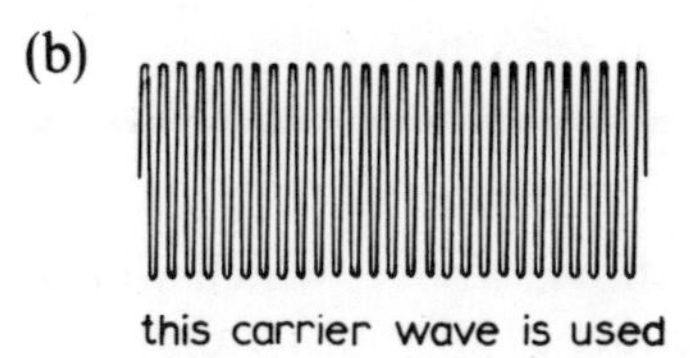

this carrier wave is used

(c)

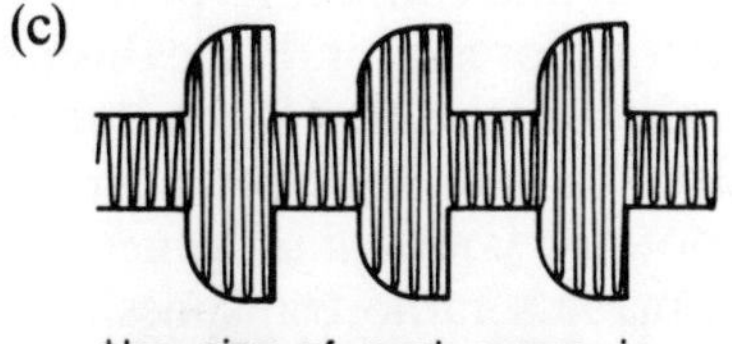

the size of each wave is changed to give the correct shape

(d)

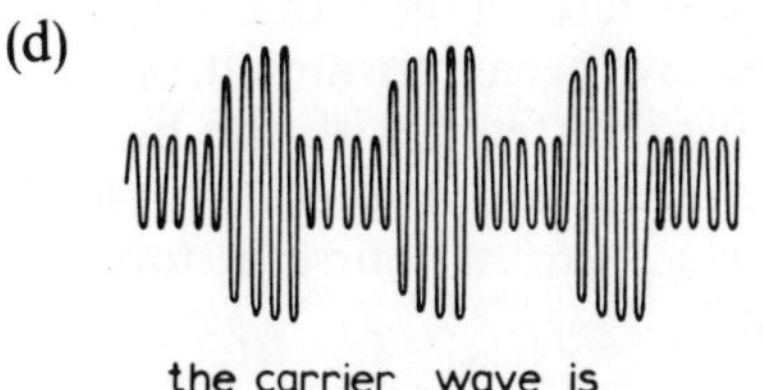

the carrier wave is transmitted like this

A Simple Radio Transmitter

Your experiments on this page will enable you to put into practice the theory of the previous page—you are to make a simple radio broadcast. This will occur in two stages. First you will produce radio waves using a radio frequency (r.f.) oscillator and then you will modulate these, using your own signal generator.

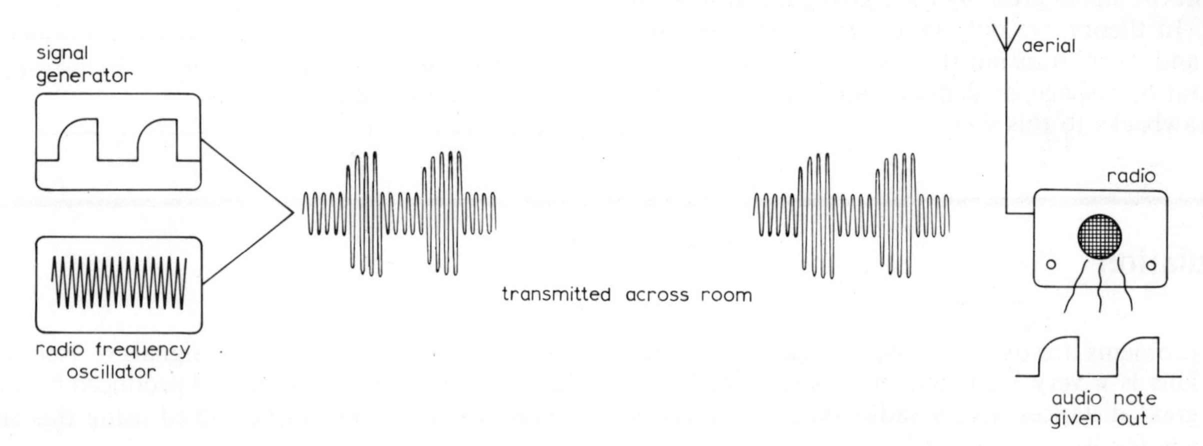

Transmission and reception

Producing Radio Waves

Construction

The circuit for your r.f. oscillator is shown. It is called a *Hartley oscillator*. The numbers for assembly on S-Dec are also shown. Take care with the construction of the coils. They should be close wound and L_1 and L_2 should not be more than 3 mm apart. Make sure they are wound in the direction shown and remember to remove the insulation where connections are needed.

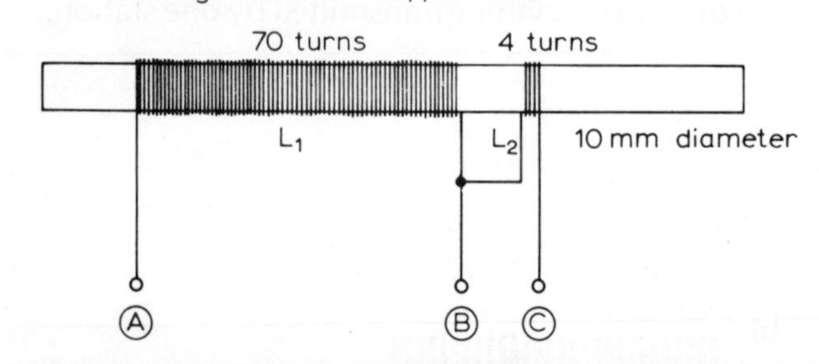

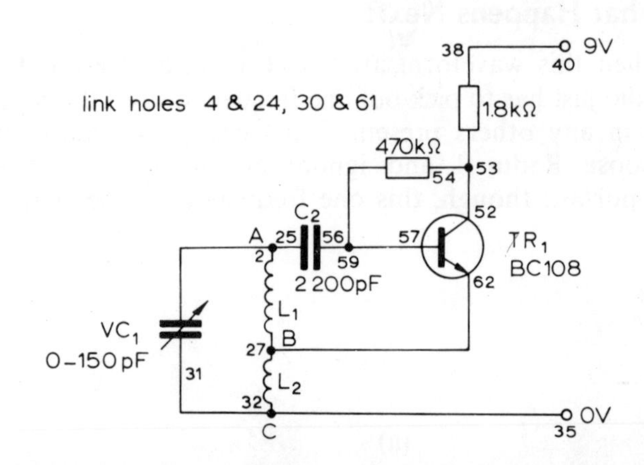

Testing

Assemble the oscillator shown above, which will produce radio waves of frequency of many hundreds of kilohertz. Place this near to a radio tuned to a station on the medium wave band and adjust VC_1.

You will find that, for one setting of VC_1, you will hear a high pitched whistling on the radio. This means that your radio waves have been transmitted, picked up by the radio and have interfered with the broadcast station. It is this interference which you heard, not your actual radio waves, which are, of course, at too high a frequency to be heard.

Modulating the Radio Waves

Now that you have successfully transmitted radio waves, you can modulate these to transmit the note of your project signal generator.

Disconnect the battery from your project signal generator and connect its power rails to the power rails of your r.f. oscillator. Place this near to your radio and tune the radio so that it is not receiving a station. Now adjust VC_1 and, for one setting of VC_1, you will hear the note of your signal generator on the radio. You have made a radio transmission!

4 CAPACITORS
Introduction

Function

The function of a capacitor in a circuit is to *store electricity*. It is, in effect, an electrical bucket that is filled and emptied by the rest of the circuit.

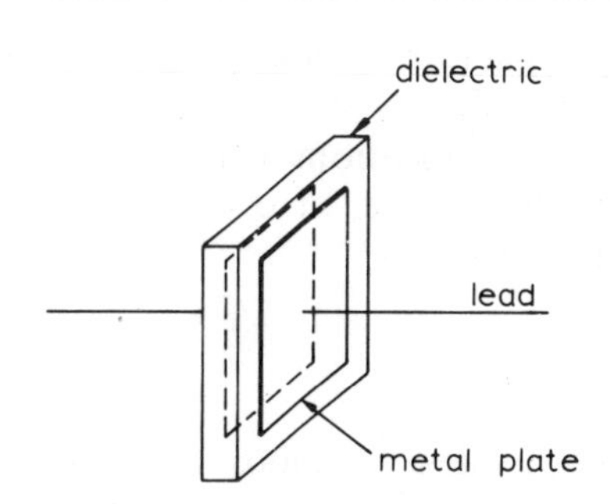

Structure

Structure

A capacitor is made of two parallel metal plates, separated by an insulator called a *dielectric*. The external leads are attached to the two metal plates.

Operation

A flow of electrons arriving at one plate is stored there and these electrons force away an equal number of electrons from the opposite plate.

For a perfect capacitor, no electricity flows across the gap.

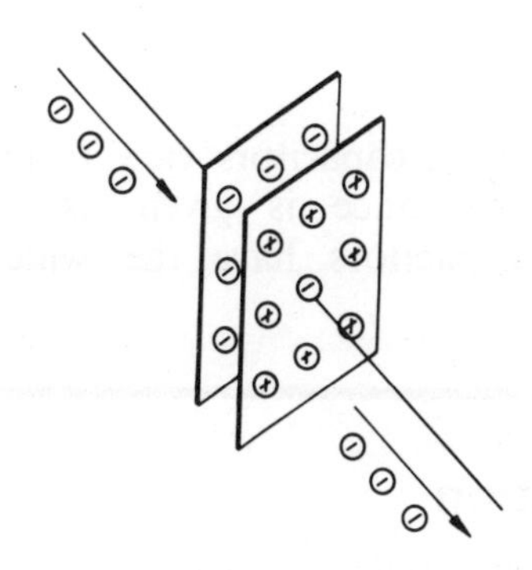
Operation

Capacitance

The important property of a capacitor is how much electricity it will store for each volt trying to push the electricity in. This is its capacitance.

The unit of capacitance is the farad.

A two farad capacitor will store twice as much electricity as a one farad capacitor, if the same voltage is supplying electricity to both of them. However, the farad is a very large unit. You will certainly not meet a capacitor as large even as one farad during this course and it is likely that the largest one you will see will be about one-hundredth of a farad. Obviously then you will be using the prefixes you learnt earlier: the microfarad (μF) and the picofarad (pF) are the units you will need.

Construction

On the next page you will see some of the faults and limitations of the capacitors you will use. There are many different methods of construction and kinds of material used in the manufacture of capacitors, but there is no such thing as the perfect capacitor. One type of construction may nearly eliminate one fault but leave another glaring fault in the finished product. You have to choose the capacitor that most suits the job you have in mind.

Capacitor Specifications

When using capacitors, you will normally be told by the circuit designer what type of capacitor to use. Here are the capacitor specifications which he has to bear in mind, along with price, of course, when choosing a capacitor for a job.

Capacitance

Obviously, you have to choose a capacitor with the right electrical size for the job. If you need a large value capacitor, say above 25 μF, then you will probably use an electrolytic type.

Tolerance

As with resistors, capacitors vary from their stated value. This tolerance is given as a percentage. Electrolytic capacitors have the widest variation. If you need a capacitor with an accurately known value then a silver mica one will probably suit.

Working Voltage

Working voltage is, as it sounds, the maximum voltage the capacitor should operate at. This figure is usually the one you find stamped on the case, immediately after the capacitance value. Some ranges, such as the Mullard C 280 series, have the same working voltage for all capacitors in the range so the value is not stamped on.

Leakage Current

The perfect capacitor does not leak. However, there is always some current passing through the dielectric. The silver mica capacitor probably gives the lowest leakage current whilst the electrolytic has the highest.

Identification

Manufacturers' catalogues will tell you how the different capacitors perform in the above respects and you should consult these for information.

The photographs show those capacitors you are most likely to meet in this course and as a constructor.

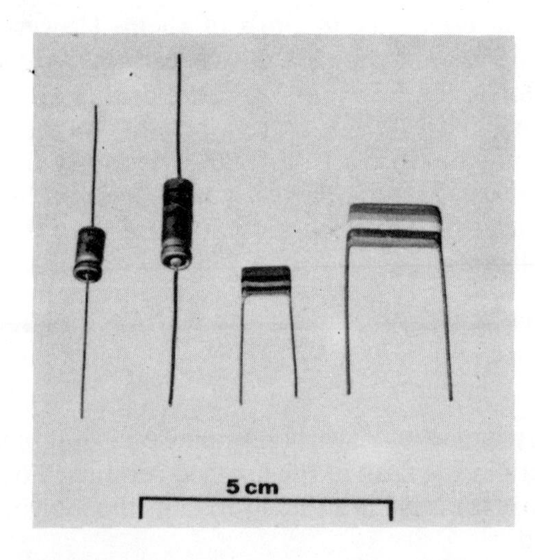

Mullard C280 series (right) and Mullard miniature electrolytic (left)

Trimmers are variable capacitors which give finer adjustment of capacitance than those shown in Chapter 1.

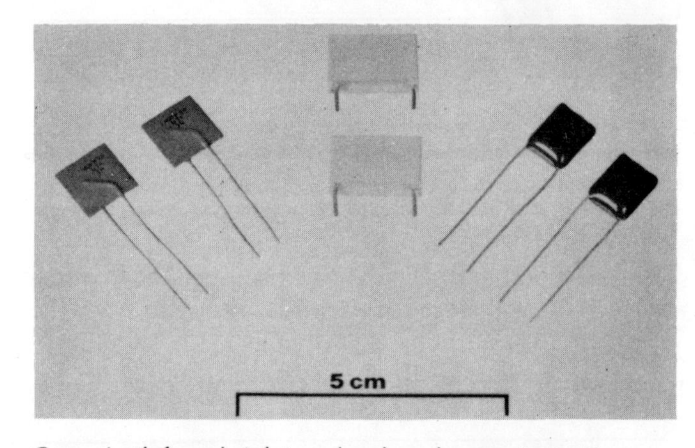

Ceramic (left and right) and polycarbonate (centre)

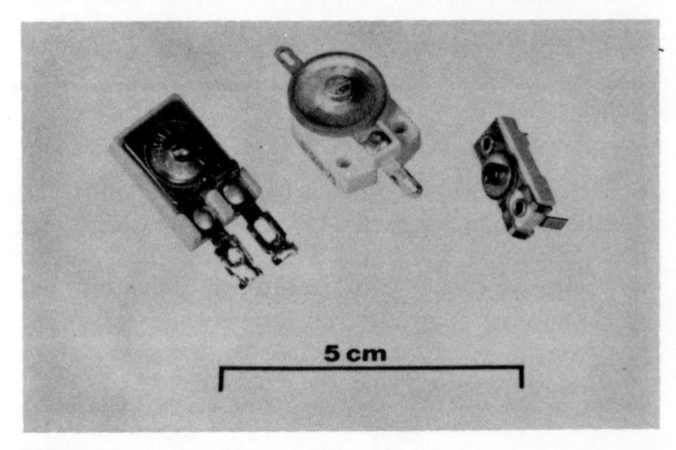

Trimmers

The Capacitor in Action

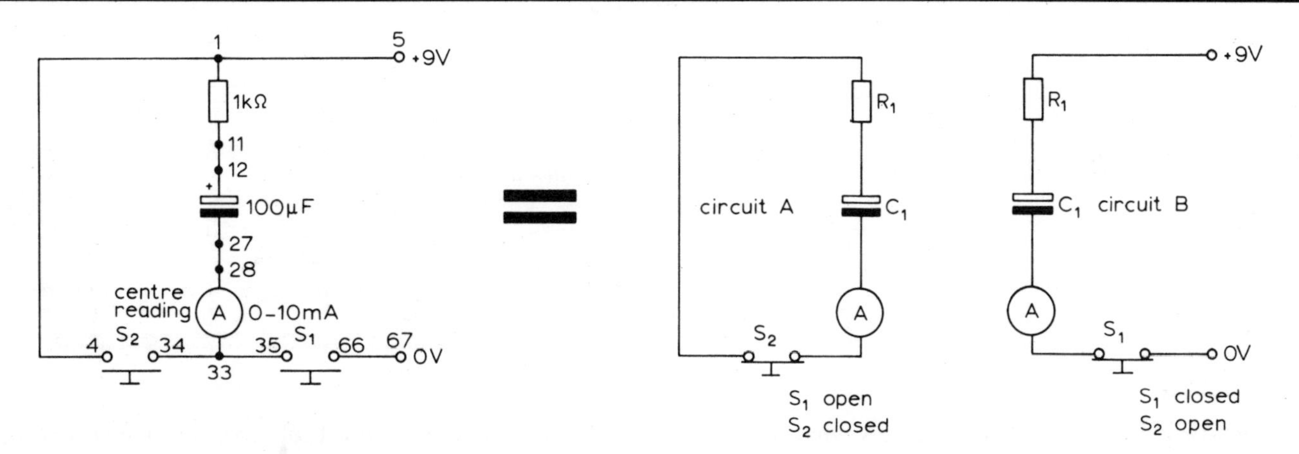

These circuits show you how a capacitor charges and discharges in a simple d.c. circuit.

When switches S_1 and S_2 are pressed independently, two part circuits are formed, labelled A and B. When switch S_1 is pressed, it completes circuit B and when switch S_2 is pressed it completes circuit A. You should be sure you understand that circuit A contains no battery and why you must not press S_1 and S_2 at the same time.

Instructions

Assemble the above circuit, using the S-Dec numbers. Press switch S_1 and watch the reading on the ammeter. We say that the capacitor is charging. Now release S_1 and press S_2. Now the capacitor is discharging. You can now replace the meter and resistor with a bulb, as shown below, and repeat the experiment. The bulb will show you the charge and discharge processes again. Find out what happens for different values of capacitor. Use capacitors of values 100 μF, 500 μF, 1000 μF, 5000 μF, 10000 μF.

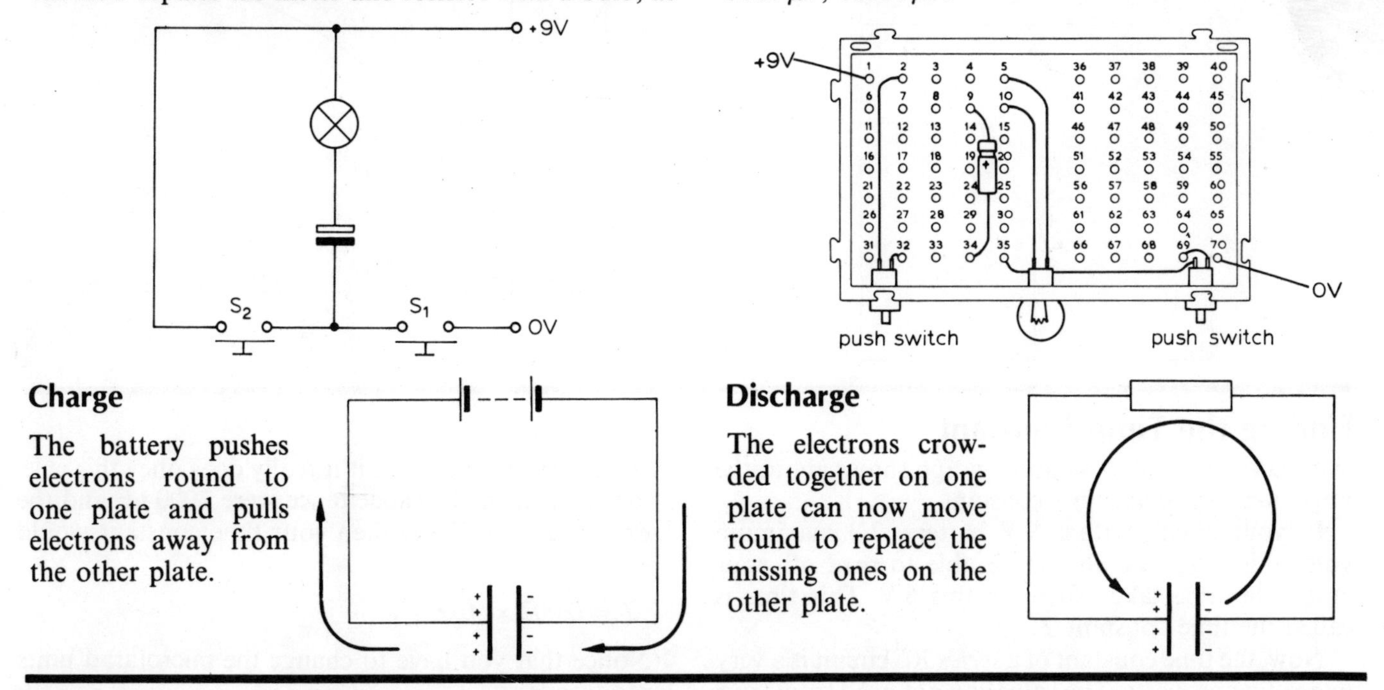

Charge

The battery pushes electrons round to one plate and pulls electrons away from the other plate.

Discharge

The electrons crowded together on one plate can now move round to replace the missing ones on the other plate.

Results

The largest current flows at the beginning. The largest current flows with the largest capacitor.
A capacitor stores electricity then releases it. A capacitor stops direct current flowing when it is fully charged.
The charge and discharge currents flow in opposite directions.

Time Constant

How Quickly does a Capacitor Charge Up?

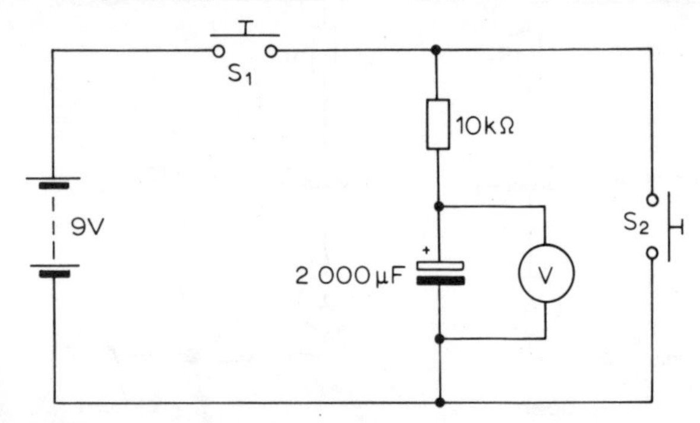

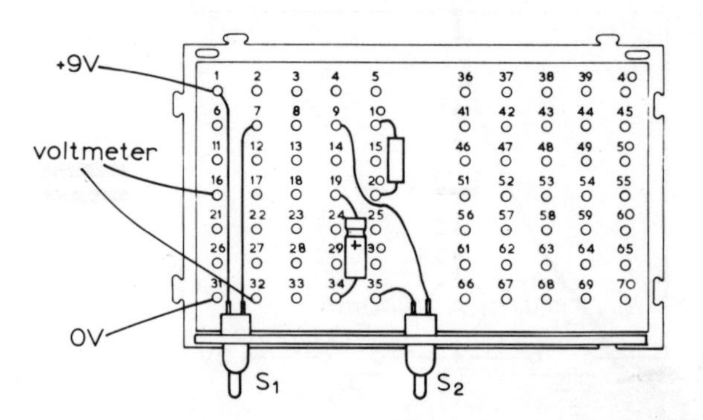

You are now going to find out accurately what happens when a capacitor charges up. You will use the resistor capacitor (*RC*) circuit above to find how the voltage across a capacitor in a circuit changes.

Instructions

Start the stop clock at the same time as closing the switch S_1. Watch the voltmeter and stop the clock when the reading reaches 1 V. Release the switch and discharge the capacitor, using switch S_2. Repeat the experiment until you obtain a consistent reading for the time. Now move on and find the time to reach 2 V, then 3 V and so on.

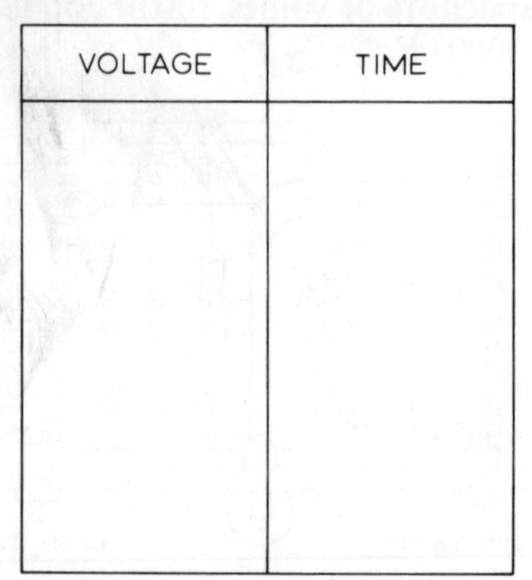

VOLTAGE	TIME

Record your time in a table like the one below and then plot a graph of voltage against time. Compare your graph with the one below. See page 172 for the problems of using a voltmeter in this circuit.

> The time constant is the time taken for the voltage to reach two-thirds of its maximum.

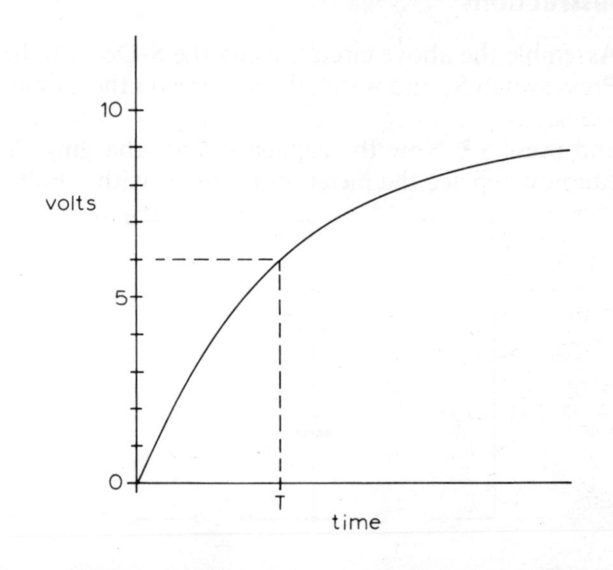

Finding the Time Constant

We shall now look at what the graph shows about the important property, time constant.

In your circuit with a 9 V battery, 2/3 maximum voltage is 6 V. You should be able to read off your graph the time taken to reach this 6 V. This time is called the time constant T.

Now, the time constant of a *series RC* circuit is a very interesting property. The value you obtained from your graph was obeying a rule and this applies to all series *RC* circuits:

$$T = C \times R$$

(T in seconds, C in farads, R in ohms).

Check your result to see if it really does obey this rule. For instance, if the capacitance were 2000 μF and the resistance were 10 kΩ then your time constant would be:

$$T = 0{\cdot}002 \times 10\,000$$

(Notice that you have to change the microfarad units into farads)

$$T = 20\,\text{s}.$$

(*Note.* If your result is a long way from your calculated value, you are probably suffering from this capacitor's major fault—its large tolerance.)

Checking Time Constants

Using the Equation

It was suggested to you on the previous page that the time taken for the voltage across a capacitor to reach two-thirds maximum was governed by the rule:

$$T = C \times R$$

You can now check that rule by finding T for a few different combinations of C and R.

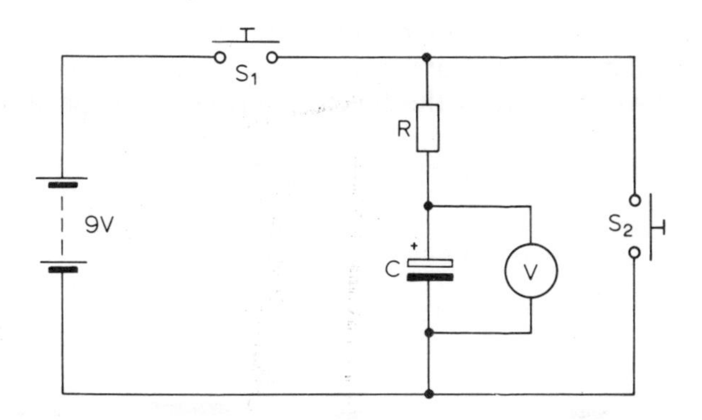

Instructions

To simplify the experiment do not plot a graph for every circuit. Choose a value for R and C and write these down in a table, like the one below. Put these two components in the circuit, close the switch and find the time taken to reach two-thirds of the maximum voltage, 6 V in this case. Discharge the capacitor and repeat till you are sure that you have an accurate value. Put this value in the table and work out what value the equation predicts. (Remember that you must use C in farads in the equation.) Now compare the value from the experiment and the value from your equation. Choose a different value for R and C and repeat the experiment. Again, you may encounter problems if you do not have accurate voltmeters to use.

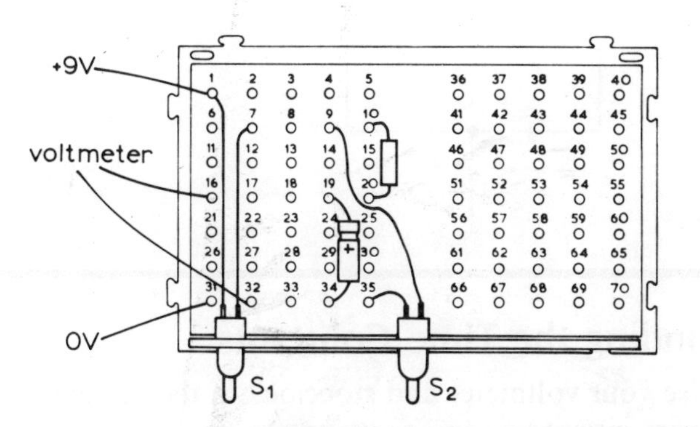

R(kΩ)	C(F)	C(μF)	T(s) by experiment	T(s) by calculation
1	0·001	1000		
47	0·000 1	100		
100	0·001	1000		
100	0·000 1	100		
100	0·000 047	47		
1·5	0·01	10 000		

The last two columns in your table should give similar results, thus proving that the equation is correct. However, electrolytic capacitors can have such a large tolerance (variation from their stated value) that your results may not be convincing. Unfortunately, there is nothing that can be done about this, since capacitors of this size have to be electrolytic.

Using the Time Constant

The series RC circuit you have been investigating is a very useful way of controlling the voltage at a point. For instance, your experiments have shown you:

1. The voltage at point A gradually rises from 0 to maximum.
2. You can choose how quickly you want point A to reach a particular voltage by choosing the value of R and C.

You will see this in action in your next constructional project.

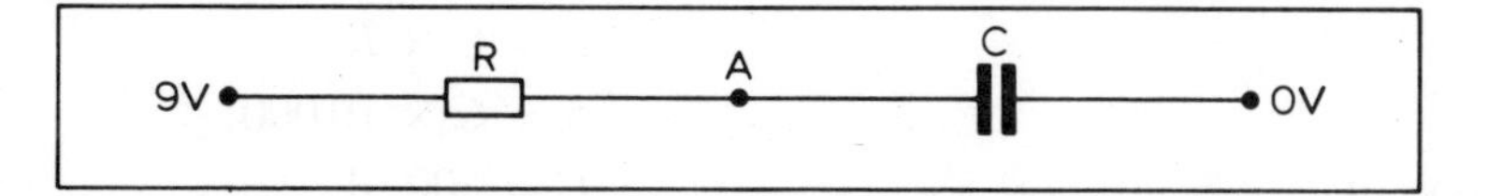

Series and Parallel

In a previous chapter you investigated the effect of putting two (or more) resistors in series and parallel. We will now study the effect of putting two capacitors in series and parallel. You want to know the size of a single capacitor which exactly replaces the combination of two capacitors.

1000μF 1000μF ?μF

Remember that you have two ways of finding out how large a capacitor is—by measuring how much current flows or by finding the time constant in an *RC* series circuit. We shall use the second one to find the capacitance of the above combinations.

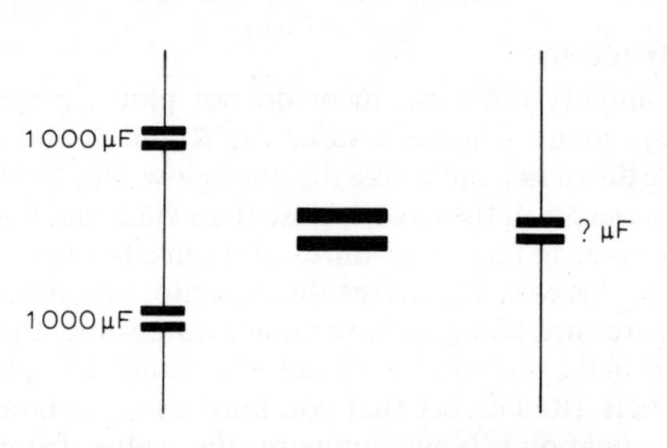

Finding the Time Constants

Use your voltmeter and stopclock in the same way as before to find the time constant, for both the parallel and series circuit.

Now calculate what *single* capacitor would have the same time constant if placed in this circuit. Typical calculations might be:

Parallel

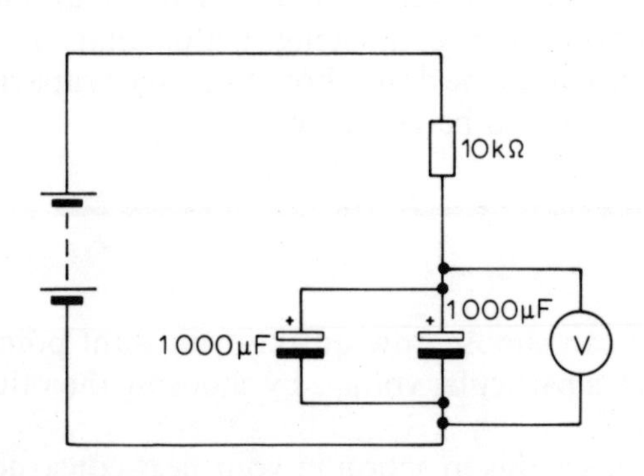

Time constant from experiment = 20 s
$R = 10\,000\,\Omega$

$$T = C \times R$$
$$20 = C \times 10\,000$$
$$C = 0{\cdot}002 \text{ F or } 2000\ \mu\text{F}$$

This means that:
Two 1000 μF capacitors in parallel have the same capacitance as one 2000 μF capacitor.

$$C_{total} = C_1 + C_2$$

Series

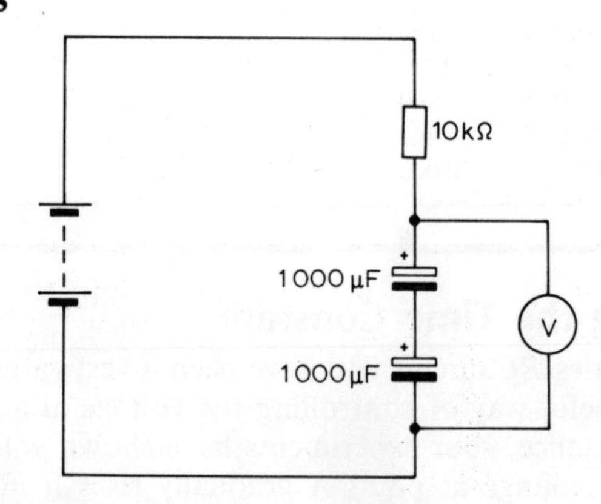

Time constant from experiment = 5 s
$R = 10\,000\,\Omega$

$$T = C \times R$$
$$5 = C \times 10\,000$$
$$C = 500\ \mu\text{F}$$

This means that:
Two 1000 μF capacitors in series have the same capacitance as one 500 μF capacitor.

$$\frac{1}{C_{total}} = \frac{1}{C_1} + \frac{1}{C_2}$$

Continuity Tester (Project)

This project gives you quite a useful piece of test equipment. You can use it, for instance, to test a fuse, find if there is a break in a cable, test for shorting between a component and its metal casing and so on. It will give an audible tone whenever its probes make a complete circuit, up to about 300 Ω. It has the advantage of producing an audible signal, so that you don't have to take your eyes off the probes when using them. This is especially useful when you are working in confined quarters, whereas an ohmmeter requires you to take your eyes off the work to read the dial.

The circuit is called a unijunction oscillator which produces an audible tone if its probes are joined by a low resistance. The sound is made by a small loudspeaker and it is powered by a 9 V battery.

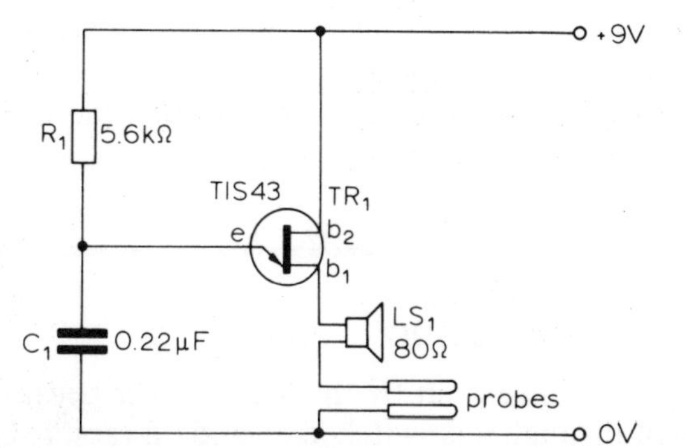

Assembly

The circuit is assembled on copper Veroboard, as shown, and no breaks are required in the strips. Any case large enough to take the board, battery and loudspeaker is suitable, but the circuit board must be mounted clear of the case if it is metal. The probes can be either specially bought or made up from brass rod, but the insulation must be foolproof, in case anyone should decide to use your unit for testing with mains equipment! The unit is definitely not intended for that purpose.

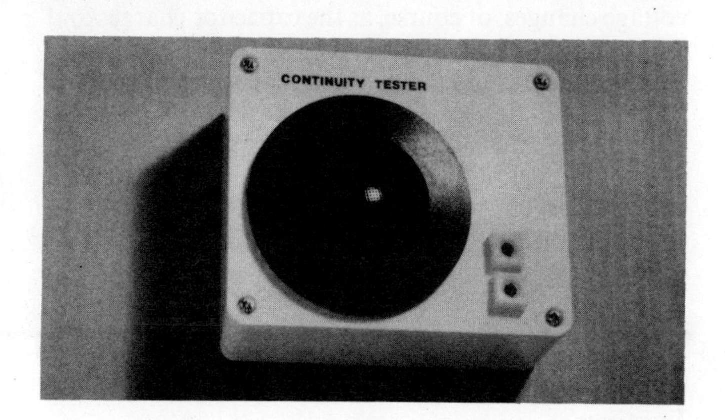

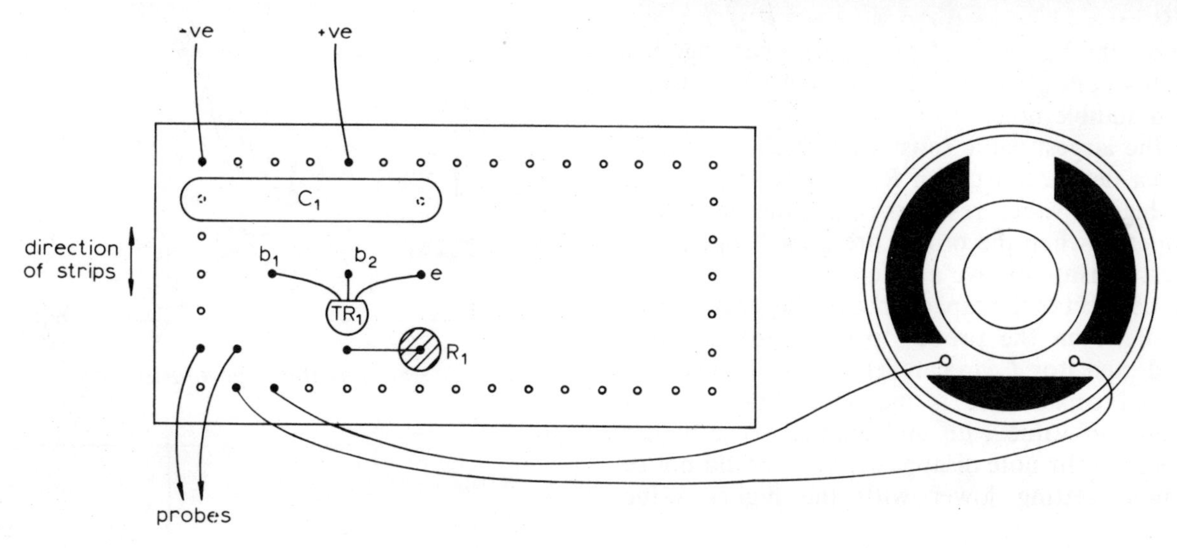

The Circuit

It is good practice to study the way components are connected in a circuit. Look at the list here and fill in the gaps by looking at the connections in the circuit diagram. Now, using the numbers from this list, see if you can mark on the veroboard diagram which connector does each job.

1. R_1 is connected to C_1 and the positive rail and e
2. C_1 is connected to . . . and . . . and . . .
3. LS_1 is connected to . . . and . . . and . . .
4. Probes are connected to . . . and . . .
5. e is connected to . . . and . . .
6. b_1 is connected to . . .
7. b_2 is connected to . . .

Continuity Tester—How it Works

Draw the part circuits which show the following currents:
1. Capacitor charging current.
2. Capacitor discharge current (through the transistor).
3. Loudspeaker current.

Circuit Operation

The note is given out by the loudspeaker because the current through it is switched on and off rapidly by the transistor. The transistor is switched *on* when the voltage at its input (emitter) rises above, about, 5 V and is switched *off* when it falls below that figure. This input voltage changes, of course, as the capacitor charges and discharges.

Temporarily solder a 2000 μF capacitor in parallel with C_1 and connect a voltmeter as shown. This large capacitor will take much longer to charge up and so you will be able to follow the switching action. Connect the battery, close the probes and watch the voltage change, as shown on the meter. When the transistor starts to conduct the capacitor discharges through it.

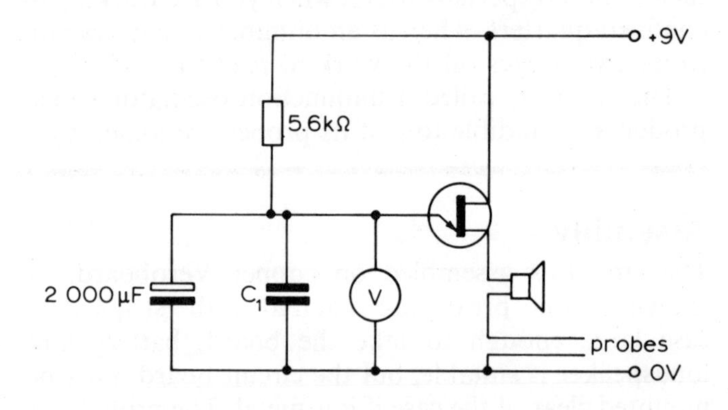

Emitter voltage at switch on = . . . V

Testing

1. Connect the battery and join the probes together with any conductor. The loudspeaker should give out an audible note.
2. Find the largest value resistance which the circuit will just detect. Start with R_x at 10 Ω and replace with higher value resistors until the note just disappears, when the probes are joined together.
   Largest resistor to give a note = . . .
3. Find the effect of the capacitance on the frequency of the note. Join the probes together and touch a second capacitor, C_x in parallel with the first so that the total capacitance is increased. Start with a capacitor of value 1 μF and gradually use higher values until the note disappears. You should notice the note getting lower with the higher value capacitors.
   Highest value capacitor to give a note = . . .

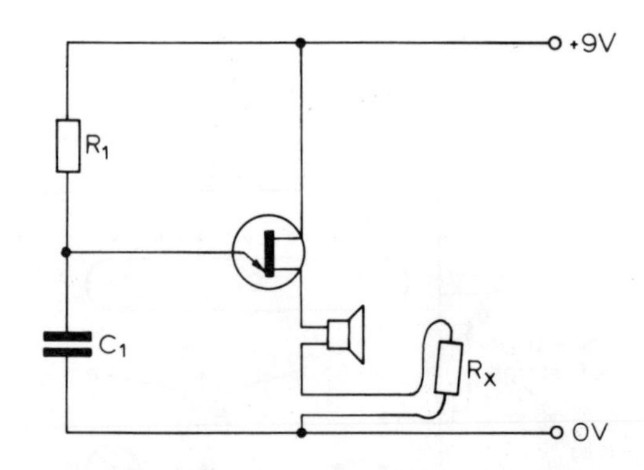

Finding the largest resistance

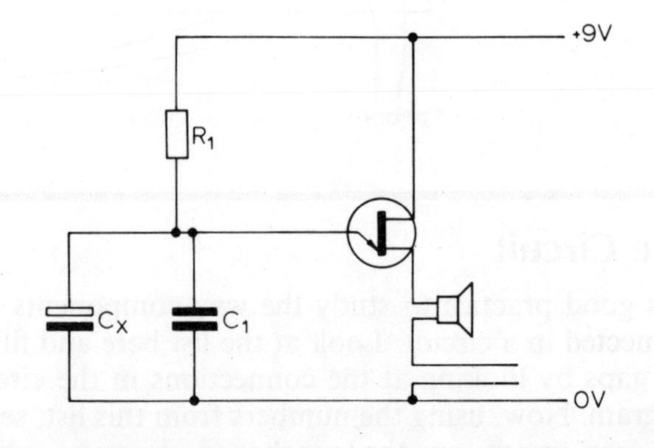

Effect of capacitance on frequency

You should now have a good working knowledge of this very useful little unit, and you should be able to put it to good use during the rest of your course.

Note: The transistor in this circuit is a unijunction type and its behaviour is not the same as that of the common transistor. Do not let this mislead you when you study the transistor chapter.

The Capacitor with Alternating Current

If you look at the bottom of page 61, you will see the diagrams of a capacitor charging and discharging. To anything else in the circuit, this looks suspiciously like alternating current—only one cycle, admittedly, but if we could charge and discharge a capacitor repeatedly then we would have alternating current passing.

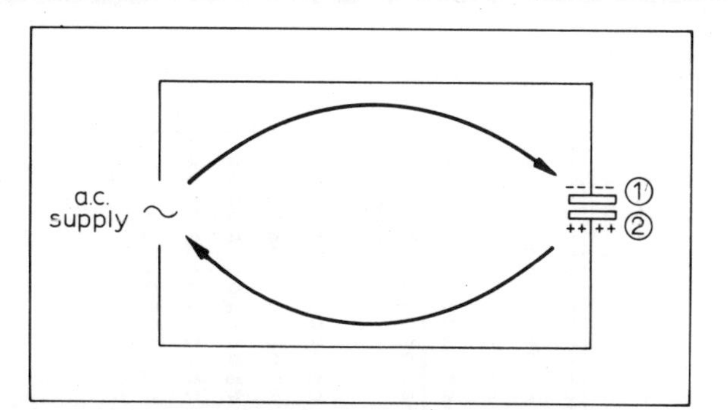

When the supply pushes electrons clockwise:
plate 1 receives electrons
plate 2 loses electrons

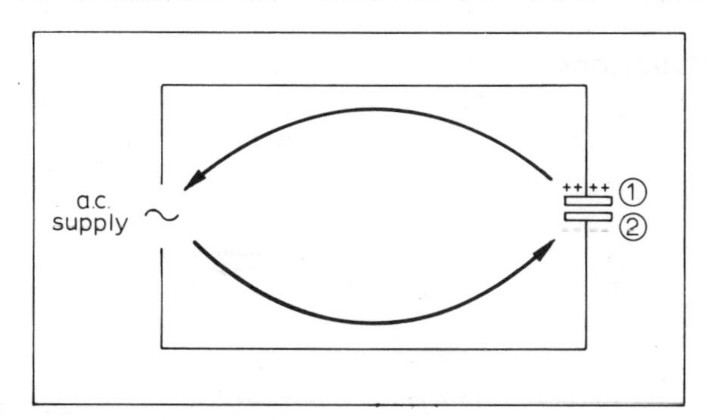

When the supply reverses:
plate 1 now has its electrons pulled off
plate 2 is happy to receive electrons

As far as the circuit is concerned, alternating current is flowing. *None passes through the capacitor*. The capacitor just *stores* the electricity which flowed, until the supply reverses, then sends it all back again!

Showing Conduction of Alternating Current

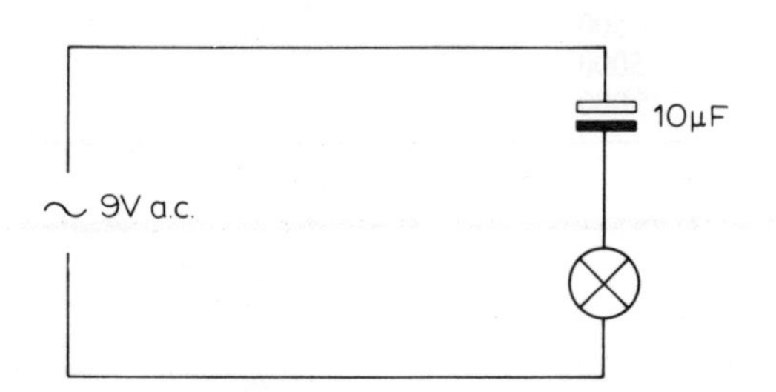

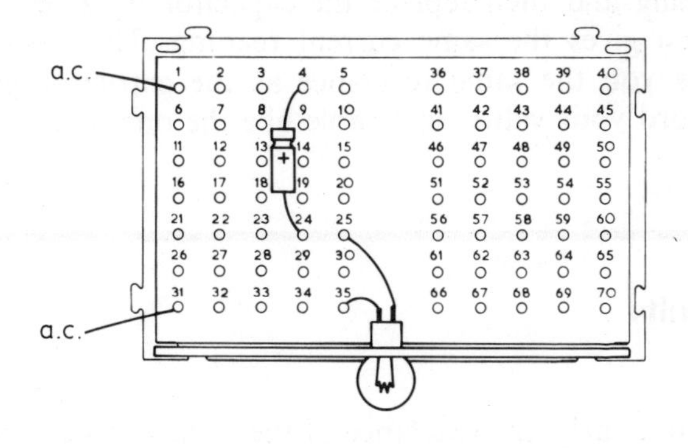

Note: Very occasionally, an electrolytic capacitor can explode with alternating current. Protect yourself and others during this experiment by using perspex safety sheets.

Instructions

Assemble the circuit and find out if a capacitor really does 'pass' alternating current. You should find that the bulb does light up, showing conduction. Now, replace the capacitor in the above circuit by a wire. You will find that the current increases—the bulb gets brighter. So, although the capacitor does allow alternating current to flow in the circuit, it clearly does have a resistance to alternating current.

Resistance in a.c. circuits is called reactance (X). Reactance is measured in ohms.

Find the approximate *reactance* of the capacitor by replacing it with different resistors until you find one which gives about the same brightness.

Write down the reactance of your capacitor in the circuit. $X = \ldots$

When talking about components, you now understand that nothing passes through a capacitor. It just allows alternating current to flow in the circuit. However, you will find it easier to say that it 'passes' alternating current to explain the conduction.

A capacitor in a circuit allows alternating current to flow.

The Effect of Frequency on Reactance

We must now look at a very important property of capacitors. This forms the basis of the tuning and tone controls in a radio. The performance of a capacitor with alternating current changes as the frequency of the current rises.

Instructions

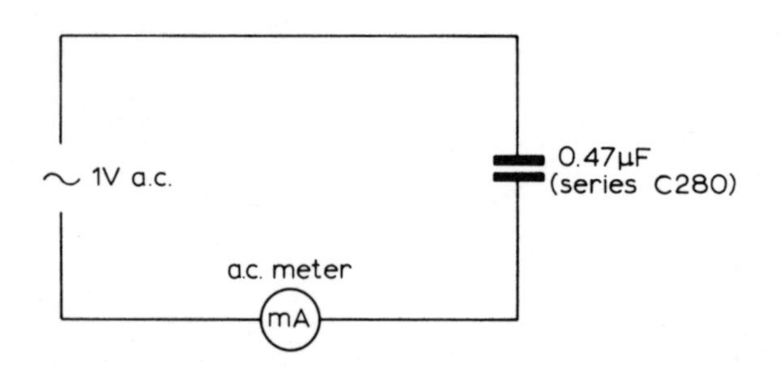

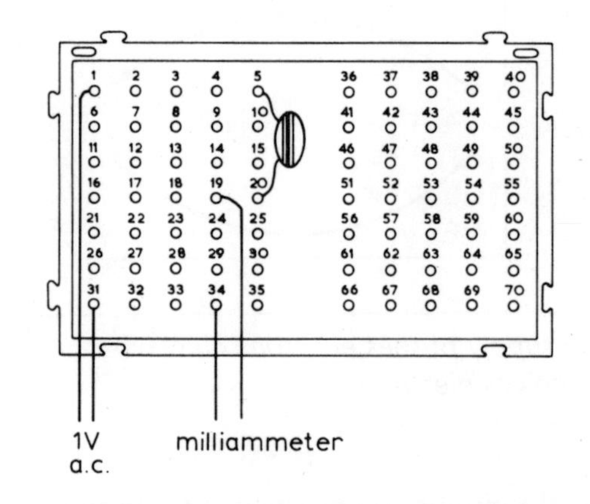

Set up the above circuit. The signal generator enables you to change the frequency of the a.c. supply. For each of the frequencies in the table here, take a current reading and then replace the capacitor by a resistor which gives the same current reading. This resistor gives you the same reactance as the capacitor had. Record your values in a table like the one here.

f (Hz)	I (mA)	Resistance, R = Reactance, X
10		
50		
500		
5000		
25 000		

Results

Quite clearly, the reactance of the capacitor decreases as the frequency increases. This is very useful since it means that, if two signals of different frequency are fed into a circuit, then the capacitor will deal with them differently. It is a *frequency-selective component*.

As frequency ↑ Reactance ↓

(The actual equation is $X = \dfrac{1}{2\pi fC}$ but you need not remember this.) Opposite is a table of values of reactance for different values of capacitance and frequency.

f (Hz)	C (μF)	X (Ω)
100	100	16
100	1	1600
1000	0·1	1600
1000	1000	0·16
10 000	10	1·6
10 000	0·001	16 000
100 000	1	1·6
1 000 000	0·01	16
1 200 000	0·001	133

Capacitors—Problems

Series and Parallel

1. What is the effective capacitance of the following pairs of capacitors in parallel?
   (*a*) 500 μF and 500 μF
   (*b*) 1000 μF and 100 μF
   (*c*) 50 μF and 100 μF
   (*d*) 2000 μF and 1000 μF

2. What is the effective capacitance of the following pairs of capacitors in series?
   (*a*) 500 μF and 500 μF
   (*b*) 1000 μF and 100 μF
   (*c*) 50 μF and 100 μF
   (*d*) 2000 μF and 1000 μF

3. You have a large collection of 100 μF and 500 μF capacitors. How would you combine them to obtain the following values?
   (*a*) 1000 μF
   (*b*) 50 μF
   (*c*) Approximately 80 μF
   (*d*) Approximately 180 μF

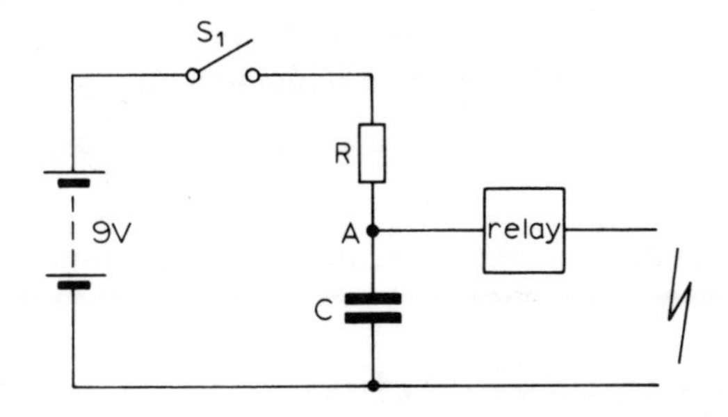

4. This circuit is used to control the time taken for the supply to the relay to reach 6 V. The rest of the relay circuit has been left out for simplicity. Assume the current through the relay can be ignored.
   Find the time taken for the voltage at A to reach 6 V when
   (*a*) $R = 10$ kΩ, $C = 1\,000$ μF
   (*b*) $R = 3{\cdot}3$ kΩ, $C =$ 500 μF
   (*c*) $R = 100$ Ω, $C =$ 100 μF

Time Constants

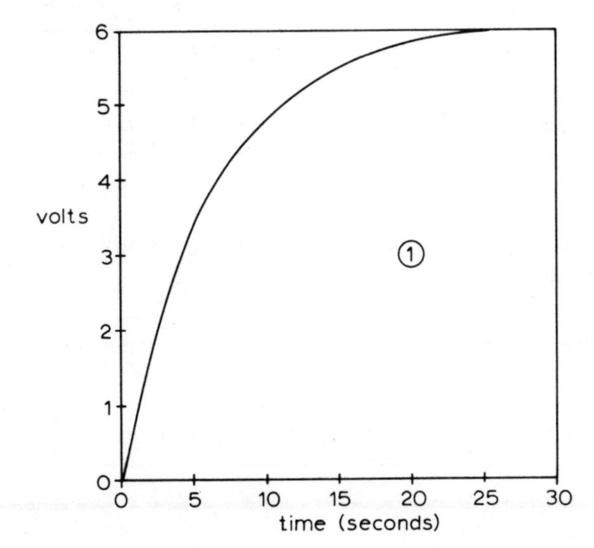

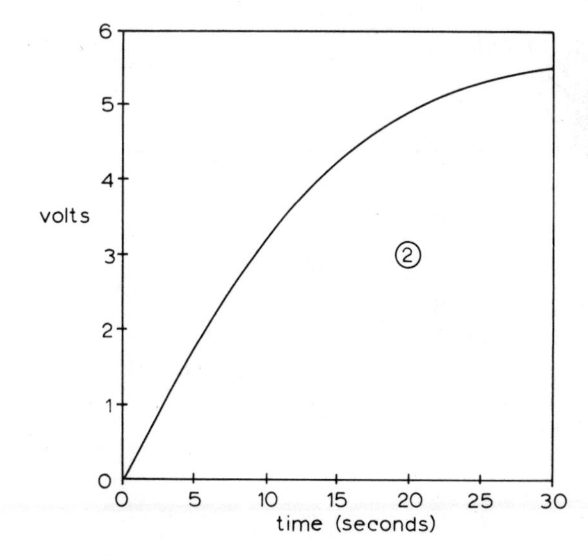

The two graphs show how the voltage across a capacitor changed with time in an *RC* series circuit, when the 6 V supply was switched on. The circuit for graph 1 contained a 6·8 kΩ resistor and capacitor C_1, whilst the circuit for graph 2 contained the same resistor but a different capacitor, C_2.

1. Use the graph to find the time constant of:
   Circuit 1
   Circuit 2

2. Use the equation $T = CR$ to work out the value of:
   C_1 in farads
   C_1 in microfarads
   C_2 in farads
   C_2 in microfarads

3. What voltage was reached after 5 s in each circuit?
   Circuit 1
   Circuit 2
   Which circuit had the fastest charging rate?

Filter Circuits

Filtering is a process applied to a.c. signals which are a *mixture of frequencies*. It involves removing some of the frequencies and leaving the remainder in.

Uses

You come across these circuits most commonly in the base and treble controls in, for example, a record player. Turning up the treble control allows more of the high frequencies to pass to the loudspeaker and cuts down the lower frequencies. You will also need a filter circuit in the radio you build. It will filter out the high-frequency (radio) waves and leave in the low-frequency (audio) waves.

Filter Circuits in Action

There are two circuits shown below and both are *RC* arrangements. One is designed to output more of the high frequencies whilst the other will give more of the lower frequencies. An input, for example a microphone or record player, can be connected to A and B and the output, perhaps another circuit or a loudspeaker, can be connected to C and D.

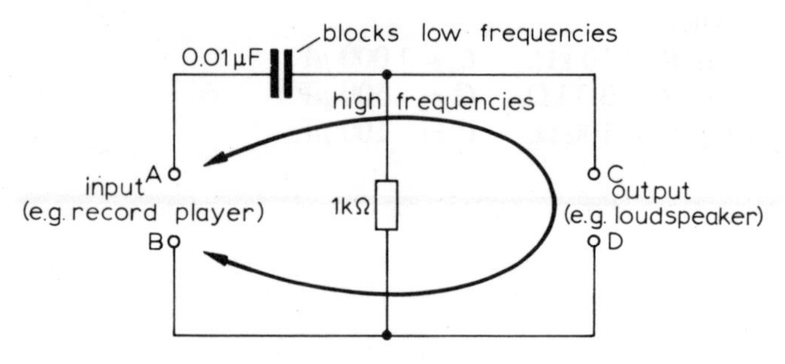

High-pass filter

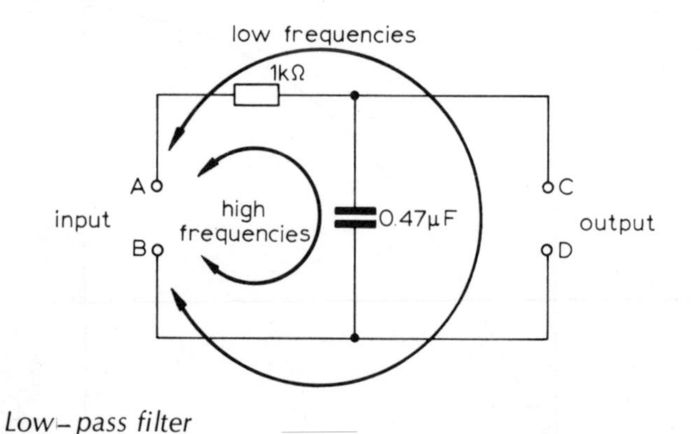

Low-pass filter

Instructions

Your teacher may demonstrate this experiment for you.

Connect a signal generator to A and B and an oscilloscope to C and D. The oscilloscope will show you the size of the output you are getting, but remember that a signal generator only gives *one frequency at a time*, so you are not yet looking at a mixture of frequencies. Find the voltage output for each of the frequencies shown below. Set your generator at 5 V for the whole experiment.

High-pass filter	
Frequency given by signal generator (Hz)	Voltage shown by oscilloscope
100	
500	
1 000	
10 000	

Low-pass filter	
Frequency given by signal generator (Hz)	Voltage shown by oscilloscope
100	
500	
1 000	
10 000	

Result

The experiment shows quite clearly the opposite effects of these two.

The high-pass filter blocks the low frequencies and outputs the high frequencies.

The low-pass filter outputs the low frequencies.

Explanation

These circuits depend on the fact that a capacitor offers less 'resistance' to alternating current as the frequency of the current rises.

High-Pass Filter

Low frequencies meeting the capacitor are blocked by it whilst the high frequencies 'pass' through it and appear at the output.

Low-Pass Filter

Low frequencies, again on meeting the capacitor, find it blocks their path so they take the alternative route to the output. The high frequencies, on reaching the capacitor, find it offers little impedance so they short circuit this way and do not appear at the output.

Capacitors—Summary

Charging

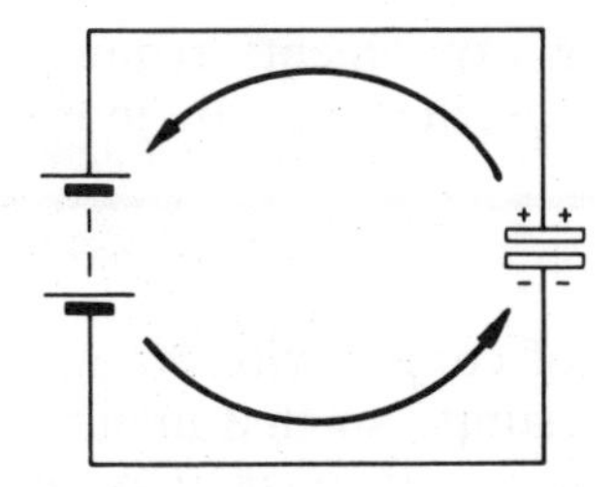

Charging

During charge the capacitor stores electricity

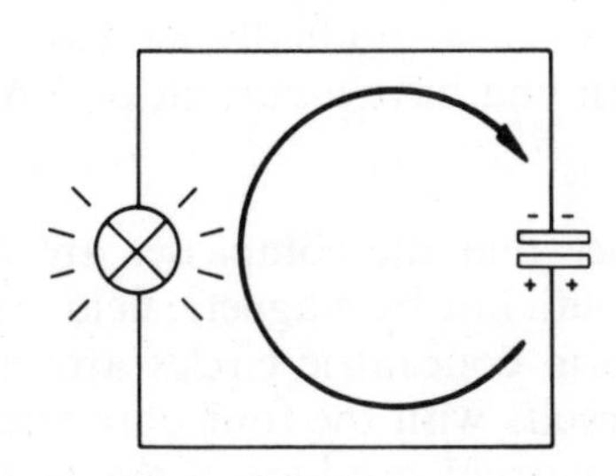

Discharging

During discharge the capacitor releases electricity

Time Constant

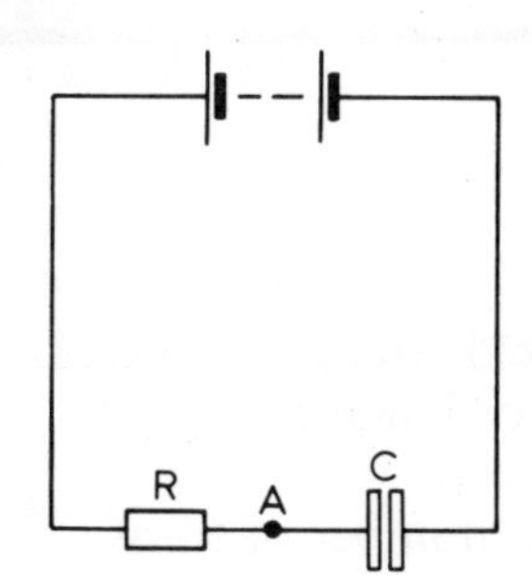

The time taken for the voltage at A to reach two-thirds maximum is equal to T

$T = C \times R$

Blocking of Direct Current

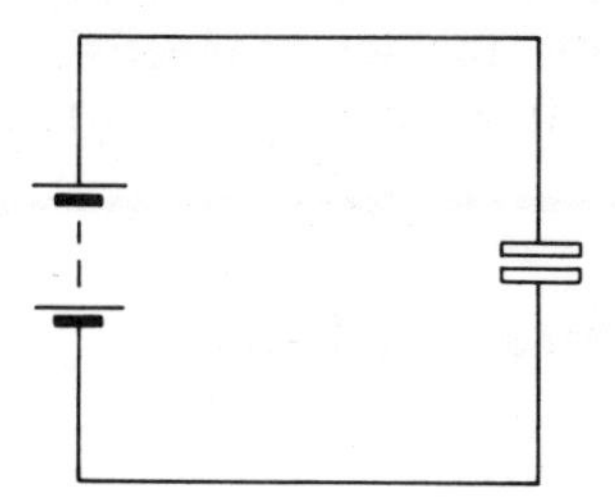

The current falls to zero when the capacitor is charged up

Voltage Control

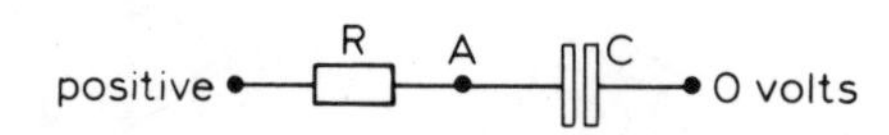

The speed of voltage change at A is controlled by the values of R and C

With Alternating Current

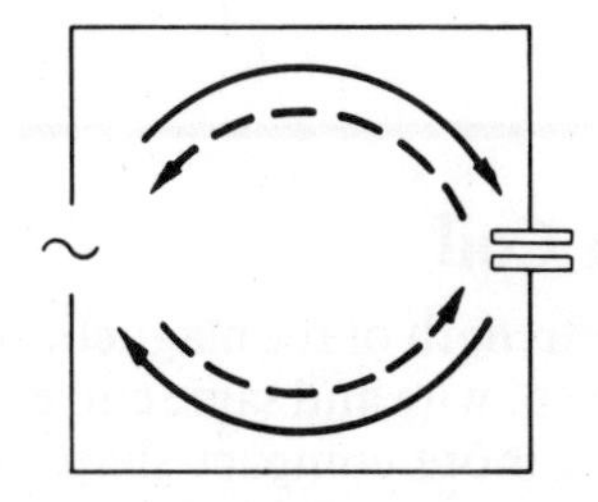

Charge and discharge with alternating current give the appearance of the capacitor conducting

> A capacitor offers a resistance to alternating current, called reactance.

> The reactance of a capacitor decreases as frequency increases.

5 ELECTROMAGNETIC EFFECTS

Magnetism Produced by an Electric Current

It may be that, so far, the only source of magnetism that you have met is the iron magnet. A source which is probably more important is the production of magnetism by electricity. In fact, magnetism is remarkably easy to produce from an electric current and, as you will see below, it is easy to control its strength and even switch it on and off, feats which are impossible with the ordinary bar magnet.

The following experiments require quite a large current and can overheat so care must be taken.

Current in Straight Wire

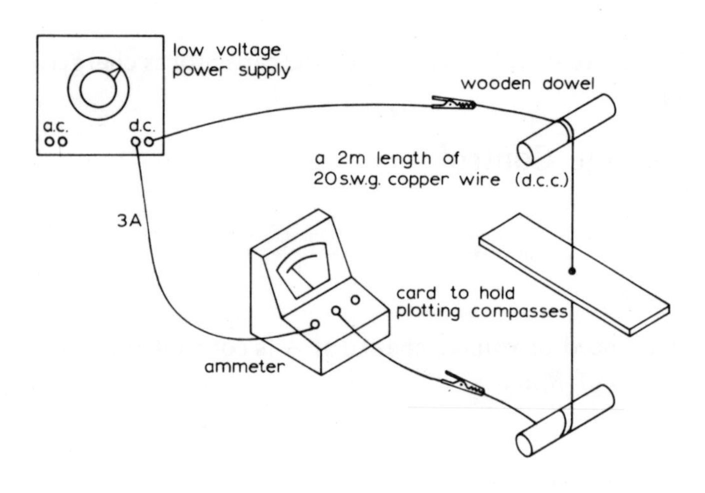

Fix the length of copper wire around two pieces of dowel, held in clamps, so that about $\frac{1}{2}$ metre is held vertically and passes through a piece of horizontal card. Spread half a dozen plotting compasses on the card around the wire and connect the wire to a d.c. supply. The plotting compasses should all be pointing north–south, so now gradually increase the voltage from zero until you have a current of 3 A.

Results

You should see that the compasses are pulled from their north–south line by magnetic field produced and the needles form concentric circles around the wire. Check your results with the four observations below:

1. An electric current produces a magnetic field.
2. The magnetic field is circular in shape.
3. The strength of the magnetic field increases as the current increases.
4. The magnetic field is only present whilst the current flows.

Current in a Coil

To increase the strength of the magnetic field produced by the same piece of wire and same current, you need to use the wire in a more compact shape. Compare the strengths of the magnetic fields produced by the four coils shown, using the same 2 m length of copper and a 3 A current.

Results

The magnetic field strength is increased by

1. the compact coil shape
2. the soft iron core
3. increasing the number of turns.

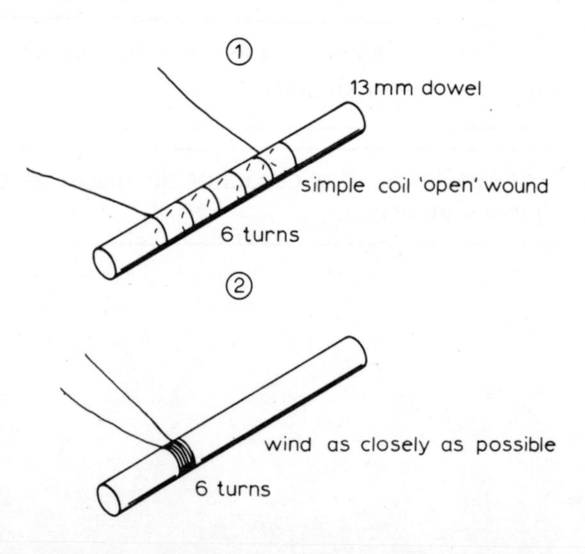

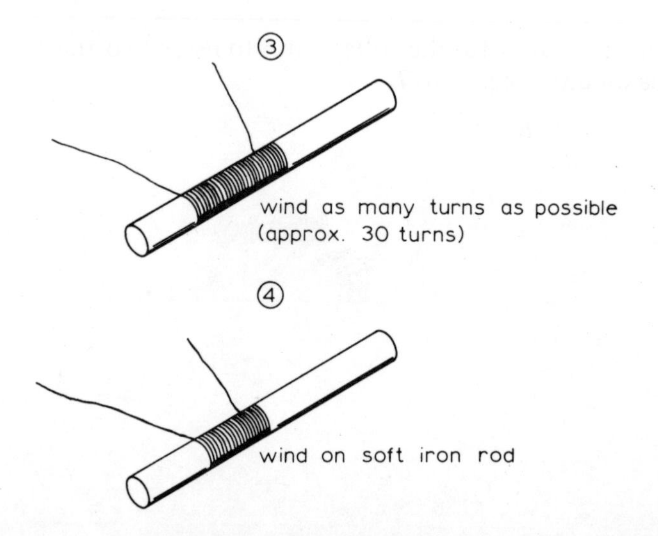

The Coil

Shape of Magnetic Field

The rest of this chapter is devoted to studying the coil, since the four devices we are interested in, the relay, inductor, transformer and loudspeaker, are all coil wound components. First do a more careful plot of the shape of the magnetic field produced by the coil.

Instructions

Use the last coil you made and again supply a 3 A current. Place the coil on a sheet of card and use plotting compasses. Mark in the direction of the compass needle in a variety of positions around the bar magnet and coil. You should find that the needles always point along the lines shown in the patterns.

These lines are called *lines of force* and they show the direction of the magnetic field at any point and they always run from a north pole to a south pole.

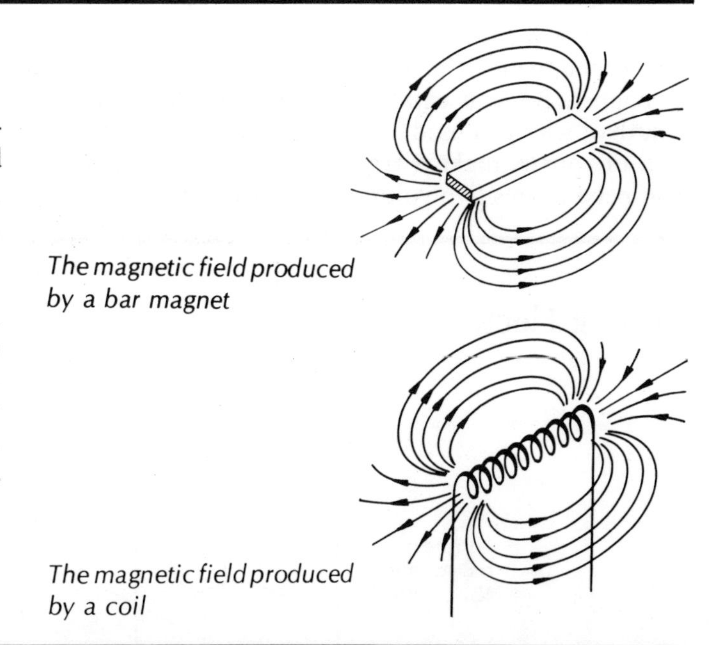

The magnetic field produced by a bar magnet

The magnetic field produced by a coil

The Coil with Alternating Current

The problem with using alternating current in magnetic experiments is finding a supply with a low enough frequency for you to be able to detect the changes as they occur. Perhaps the easiest way to obtain a crude sort of alternating current is to use a reversing switch with direct current, so that you may control the frequency of the supply. This will give you a square-wave alternating current and not the familiar sine wave that we obtain from the mains. The supply to the coil will reverse in direction every time the reversing switch is operated.

Instructions

Use the coil from the last experiment and connect it as shown, so that you use a 3 A current. Again observe the positions of the compass needles and then use the reversing switch to find the effect of a change in the current direction.

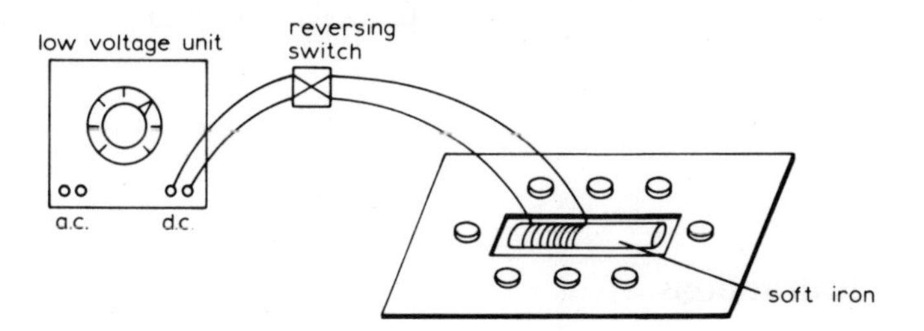

Results

When the current drops to zero the field disappears and when the current reverses in direction the field reappears in the *opposite direction*. However, what is more difficult to show is the change in the magnetic field as the current is increasing or decreasing. This change is shown in the three diagrams.

The lines of force move inwards and outwards as the current changes.

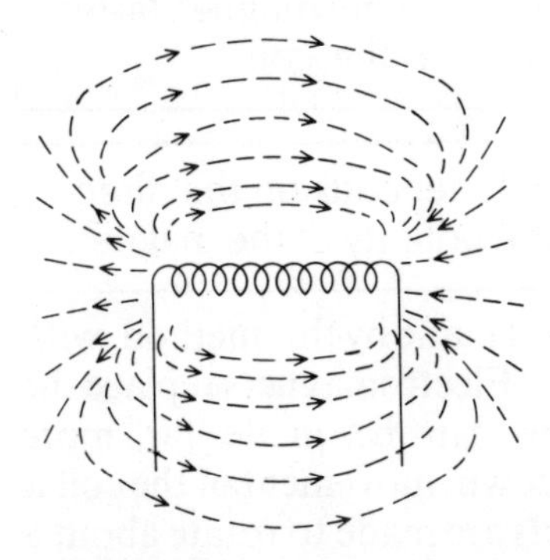

A large current produces a large field

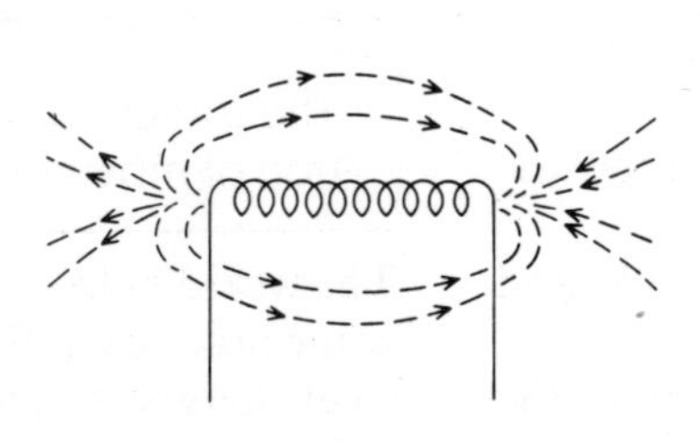

The field produced by a smaller current. The lines of force have moved inwards

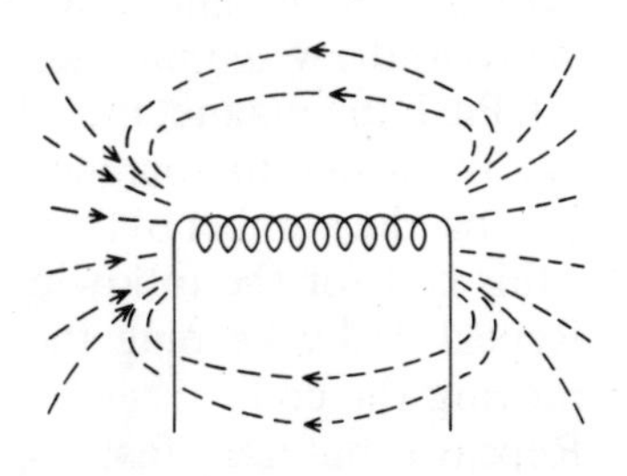

With the current in the opposite direction the compasses also point in opposite direction

Producing Electricity

You can now see how easy it is to produce electricity, using a magnetic field. You need only a coil of wire and a magnet. So you can use the bar magnet and the length of wire you used for the previous page. You will not, of course, make a large quantity of electricity so you need a sensitive voltmeter to detect it.

Making Electricity is Easy

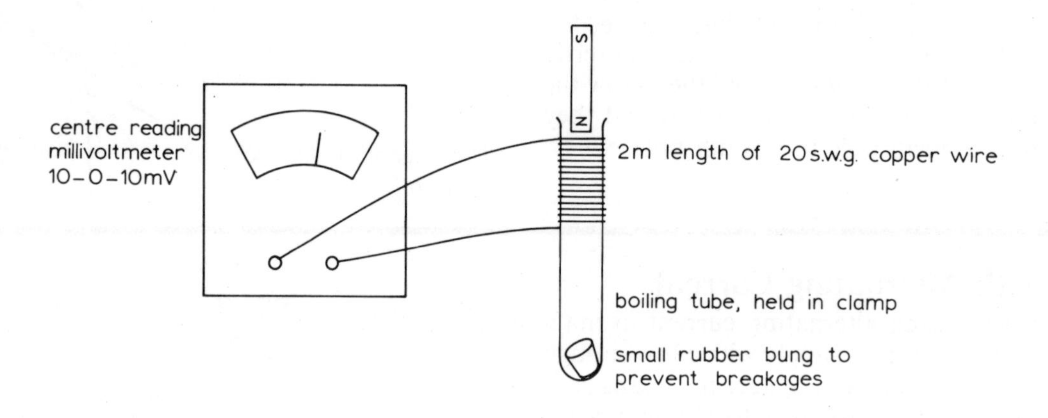

Instructions

Wind the copper wire into a coil as shown above and connect the ends to a galvanometer. This sensitive meter will show you when you have produced electricity and also, being a centre reading type, it will tell you the direction of the electric current produced. A positive reading on the meter indicates an electric current flowing into the positive terminal and a negative reading indicates a current flowing out of the positive terminal.

For each of the following instructions write down the direction of the current produced and the reading on the meter.

1. (*a*) Push the magnet into the coil, north pole first.
   (*b*) Withdraw the magnet
   (*c*) Push the magnet into the coil, south pole first.
   (*d*) Withdraw the magnet.
2. Double the number of turns on the coil and repeat 1.

Use this coil for the following:

3. Repeat 1, by keeping the magnet stationary and moving the coil.
4. Repeat 1, but use a faster speed of movement of the magnet.
5. Make a stronger magnet by sellotaping two magnets together, north pole to north pole. Repeat 1, using this double magnet.

Results

All these experiments show quite clearly that electricity is produced, in varying amounts. An electricity supply of this kind is called an *induced e.m.f.*

An induced e.m.f. is always produced when the lines of force of a magnetic field cut a conductor.

In particular, the results should have shown the following rules for the production of an induced e.m.f.

> An e.m.f. is produced only when there is movement of the coil or magnetic field.

> A larger e.m.f. is given by a stronger magnet, faster movement and more turns on the coil.

> The direction of the e.m.f. depends on the direction of movement and the polarity of the magnet.

The world's electricity is still made by this method, with some practical refinements. Electromagnets replace the steel magnets, since they can be made far more powerful and the 'up and down' movement of the coil is dispensed with and the coils are made to rotate about a shaft. The principle is, however, the same and accounts for practically all of the electricity produced.

The Transformer—for Changing the Voltage

You have seen the essential elements of the production of electricity. The lines of force of a magnetic field must cut through a conductor, usually a coil of copper wire. However there is another way of getting the magnetic field to cut the coil and it depends on the behaviour of alternating current in a coil (page 73).

> Alternating current in a coil produces a moving magnetic field.

This is put to use in the *transformer*.

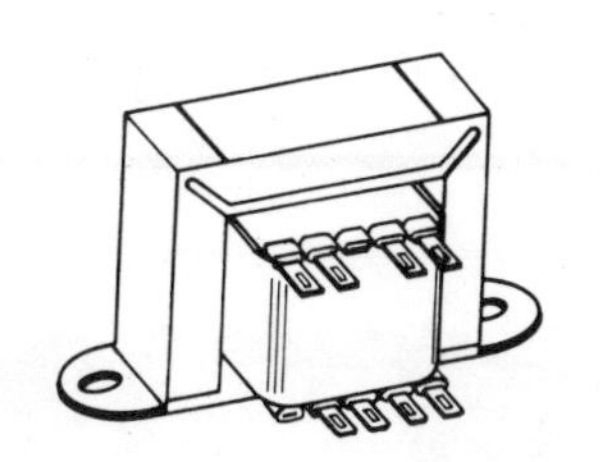

External appearance of a transformer

Construction

A transformer consists simply of two coils of wire, like the ones you have been using. Alternating current is put into one coil (the primary) and this produces a *moving* magnetic field which continually cuts the other coil (the secondary), thus producing electricity. Obviously, in this case, electricity is being produced only when a supply of alternating current is already available, so this is not a primary source of electricity. However, the transformer is still a very useful piece of equipment, chiefly because it can produce electricity at a different voltage from the input voltage.

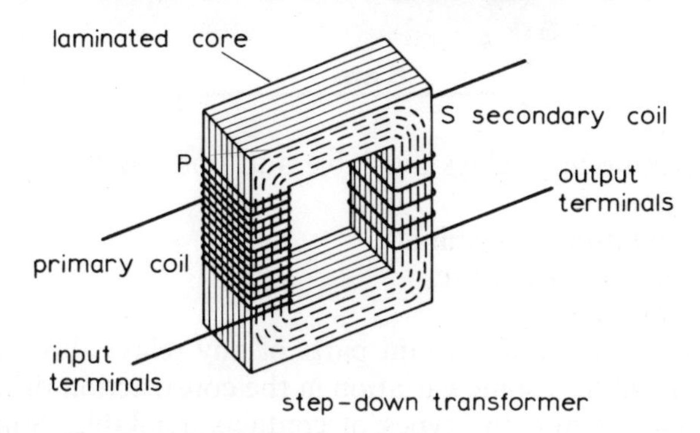

Internal construction of a transformer

Investigating the Transformer

The experiments you perform with transformers will depend very much on the apparatus which you happen to have. You may be able to wind simple coils yourself to make up a transformer or you may have commercial models to work with. Remember during these experiments that *no* electricity flows between the two coils since the copper wire is insulated. The connection between the two coils is only *magnetic*. The magnetic field caused by alternating current in the primary 'cuts' the wire of the secondary coil and causes electricity to flow there. If possible, measure the input and output voltages and investigate the effect of:

1. Using both alternating and direct current input, up to the maximum voltage given by your teacher.
2. Changing the number of turns of wire on the two coils.
3. Using a soft iron core for the two coils.
4. Changing the separation of the coils.

Your experiments should show you the following:

> Alternating current fed into the primary of a transformer generates alternating current in the secondary.

> The change in voltage caused by a transformer depends on the number of turns on the coils.

Uses

Because of this performance the transformer has a few uses.

Voltage Change

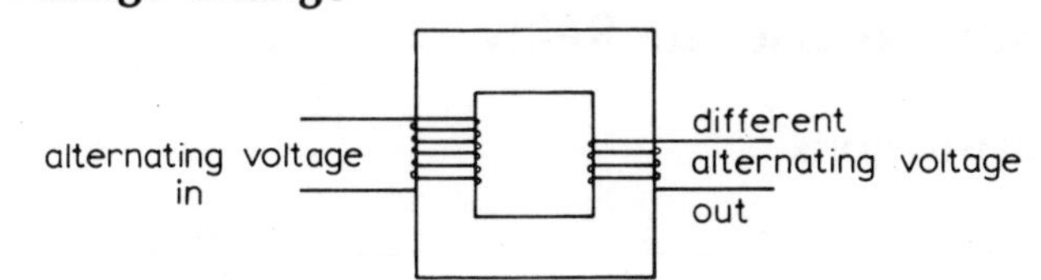

Isolation

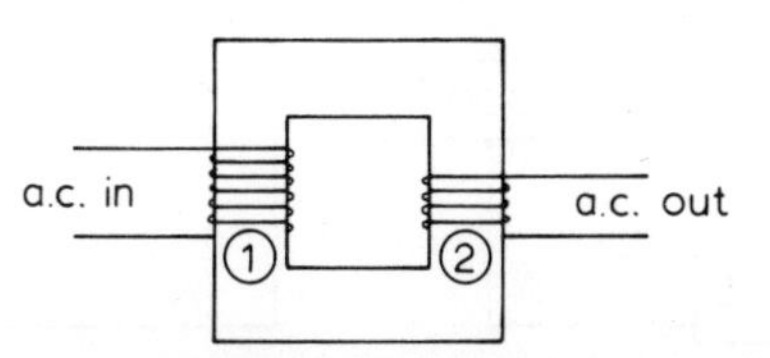

Alternating current is 'passed' from circuit 1 to circuit 2 but there is no electrical connection between them.

D.C. Block

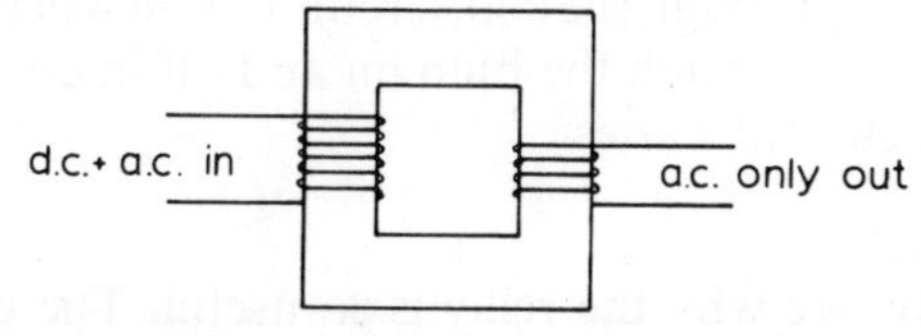

Alternating current is 'passed' into the secondary but direct current is blocked.

The Relay

You are now to look at the second of the four devices which depend on the electromagnetic effects which you have studied. The heart of the relay is just a coil of wire, wound on a soft iron core, like the ones you have made.

Construction

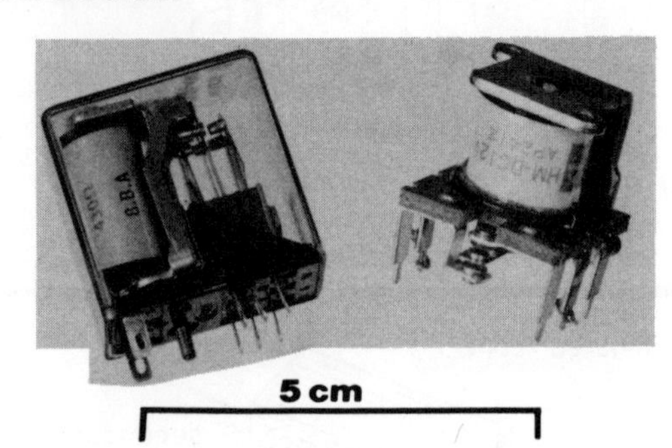

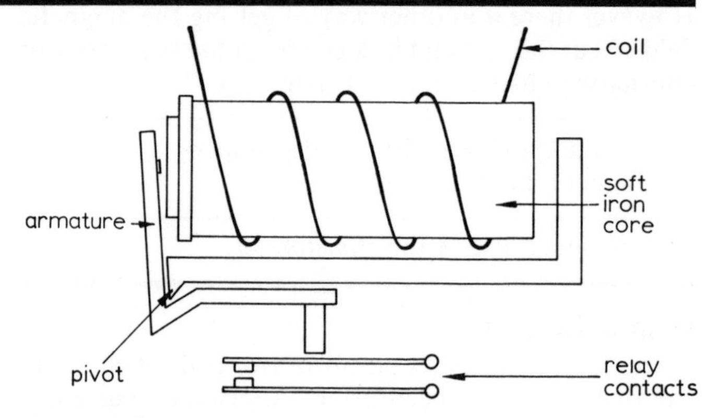

Examine any relays available and find the following parts:

1. coil and its terminals;
2. moving armature;
3. contacts.

These are the important parts of any relay, although you will find some variation in the construction of the armature and the types of contacts available. Some relays are completely enclosed in a case, so that their contacts stay clean and do not corrode and you may find these useful in any projects you build which have to operate under difficult conditions.

Operation

Current through the coil of the relay creates a magnetic field and when this is strong enough it will attract the armature and move it on its pivot. The movement of the armature opens and closes the contacts and you can easily see this by moving the armature with your fingers. These contacts are switches which control any circuit for you.

> The relay is a switch controlled by the current through its coil.

How to Use the Relay

Instructions

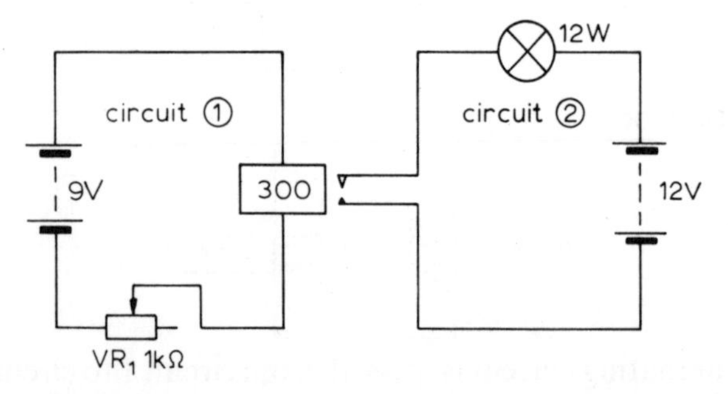

Control circuit *Switched circuit*

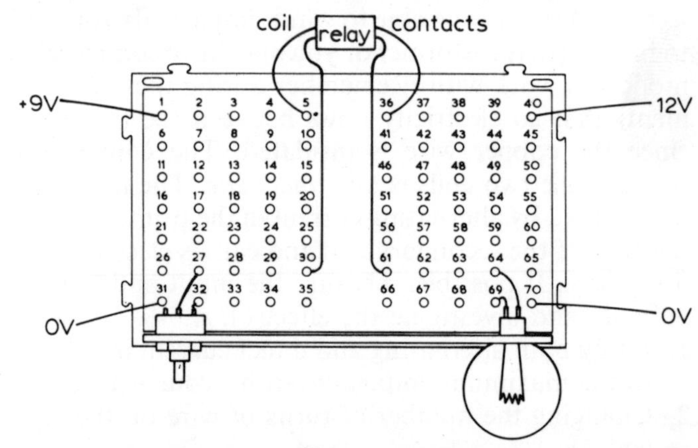

Assemble the circuit as shown and use VR_1 to control the current through the coil, circuit 1. You will find that the relay will switch the bulb on and off in circuit 2 as the coil current varies.

Results

You now see why the relay is so useful. The currents and voltages in the two circuits can be completely different since the two circuits are separate, so you can use low power devices to control high power devices. In the above circuit, the bulb could have been a mains driven type and could easily and safely be controlled by a 9 V circuit.

Uses

You can use a safe low voltage circuit to switch a high voltage or high current circuit.

You can keep a switched circuit in the 'on' or 'off' condition even though the current in the control circuit varies gradually.

Relay Specifications

Coil Resistance

Coil resistance is measured with the ohmmeter. The relay need not be connected into a circuit, of course; merely connect it to the meter and take the resistance reading. This measurement is the figure often found on the symbol for a relay.

Operating Current

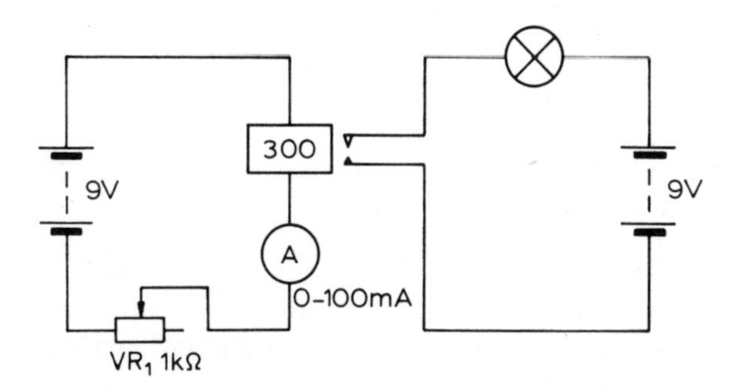

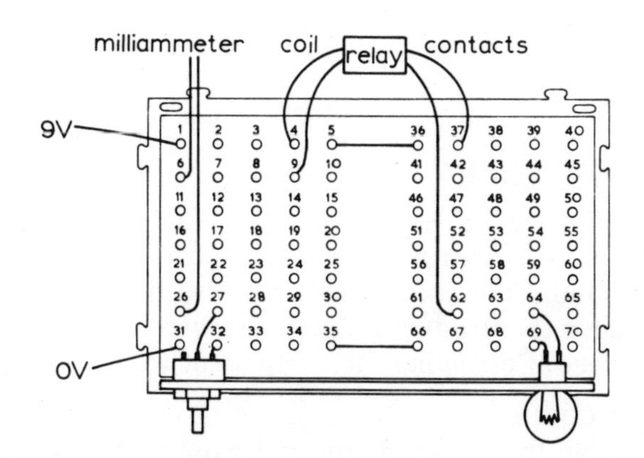

Instructions

Use the circuit from page 76 (you may find it easier to change the 12 V circuit for a 9 V one, as shown) and connect a milliammeter into the coil circuit. Use VR_1, to vary the current and find the two measurements:

1. Energising current to switch on.
2. Current to switch off.

Once you have these figures for a relay you can decide whether it is suitable for the control circuit you have in mind.

Controlling the Relay—a Light Operated Switch

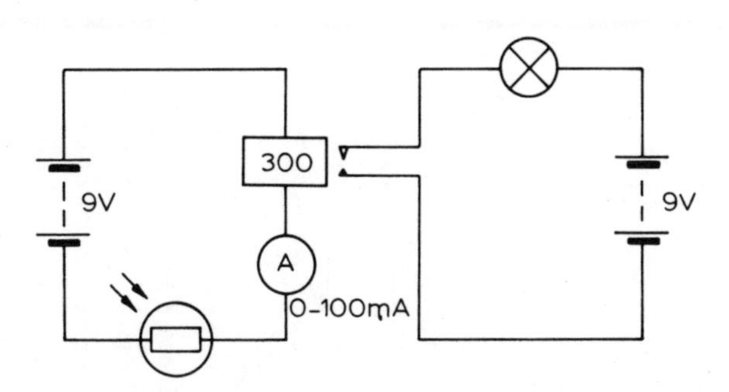

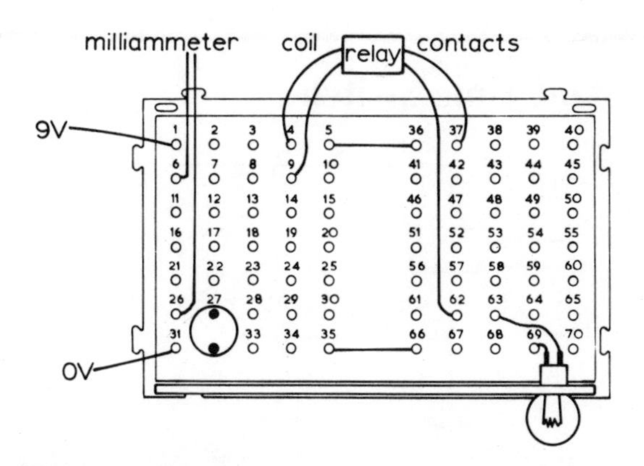

Any device that can control current can be placed in the control circuit of a relay and used to switch the contacts, as long as a suitable relay is chosen. In this case, a light dependent resistor (l.d.r.) is to be used and its properties are:

Mullard ORP 12

In darkness it has a high resistance (about 1 MΩ)

In bright light, it has a low resistance (about 200 Ω)

Assemble the circuit shown and vary the light falling on the l.d.r. Your milliammeter should show the current varying as the resistance of the l.d.r. changes. With a suitably chosen relay the l.d.r. will switch the bulb on and off. Take the current readings:

1. Minimum current.
2. Switch on current.
3. Maximum current.

You now have a circuit which will switch for you as the level of light changes.

Notice another advantage of a relay used in this way. The switched circuit is always either *on* or *off*, even though the l.d.r. controls the current in a gradual manner. If the l.d.r. controlled the light bulb directly, without a relay, then it would give continuously variable light output.

Two Uses for a Relay

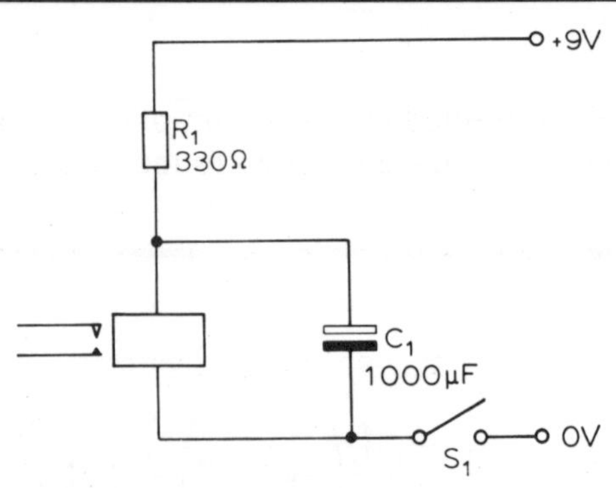

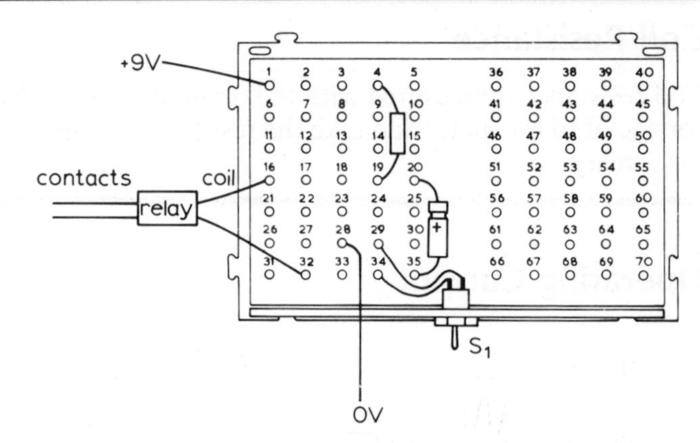

Delay Switch

Here you see a resistor–capacitor (*RC*) combination being used to control the time at which a relay switches. You should now be quite familiar with the use of a capacitor to introduce a time factor into a circuit and the effect of changing the component values.

Instructions

Assemble the above circuit and close switch S_1. Try different values for C_1 and R_1, and find the delay time in the relay operation for each combination. You may connect the relay contacts into a simple bulb/battery circuit to watch the switching action. Ensure that the capacitor is discharged between each timing.

Result

You should have observed the familiar result that the larger the *RC* combination, the longer is the delay in the circuit operation. Also, you will see that there is a maximum value for R_1 that will allow the relay to switch on, and above this value there is not sufficient current to energise the relay.

The Circuit

Examine the circuit carefully and draw out the part circuits which show the following current paths:

1. Coil current.
2. Capacitor charging current.
3. Capacitor discharge current.

Notice that C_1 is in parallel with the relay coil and so the current through R_1 is shared between them. This means that the amount of current which the coil receives depends on the state of charge of the capacitor:

When C_1 is uncharged it takes most of the current.

When C_1 is charged the coil takes the current.

Constant Time Switch

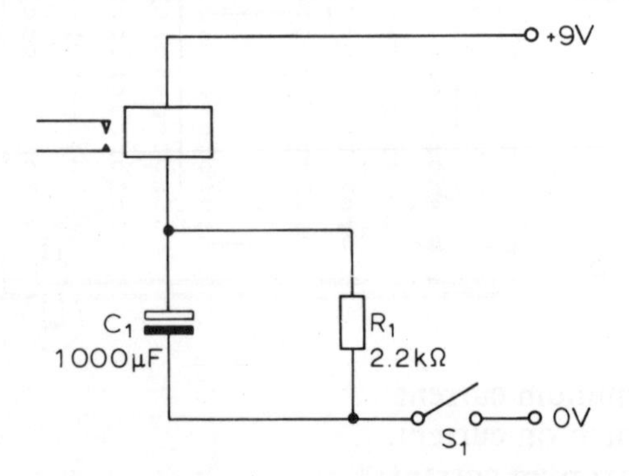

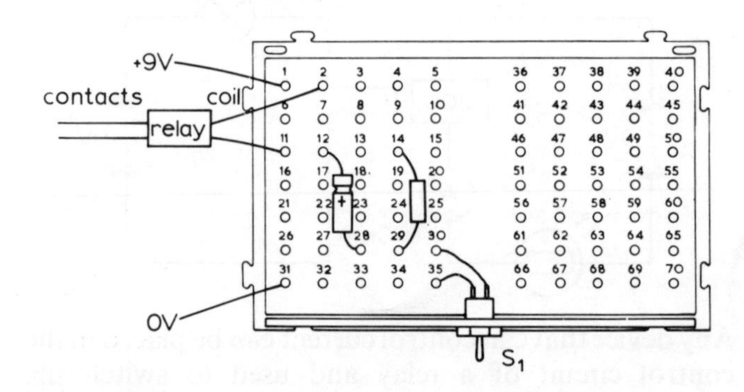

This circuit provides a switch which stays on for a constant length of time and then switches itself off. It is particularly useful if you have to make a series of operations which all require the same length of switching time.

Instructions

Close S_1 and notice that the relay stays on for a particular length of time and then goes off again. Allow the capacitor time to discharge (through R_1) between operations and try different values of capacitors.

The Circuit

Notice that the relay coil and capacitor are *in series* so the capacitor charging current is also the coil current.

The relay stays on as long as the capacitor charging current is large enough.

Burglar Alarm

Here is a design for a simple burglar alarm. The guard switches could be placed in window and door cavities so that, should a burglar try to enter, he will trip one of the switches and set the bell ringing.

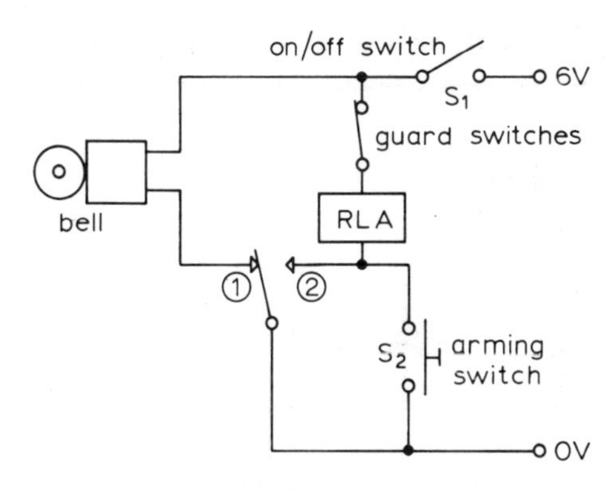

Alarm not in use

Alarm Not in Use

This is the circuit before 'arming'. No current is being drawn. Note that the relay must have a set of change over contacts. Only one guard switch has been shown, but any number, in series, may be used. The normal position for the relay contacts is position 1, as shown, when the relay is off.

Armed—Relay On

To arm the alarm, close switch S_1. This will set the bell ringing, via relay contact 1. Now press switch S_2 for a moment. This will allow relay current to flow and switch on the relay. The contacts will move to position 2, as shown, and now the relay coil current flows through its own contacts, and the bell stops ringing. The alarm is now ready for use.

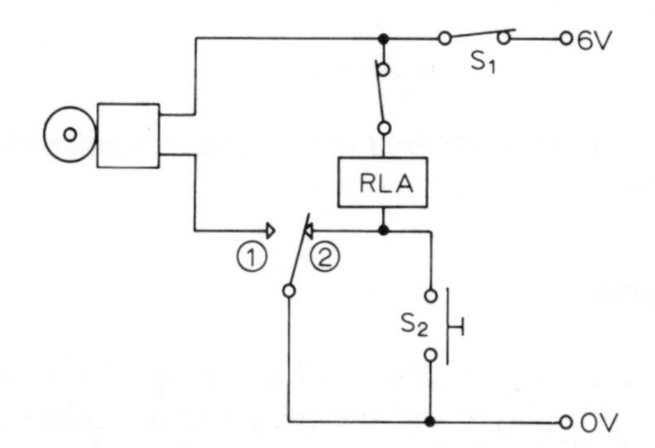

Armed—relay on

Tripped—Bell Ringing, Relay Off

If an intruder should now enter and open a guard switch, this switches off the relay current. The relay contact will now move back to position 1 and so switch on the bell. Even if the burglar now closes the guard switch the bell keeps on ringing, because the relay cannot be switched on again until S_2 is pressed again.

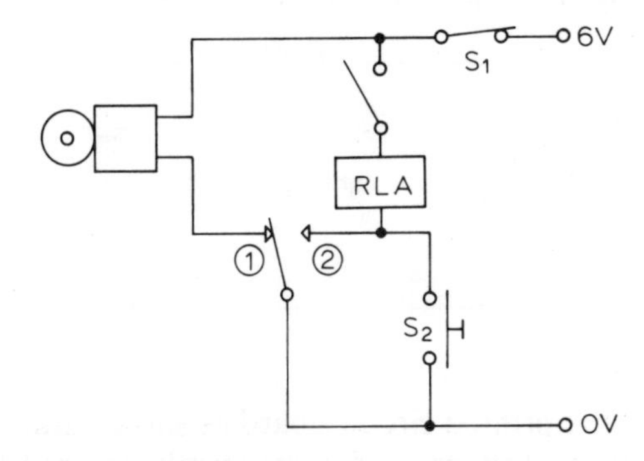

Tripped—bell ringing, relay off

Construction

Although this circuit does not warrant the use of a circuit board of the electronic type which you have used already, you will need an orderly layout to make sure that you have made the correct connections to the switches and contacts. You may wish to assemble a working model first before considering a full scale version, and it would be worthwhile to mount the components on a large board and label all your connections and wiring very clearly.

The guard switches should be constructed so that they spring open when released. Lengths of springy brass could be adapted to make a large number of switches quite cheaply.

The main drawback to this circuit is its high current consumption. It draws current all the time it is armed so a dry battery would be unsuitable. A rechargeable battery, such as a motorbike battery, would be most practical in the long run.

The Circuit

Draw out the following part circuits:

1. Current for bell.
2. Current to switch on the relay.
3. Current to hold on the relay.

Sound to Light Unit

This sound to light unit will produce a flashing lights effect, such as is seen in discotheques and television music programmes. The flashing is in response to the music being played and in this case is controlled by the signal at the loudspeaker terminals of a record player.

Construction

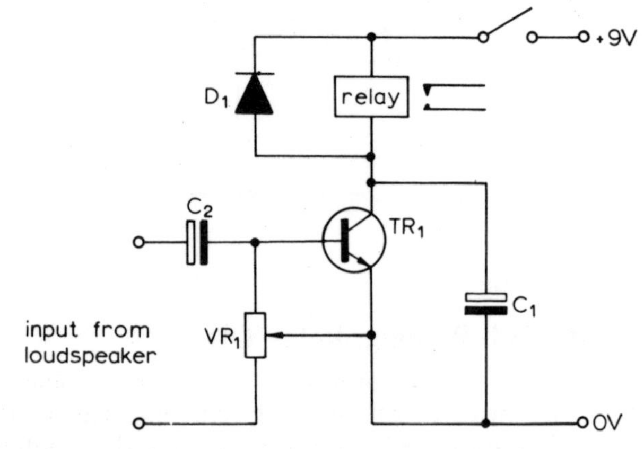

$VR_1 = 500\,\Omega$
$C_1 = 100\,\mu F$
$C_2 = 47\,\mu F$

TR_1 : BC108
Relay coil resistance = $300\,\Omega$
D_1 : 1N4004

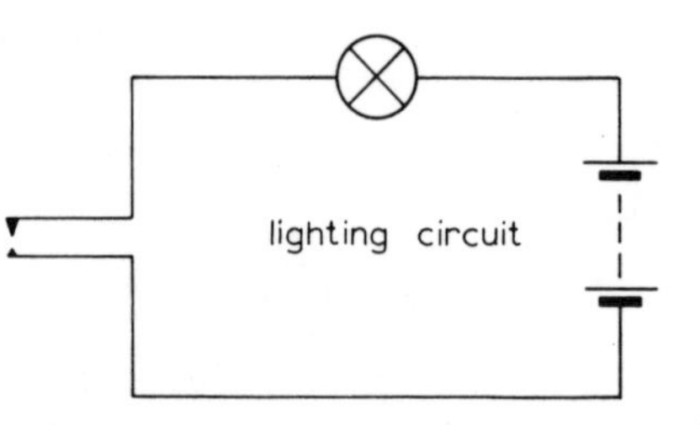

Lighting circuit

Prepare the printed circuit board as shown and ensure that your battery and relay are firmly secured to the board, after the other components have been soldered in place. Mount the unit inside a small case, with cut-outs (and grommets) for the switch cables, relay contact cables, loudspeaker cables and cables for VR_1.

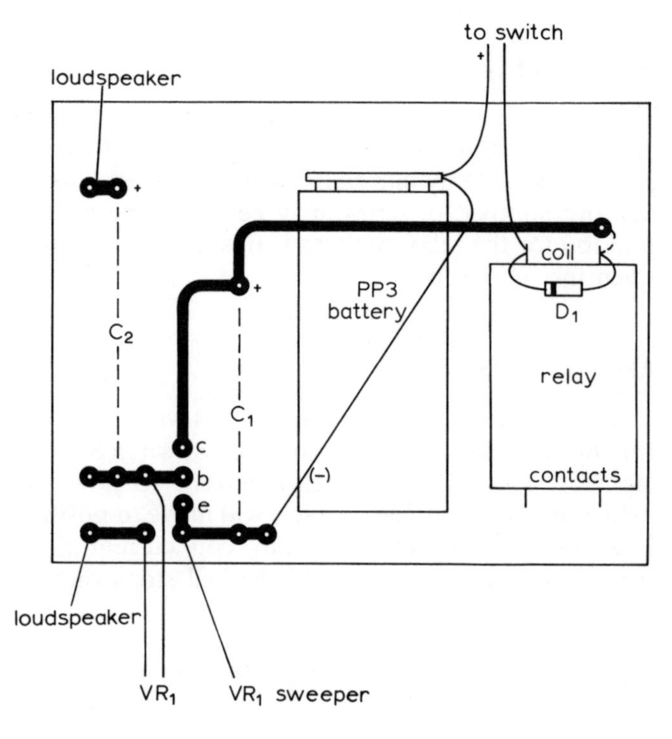

Battery and relay are fixed to non-copper (underside) of board

Testing

Connect the input cables to the loudspeaker terminals in a record player and the relay contact cables into a simple bulb and battery circuit and then play a record. You should find that the relay switches the bulb on and off, in response to the music. Alter VR_1 and you should find that this controls the volume level at which the unit starts to operate.

In Use

The case is fixed inside the record player and the loudspeaker cables soldered in place. Holes are drilled in the record player case for the switch, VR_1 and for a plug and socket arrangement for the light circuit.

The lighting circuit presents a problem, since, in the interests of safety, a low-voltage circuit is recommended and yet this will be very expensive on batteries. A 12 V circuit, using car bulbs, but powered by a 12 V transformer may prove to be a practical unit for you.

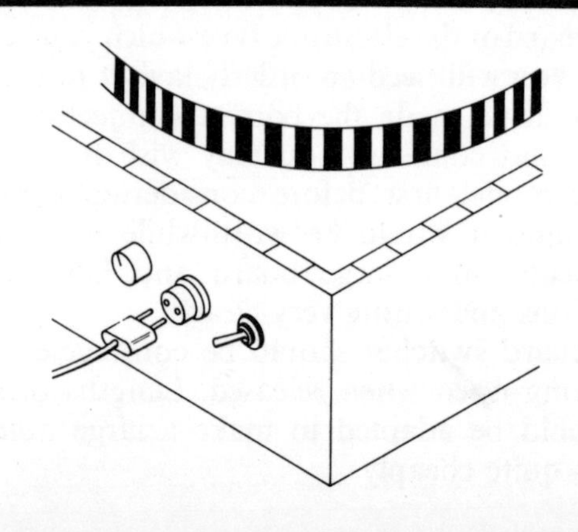

The Inductor

The inductor is the third of the devices that depend on the electromagnetic effect of a current for their operation and is in fact the simplest of the three. The inductor consists, quite simply, of a coil of wire, usually insulated copper, wound on a former.

An inductor has a property, inductance, which is measured in henries (H).

The action of an inductor is described below.

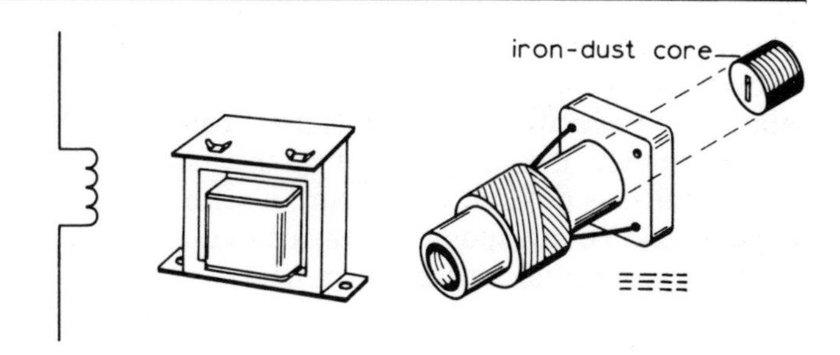

Symbol *Typical appearance*

Operation

Remember two important results from your earlier work:

Changing current in a coil produces a moving magnetic field.

A moving magnetic field cutting' a coil produces an e.m.f. in the coil.

You saw these two effects in operation, quite clearly, in the transformer. The changing current in the first coil produced an e.m.f. in the second. In the inductor, both these effects occur in the same coil and the e.m.f. which is produced is opposite in direction to the one causing it.

The magnetic field produced by changing current in an inductor 'cuts' the inductor itself and produces an e.m.f. in the inductor.

Changing current in an inductor produces an e.m.f. in the inductor which opposes the current change.

An inductor always opposes changes in current.

The Inductor in a Circuit

For the reasons given above, the inductor only affects a circuit (in theory, at least) when the current is changing. The diagrams show what happens at the moments when a current is switched on or off.

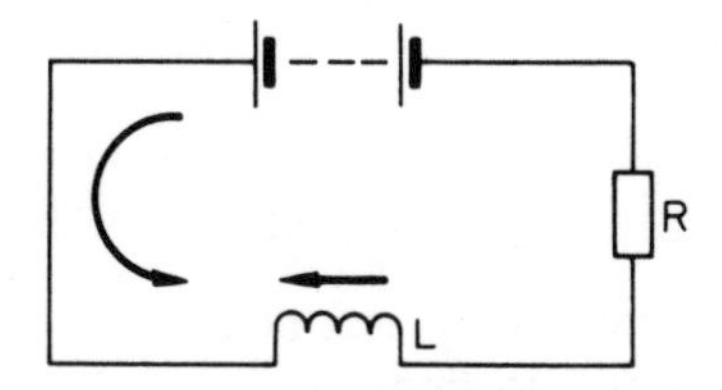

At switch on, the current is rising. The inductor tries to push the current in the opposite direction.

When current starts to increase in an inductor, it produces an e.m.f. to oppose this current increase.

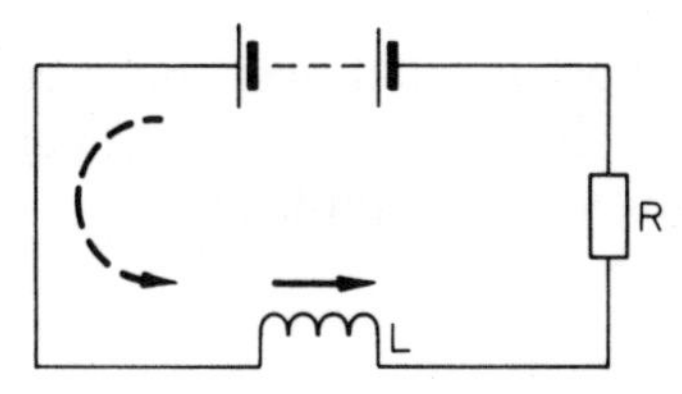

At switch off, the current is falling. The inductor tries to keep the current high.

When current starts to decrease in an inductor, it produces an e.m.f. to try to keep the current high.

Thus, when there is an inductor in a circuit, it always takes longer for the current to rise to its maximum or fall to zero.

This behaviour is called *self-inductance* and its unit is the *henry*.

Current Changes with an Inductor

Having been told that an inductor slows down changes in current, it would be natural for us at this stage to find out how effectively an inductor can do this. Unfortunately, the delay caused by the inductor is so small that it is difficult to measure it with the simple apparatus at your disposal. By way of an example, look at the experimental set-up shown here.

The results you should get from this experiment are given in the graph. The time taken for the current to rise from 0 to 2/3 A is 1/10 s and this is too small for you to measure easily. Even by using a different value inductor you could not achieve a delay that you could accurately measure with school apparatus. This means that you will have to be content with being given the information on the performance of the inductor.

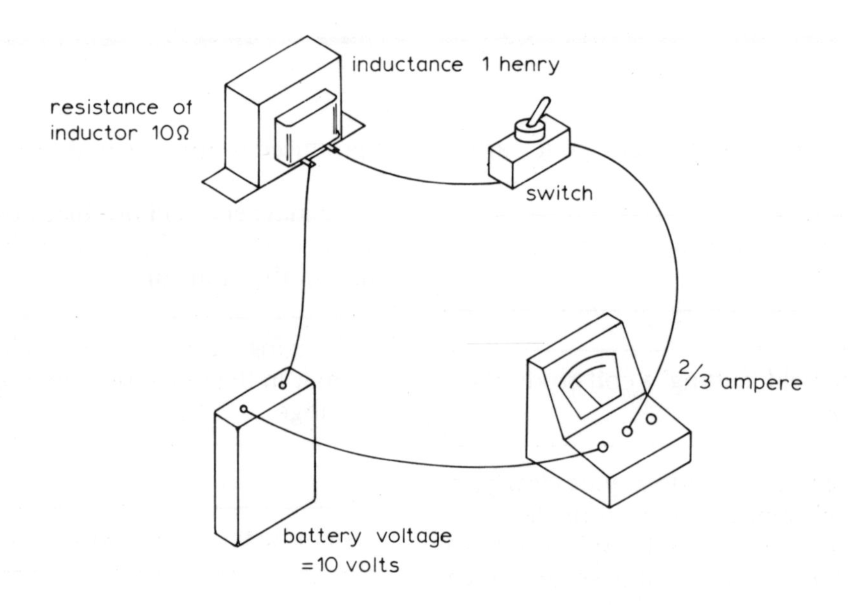

Time Constant

In the above experiment, the current rises from 0 to 1 A (Ohm's Law). The time taken for the current to rise to 1 A is shown by the graph below:

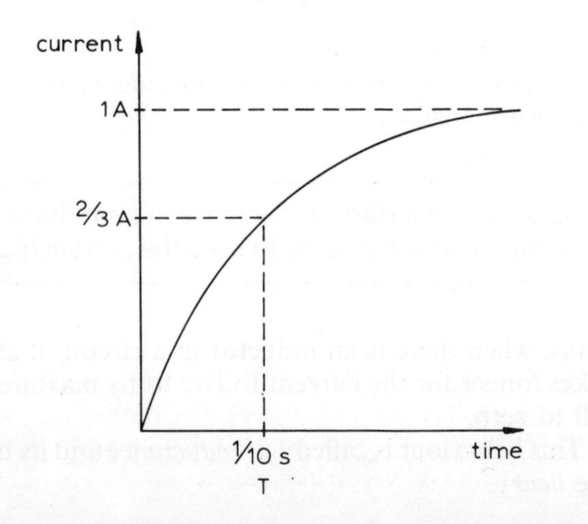

The time constant for an *LR* series circuit is the time taken for the current to rise from zero to two-thirds its maximum value.

The time constant obeys a rule:

$$T = \frac{L}{R}$$

T = time constant in seconds
L = inductance in henries
R = resistance in ohms

This result applies generally to circuits containing a resistance and an inductor. You may see the similarity to the result obtained for the change in voltage across a capacitor. The equation enables you to calculate the time taken for the current to change from zero to two-thirds its full value for any combination of inductor and resistor.

A Metal Detector

This ingenious metal detector is, in fact, a small medium wave transmitter, with a self-assembled coil, like the one you studied on page 58. It can be picked up by a portable radio, tuned to a station, because of the interference it causes. When the coil comes close to a metal object the interference note changes, showing the presence of the metal.

Construction

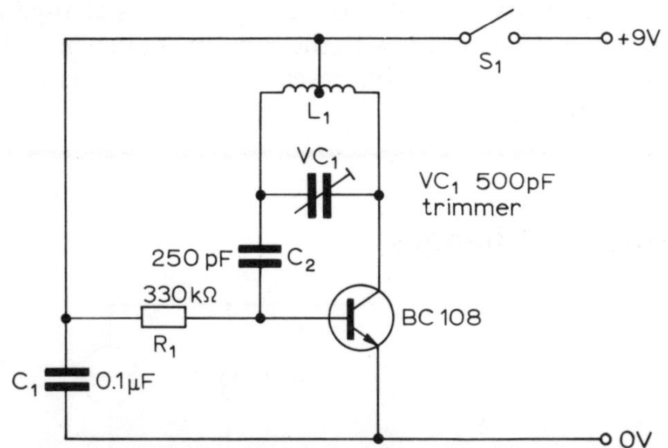

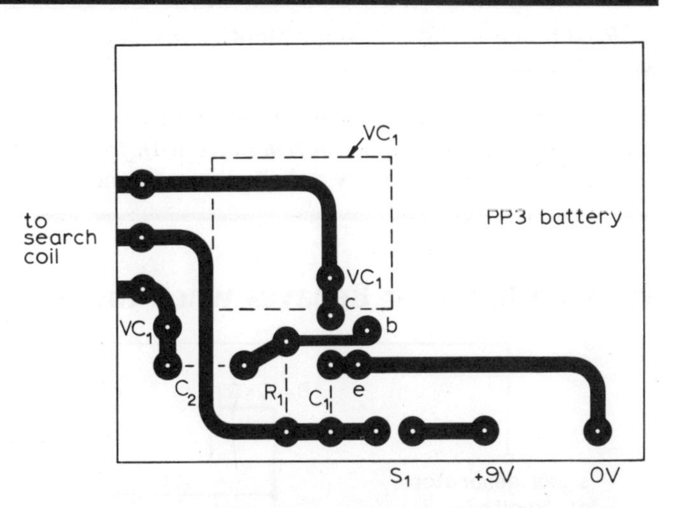

The printed circuit board is shown with space to fit in the battery, variable capacitor and switch directly onto the board, but you must check the sizes of the components you are given against the spaces allowed and also that your variable capacitor does not foul the copper track.

The search coil is made of 20 turns of 22 s.w.g. enamelled copper wire wound on a 15 cm former and centre tapped at 10 turns. The former could be made of a 15 cm diameter circle of hardboard with two slightly larger circles of, say, Formica to give a channel in which to wind the coil.

The circuit board should be mounted in a small case, with cutouts for the switch and the tuning capacitor. The circuit case, coil and the transistor radio could be assembled as shown.

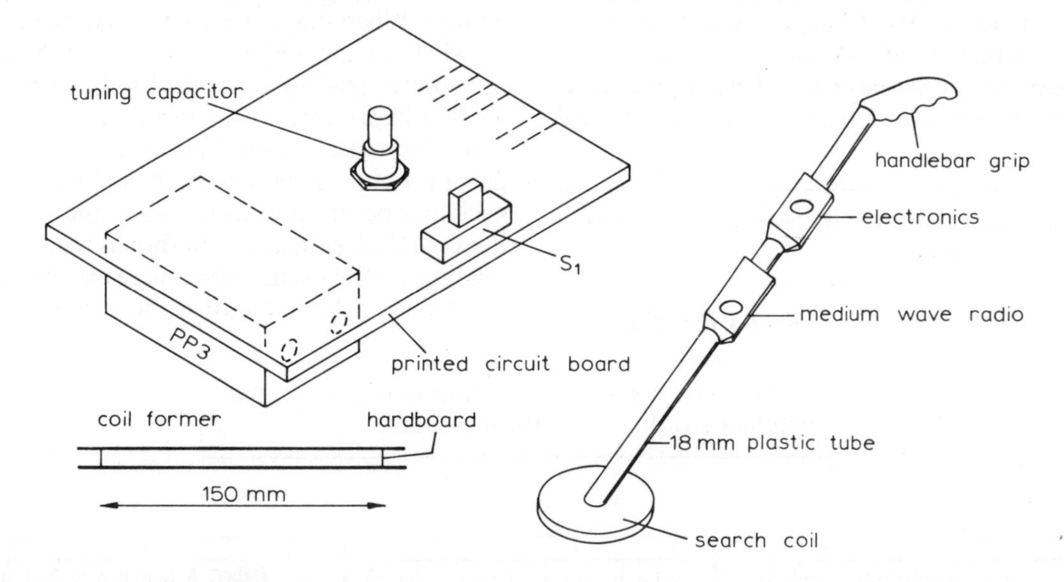

Operation

Tune the radio to a station and then adjust VC_1 till your transmitter causes whistling on the radio. This means that your transmitter is now broadcasting on the same frequency as the radio station. If the coil now comes close to a metal object the whistling will change in pitch or even disappear. The reason of course, is that your transmitter's frequency has changed since the inductance of the coil changes near metal.

The Effect of an Inductor on Alternating Current

Since an inductor opposes changes in current, in a circuit, it follows that it is going to have a reactance to alternating current. Remember

> Reactance is the a.c. equivalent of resistance.

Reactance, is of course, measured in ohms and you are now to follow the *change in reactance with frequency*. The obvious way to follow a change in reactance (or resistance) is to watch the effect on the current in a circuit. As the reactance increases, you will see the current decrease. However, you will probably find that your a.c. ammeters do not give satisfactory results.

Instead, use the method shown here, in which an oscilloscope acts as a voltmeter and the change in voltage across the inductor tells you of its change in reactance. Compare it with the experiment in which the voltage across a variable resistor changes as its resistance changes.

How an Inductor Behaves when the A.C. Frequency Changes

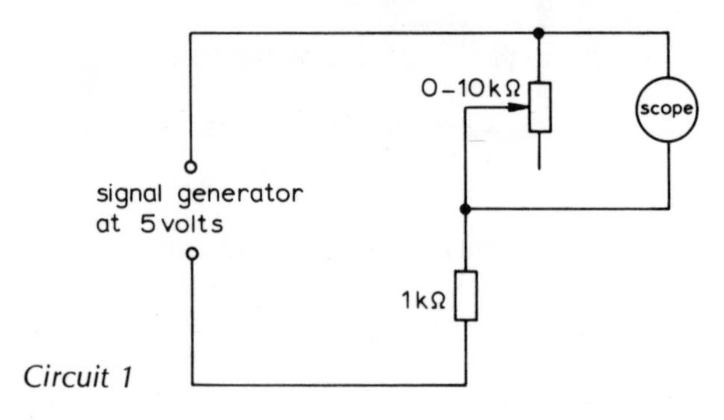

Circuit 1

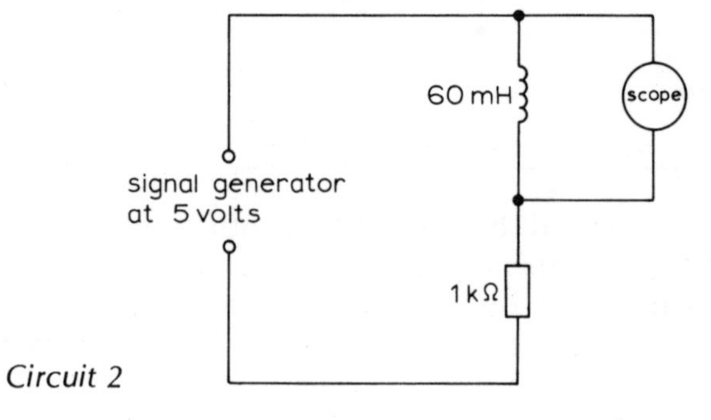

Circuit 2

Instructions

(*a*) First assemble circuit 1, with the generator at, say, 5 V and frequency 1 kHz. Change the setting of the variable resistor and you should find that the size of the signal changes on the oscilloscope. This is, of course, the potential-divider effect which you have studied already.

> As the variable resistance increases, its share of the voltage increases.

For example, in the circuit above, when *VR* equals 1 kΩ, then the voltage across it will be half of the supply voltage. When it is increased to 9 kΩ, the voltage across it will have increased to nine-tenths of the supply. Thus the voltage reading from the oscilloscope will tell you what is happening to the reactance in a circuit.

(*b*) Now use circuit 2 and set the frequency of the generator to its minimum. Read the voltage off the oscilloscope in the usual way and then gradually increase the frequency. You should find that the voltage reading on the oscilloscope increases showing that the reactance of the inductor is increasing also.

> As frequency of alternating current increases, the reactance of the inductor increases.

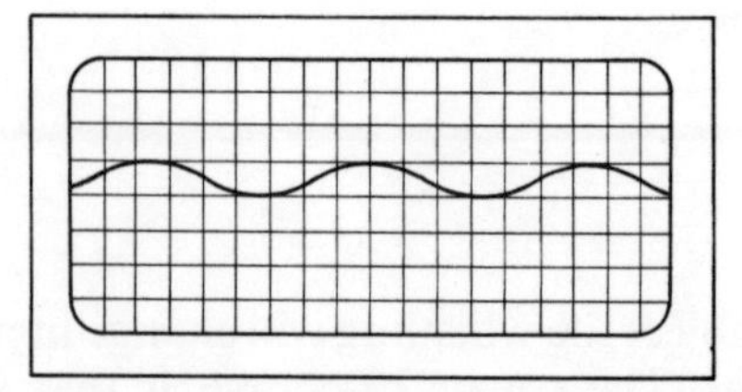

Low reactance with low frequency input.

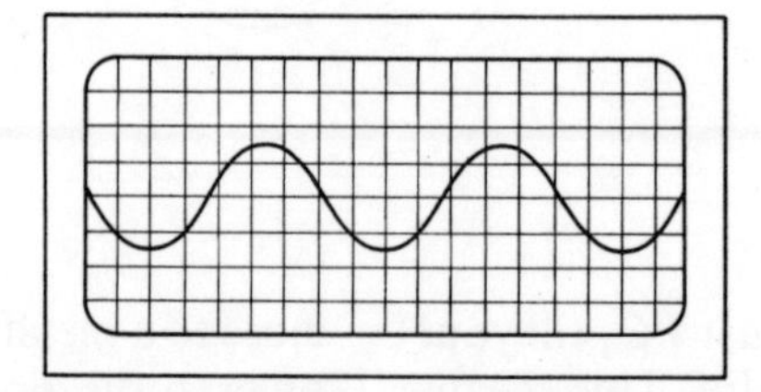

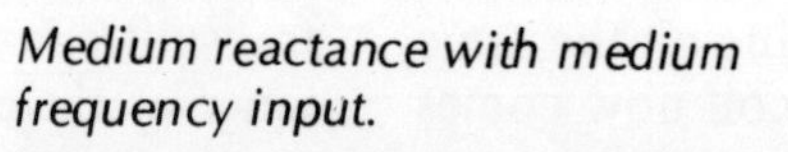

Medium reactance with medium frequency input.

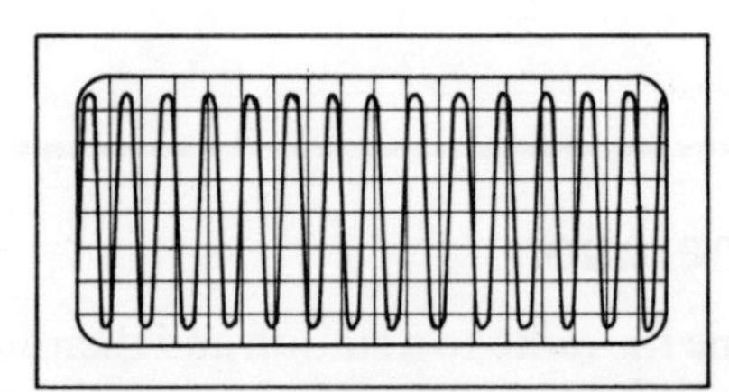

High reactance with high frequency input.

Practical Inductors

Resistance to Direct Current

The inductor is a component which, like the capacitor, strives for perfection but can never reach it. In our experiments we talked about the perfect inductor which meant an inductor which had no d.c. resistance, just a reactance with alternating current.

Since an inductor is made from a length of copper wire it is going to have a small resistance but never zero resistance. So all our inductors are a compromise.

Using thick copper wire makes an inductor which is big, heavy and expensive but keeps its d.c. resistance low, whereas using fine wire makes a small and cheap inductor but its resistance will be higher.

However, if fine wire is used instead of thick wire then far more turns can be wound in the same space so the inductor will have a higher value inductance. The inductor you choose will have a balance of these extremes of its properties.

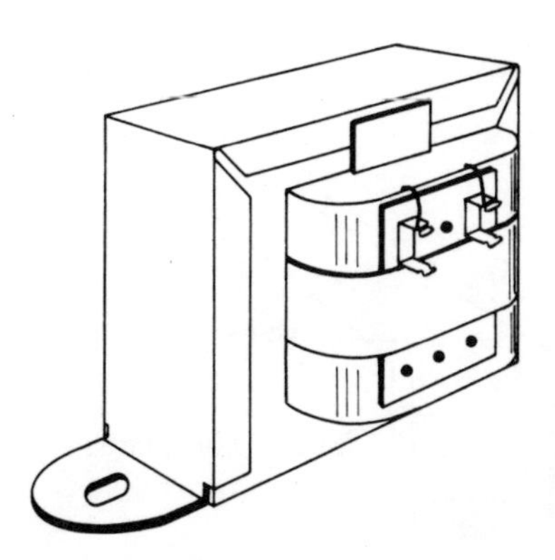

Inductance = 3 H
D.C. resistance = 68 Ω
Weight = 2·1 kg
Thick wire in the coil gives a low resistance but a high price and weight.

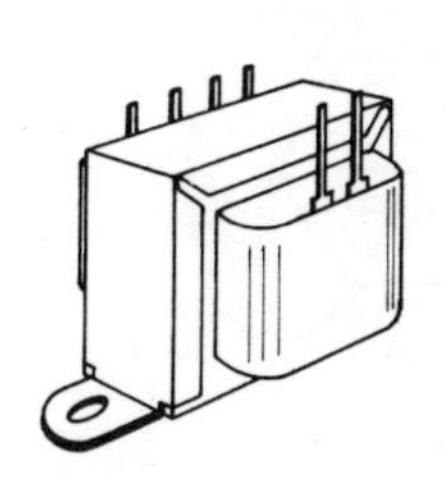

Inductance = 5 H
D.C. resistance = 290 Ω
Weight = 0·24 kg
The use of fine wire increases the number of turns and gives a high inductance at the expense of a high resistance.

The Core

The core material in an inductor is used to increase the electromagnetic effect produced by the coil and so it increases the inductance. The two materials most commonly found in the core of an inductor are air and ferrite, which is a compressed powder, and iron. When only a small inductance is required, the inductor may be air-cored.

Variable Inductor

This inductor is especially useful since its inductance can be varied. When the core is screwed in to the coil, the inductance increases.

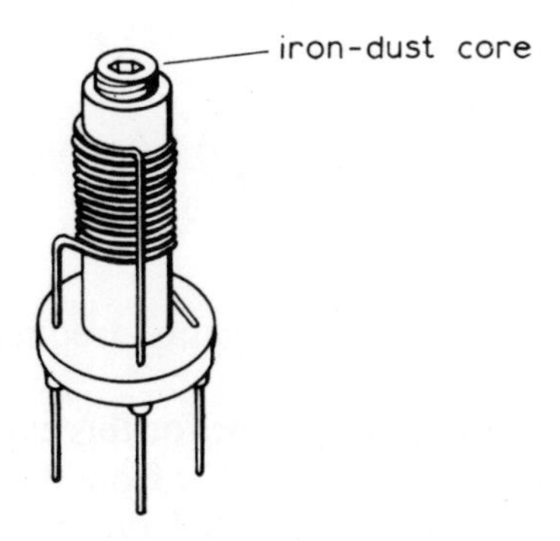

Laminations

Here the core is laminated soft iron. It is composed of plates of soft iron which are insulated from each other in a multi-layer sandwich. As you know, the current flowing in the coil will induce an e.m.f. in the soft iron and a current would flow in the core if it were made from one piece. This undesirable effect is stopped by splitting the core into many plates, all insulated from each other.

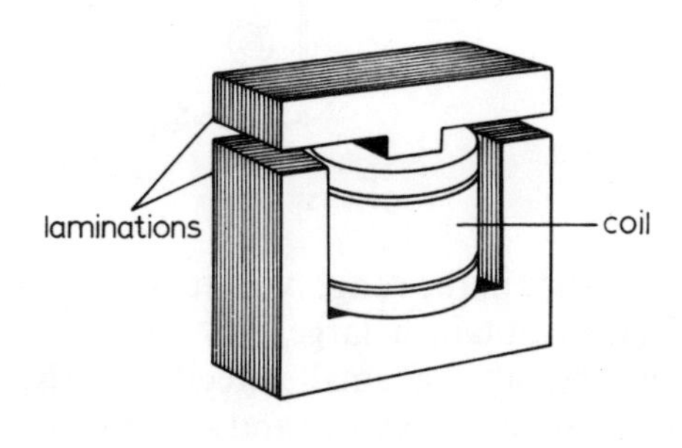

The Loudspeaker

Construction

The loudspeaker is the fourth device that depends on the magnetic effect of an electric current. Like the others, it contains a coil, where the magnetic field is produced and it also has a powerful permanent magnet. A lightweight fibre cone is fixed to the coil near its centre and is fixed to the metal frame at its outside edge.

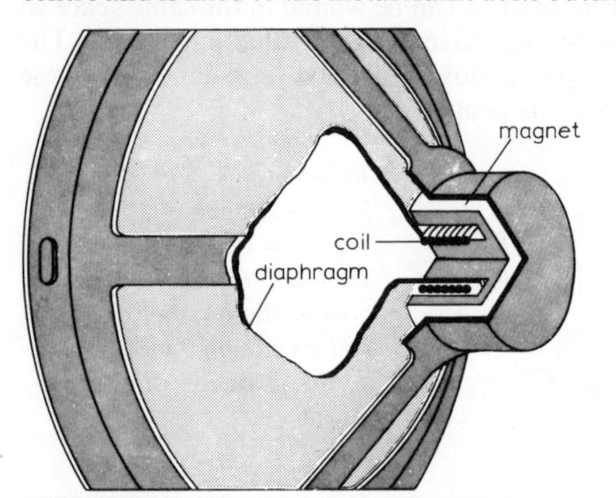

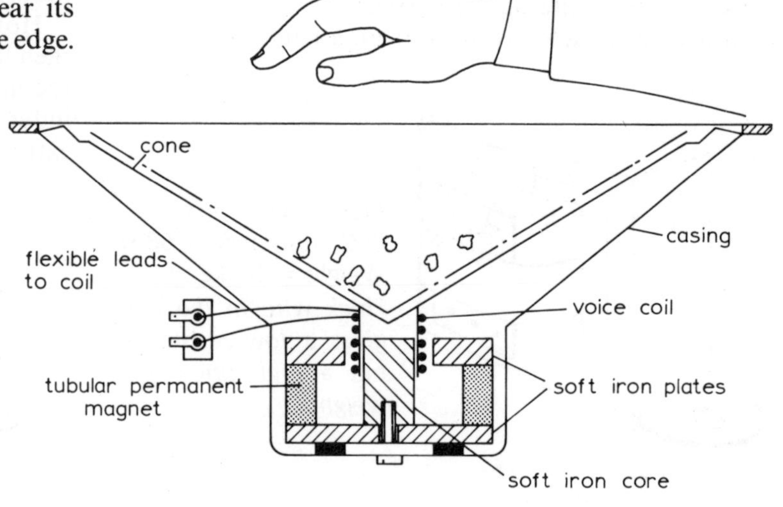

Operation

Alternating current is fed into the coil and this, of course, causes a magnetic field which constantly changes direction. This creates a 'push and pull' effect against the permanent magnet and so the cone moves backwards and forwards. This movement pushes sound waves out from the loudspeaker. It is easy to show this movement, by placing small rolled balls of paper on the cone and playing a low note through the speaker.

A hand held over the speaker will feel pulses of air produced by a low frequency note.

Frequency Range

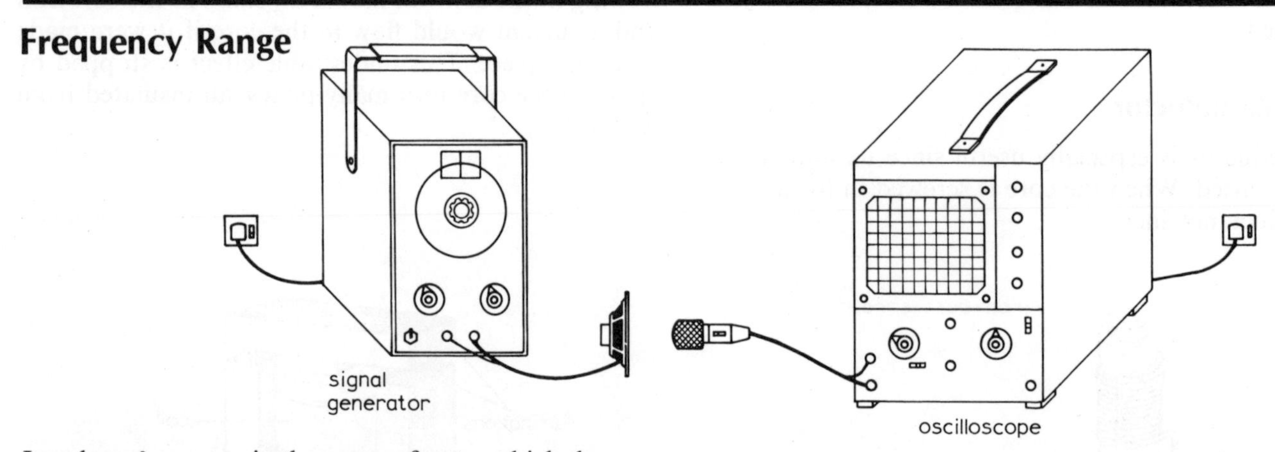

Loudspeakers vary in the range of notes which they can produce. Try to obtain a large, say 25 cm or 30 cm loudspeaker and also a small tweeter loudspeaker. Connect them, in turn, as shown and sweep through the whole range of your signal generator. Measure the voltage shown on the oscilloscope for different frequencies. You will see that your loudspeaker produces a certain range of notes quite well but its performance falls off either side of this range.

Plot a graph of voltage against frequency for your different loudspeakers.

Note: We have had to assume for this experiment that your microphone does not suffer from the same fault as your speaker—that is, a limited range. This is not of course true, but a reasonable quality microphone should be able to show you the differences between your loudspeakers.

Electromagnetic Effects—Summary

Direct Current in a Conductor

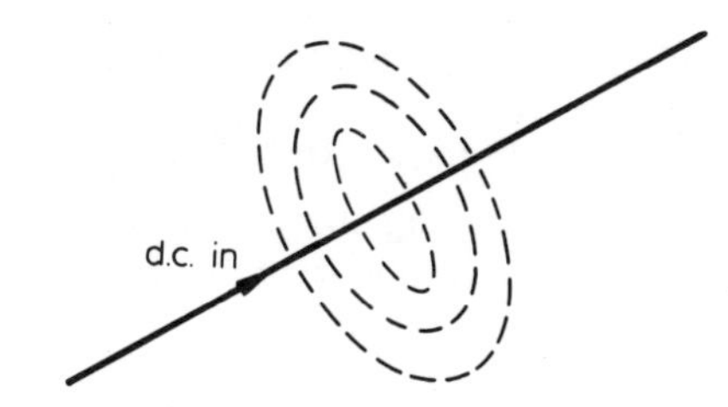

Current in a conductor produces a magnetic field.

Direct Current in a Coil

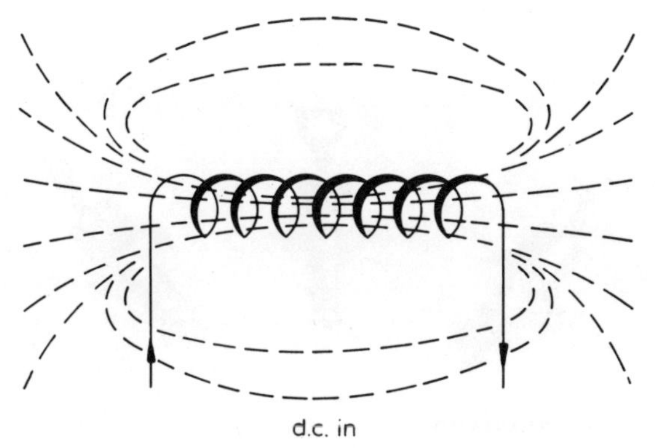

A coil wound conductor produces a stronger magnetic field.

Alternating Current in a Coil

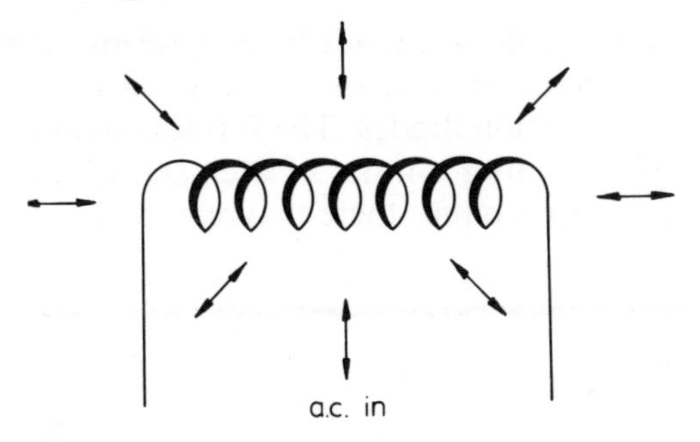

Alternating current in a coil gives a magnetic field that is constantly changing.

Moving Magnetic Field

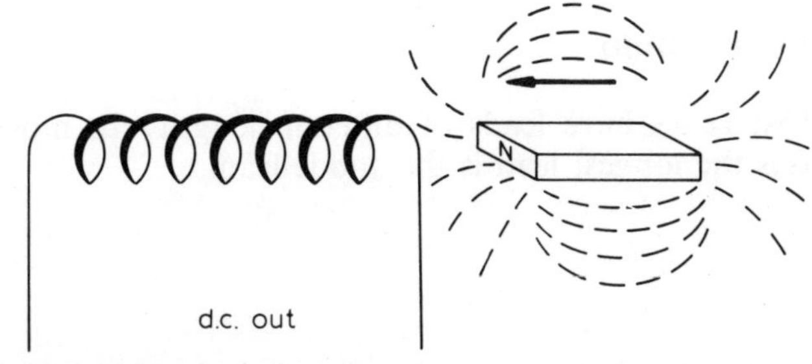

If a magnetic field moves near a coil an e.m.f. is produced.

Induced E.M.F.—The Transformer

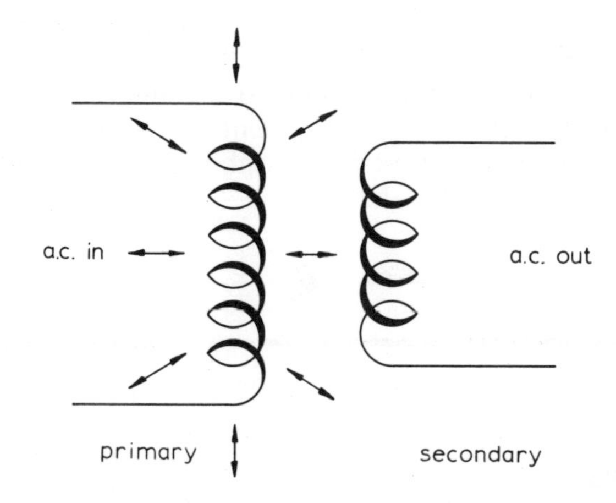

The moving magnetic field produced by alternating current results in an e.m.f. being induced in the secondary.

Induced E.M.F.—The Inductor

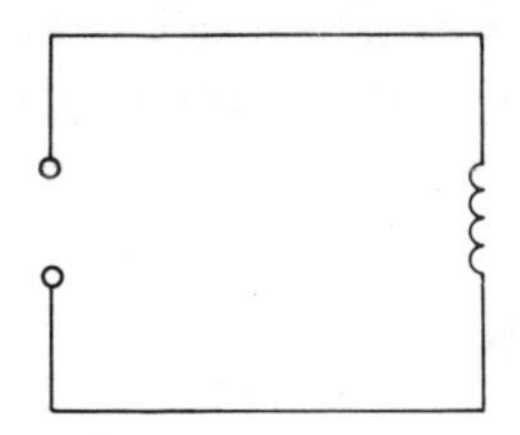

The e.m.f. produced by the current in an inductor slows down any changes in current.

Reactance

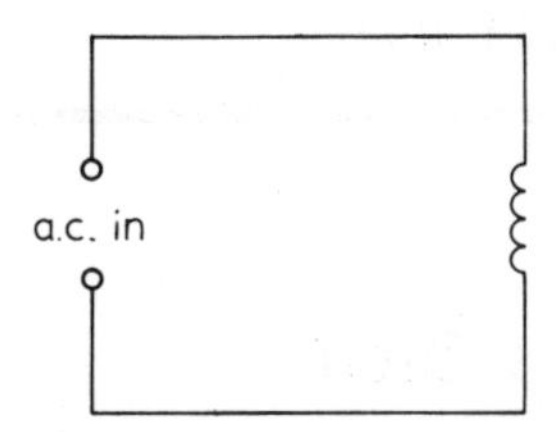

The inductor provides a resistance to alternating current, called reactance.

Reactance Increases with Frequency

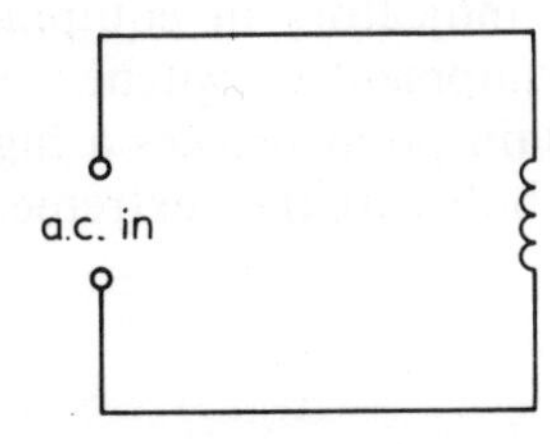

As the frequency of the alternating current rises, the reactance also rises.

6 THE DIODE
Introduction

Function

The diode is a device that allows current to flow in one direction only. It has two leads, called the anode and the cathode, and it is important to distinguish between them. The cathode is often identified by a red band round the casing or the diode symbol may be printed on the case to identify the leads.

Specifications

The most important specifications about a diode are the maximum current and voltage it can tolerate without damage and these will always be quoted for any diode. With alternating current circuits the current will be quoted on the average for the whole cycle, $I_{F(ave)}$.

Uses

Particular uses for a diode include power rectification (see page 93), switching circuits and detector/demodulation circuits (see page 98). Manufacturers' catalogues will usually state for which purposes a diode was designed. Diodes designed to carry large current may require mounting on a heat sink to keep them cool and these are usually stud-mounting types.

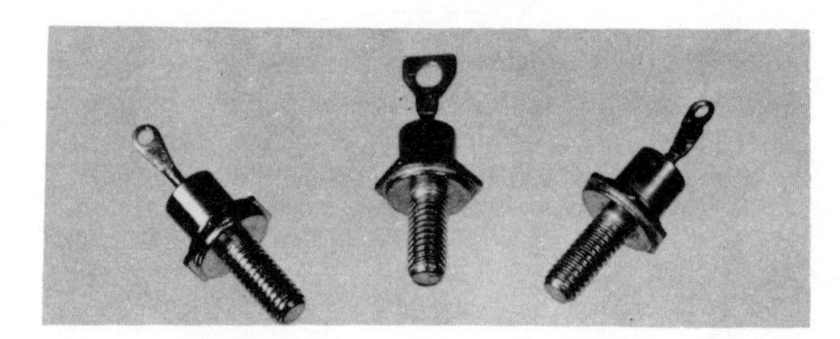

Stud mounted diodes

Construction

The diode is the first of the semiconductor devices you will study. These are devices made from the elements silicon or germanium, which have resistance values between these of the conductors and those of the insulators.

The diodes you will meet will be of two types—point contact and junction diodes. The former contain a thin metal wire in contact with a piece of silicon or germanium whilst the latter consist of a crystal of silicon or germanium with two distinct regions in it. The crystal has been treated to produce these two regions and they are called n-type and p-type.

Light Emitting Diode

The light emitting diode (l.e.d.) is a special kind of diode which does as its name suggests. However, you would never use it to help you see in the dark since it does not produce that much light. Light emitting diodes are mainly used as indicators in equipment. They can indicate when equipment is switched on or when the voltage at a certain point reaches a high value. Their advantages over bulbs are their extremely long life, low current consumption and, of course, they only emit light when the current flows in the 'correct' direction.

Identification

Most l.e.d.s have leads of different lengths. In most cases the longest lead is the anode lead.

Diodes in Circuits

Simple Circuits

Instruction

Assemble, in turn, the two very simple circuits shown here. Notice that your diode must have a rating of at least 60 mA. ($I_{F(ave)}$ must be greater than 60 mA)

To memorise this easily, remember that the diode conducts conventional current in the direction of the arrow of its symbol.

9V

current flows

Said to be forward biased

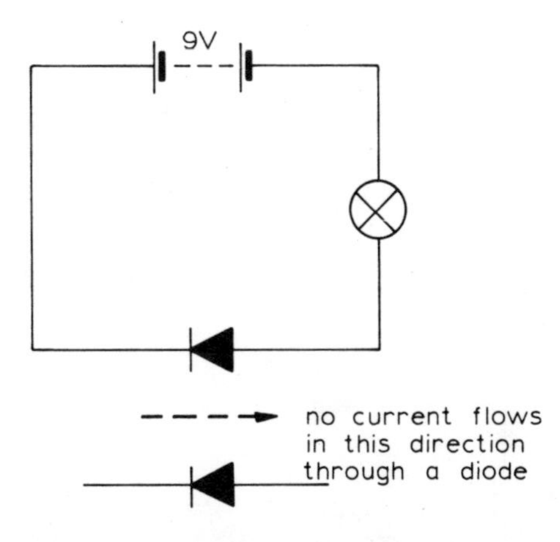

Reversed bias

Further Practice

Instructions

The two circuits here merely have the battery polarities reversed. Show what happens when the circuits are connected to the battery. Now ask your partner to connect the circuit for you with the battery terminals obscured and then identify which way round he has connected it.

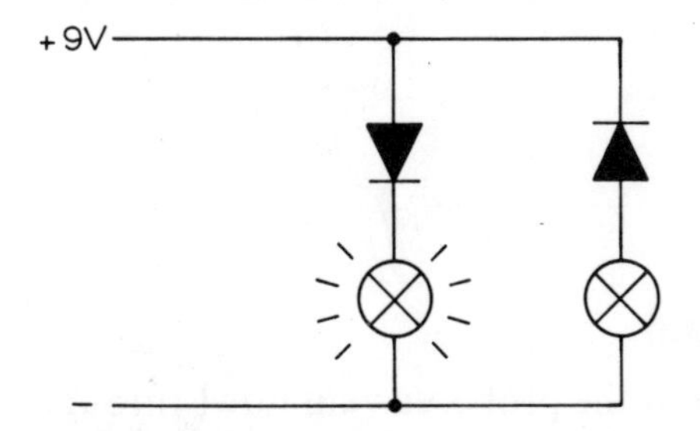

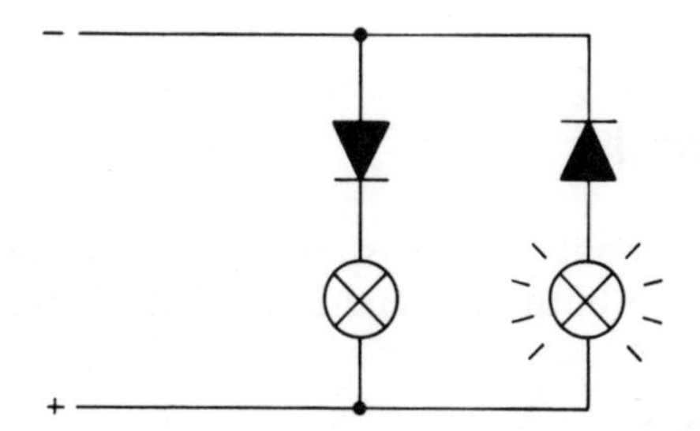

Light Emitting Diode

Instructions

Use whatever l.e.d.s you have available in the simple circuit shown and use a value of 2·2 kΩ for *R*.

Reduce the value of *R* in stages (to increase the current) and note the change in brightness of the l.e.d. Do not take *R* below 330 Ω unless you are told to do so.

Try to calculate, roughly, how much current is needed to give an easily visible glow.

Try reversing the diode in the circuit.

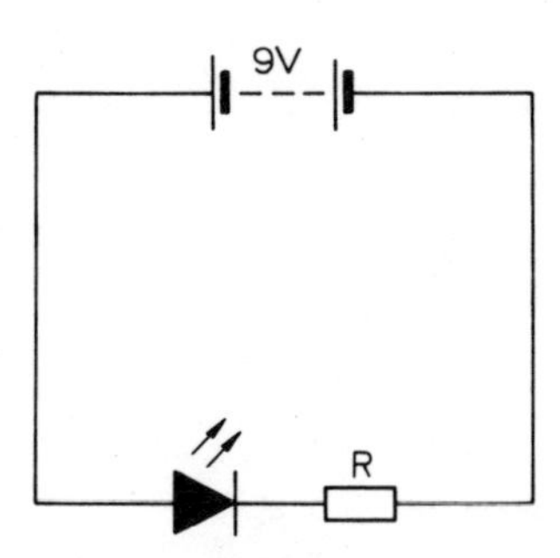

Voltage Indicator

This is a project involving the use of light emitting diodes and depends on the fact that they conduct in one direction only and emit light when doing so. It is a piece of test apparatus and will act as a simple voltmeter, so it will detect points at high and low voltage in a circuit under test. It is especially useful for testing computer circuits, as in Chapter 12.

Construction

The diagrams are self-explanatory, but the circuit must be tested before it is glued into its case. The case could be a felt-tip casing and the diodes should be a suitable size for fitting inside this: 3 mm or 5 mm would do. The small size $\frac{1}{8}$ W resistors would make construction easier.

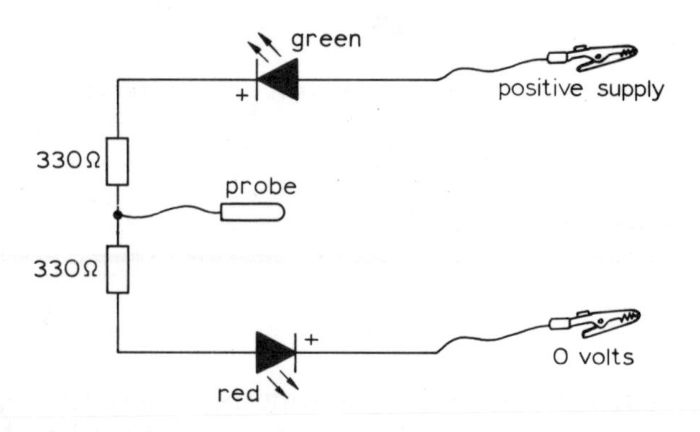

Use

The two crocodile clips attach the indicator to the battery terminals of the circuit under test. The probe is used at any point in the circuit in which you are interested. A green light indicates a point at low voltage and the red light shows a point at high voltage.

Do not use the indicator with a supply above 12 V.

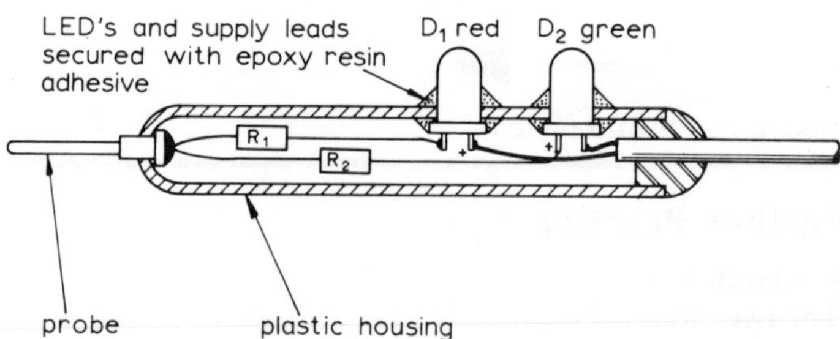

The Circuit

The two circuits below show you how the device performs for both high and low inputs to the probe.

High Input

With 9 V at the probe, current flows as shown through the red l.e.d. No current flows through the green one because both ends of it are at the same voltage (9 V).

Low Input

With 0 V at the probe, current flows through the green l.e.d. No current flows through the red one because both ends of it are at 0 V.

This is how the l.e.d.s show you the voltage at the probe.

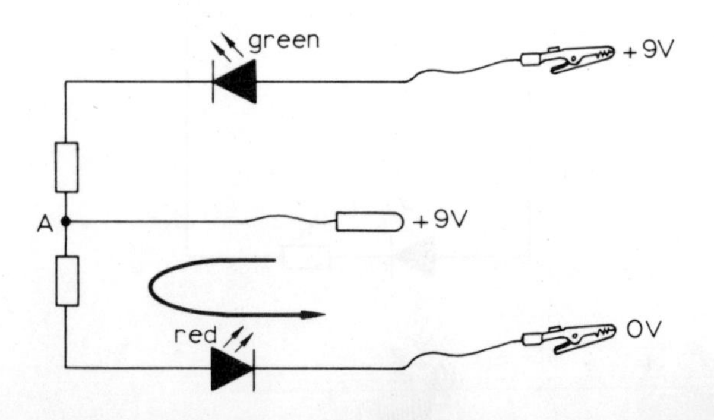

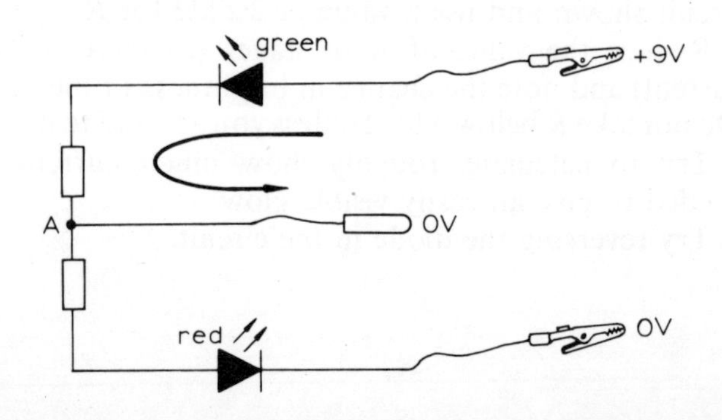

Current/Voltage Measurements with a Diode

You have seen the main effect of a diode, namely that it passes current in one direction only. This experiment will give you information about what happens to the currents and voltages in a circuit containing a diode, when it is conducting. The current is controlled by the resistor *R*. You are to find the voltage produced across the diode by this current.

This is another experiment where a low quality voltmeter will not give good results.

How are Current and Voltage Related for a Diode?

Instructions

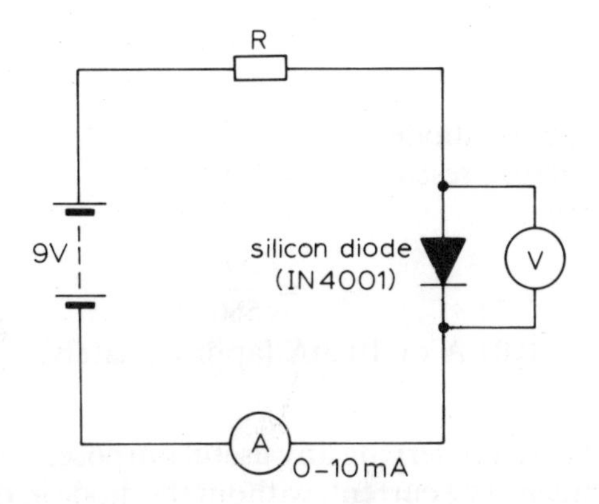

Take the current and voltage readings asked for in the above circuit for $R = 1\ k\Omega$ and repeat for the other values of *R* in the table. Record your values and plot a graph of current against voltage.

R (kΩ)	I (mA)	V
1		
2·2		
4·7		
10		
22		
100		

Results

Your graph should look like the one here. Now, what does this mean about the practical use of a diode?

Diode Protection

The diode does very little to cut down the current in any circuit. With a supply to the diode rising above 0·6 V, the current would rapidly reach a large enough value to damage the diode. This means that a diode will need a resistor in series with it to protect it from too large a current.

Voltage Supply

Since this figure of 0·6 V for a silicon diode is reasonably constant, for any current over a few milliamperes a diode is sometimes used in circuits to supply a fixed voltage. Two diodes in series, for example, would supply 1·2 V. These two figures can be very useful in transistor circuits.

Germanium Diodes

The germanium diode behaves in a similar way to the silicon diode, except that it reaches its maximum at 0·2 V.

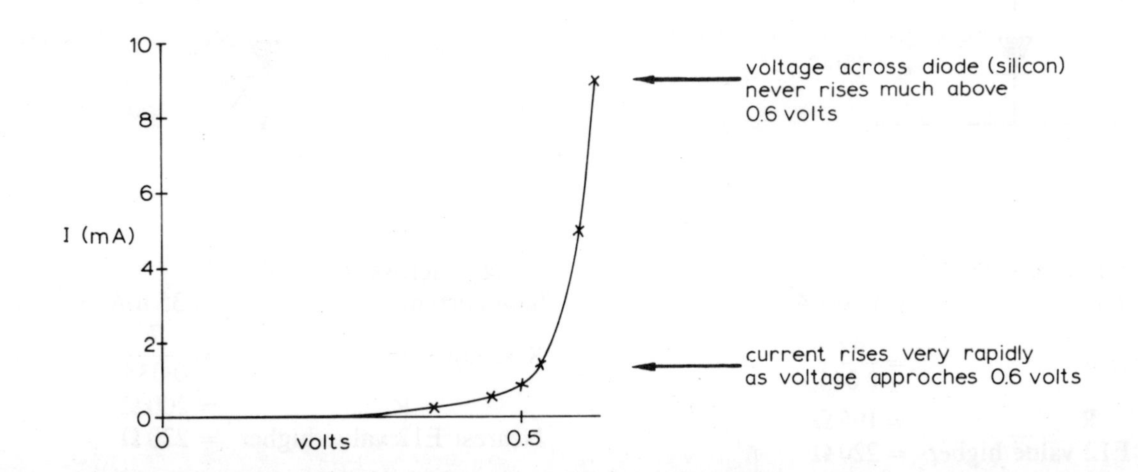

Current Calculations

The key to doing these calculations is to assume that the voltage drop across the diode is: Silicon: 0.6 V; Germanium: 0.2 V, Light emitting diode: 2 V (approximately).

Find the Current in this Circuit

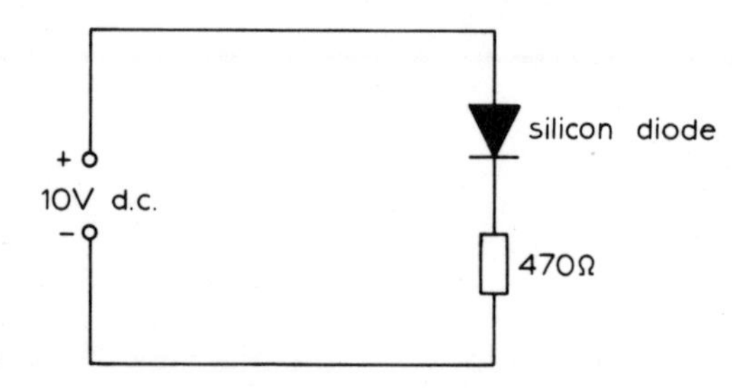

Voltage across diode = 0·6 V
Voltage across resistor = 9·4 V
Resistance = 470 Ω

Current $= \frac{9{\cdot}4}{470}$

Current = 0·02 A or 20 mA.

Find the Current in this Circuit

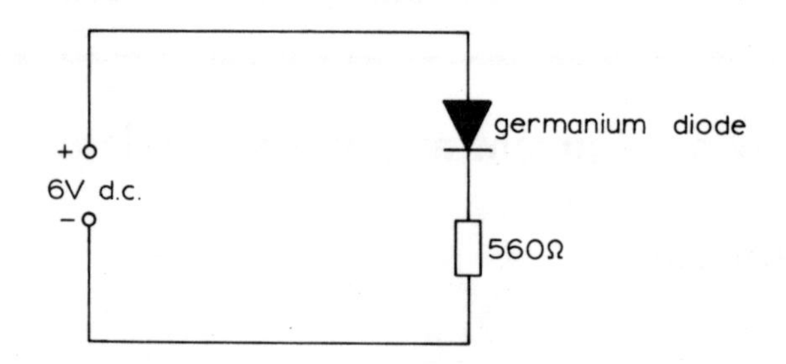

Voltage across diode = 0·2 V
Voltage across resistor = 5·8 V
Resistance = 560 Ω

Current $= \frac{5{\cdot}8}{560}$

Current = 0.01 A or 10 mA (approximately).

(These circuits are only to illustrate the calculation. The diode is not actually serving any useful purpose. You may notice that the diode does not, in actual fact, affect the current very much. The current, without the diode in the first circuit above would have been 21.3 mA)

Diode Protection

You need to be able to calculate the resistor needed to protect a diode from too large a current. Find the nearest E12 series value to allow a safe current in the part circuits here.

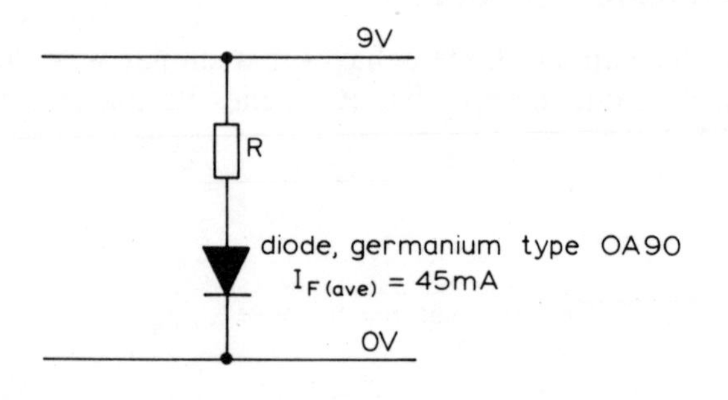

Voltage across R is 8·8 V
Safe current = 0·045 A

Resistance R $= \frac{8{\cdot}8}{0{\cdot}045}$

R = 195 Ω
Nearest E12 value higher = 220 Ω

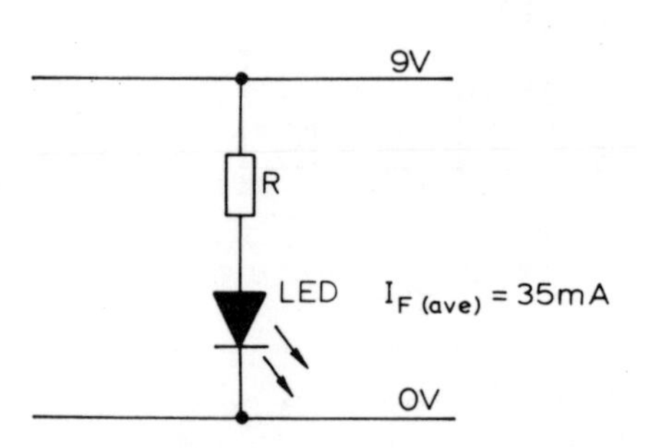

Voltage across R is 7 V
Safe current = 35 mA

Resistance R $= \frac{7}{0{\cdot}035}$

R = 200 Ω
Nearest E12 value higher = 220 Ω

Rectification

The fact that a diode passes current in one direction only is put to good use since the electricity supply in most countries is alternating current and most electronic devices operate on direct current. The process of changing alternating to direct current is the job of the diode and is called rectification.

The Diode in an Alternating Current Circuit

Your teacher will select a diode that will conduct a large enough current for the circuits shown. He may wish to do the motor experiment himself to prevent damage to the motor.

Instructions

Examine the effect of putting a diode into the three alternating current circuits here. Describe what happens to the property listed to the right of each circuit.

Results

You should see that, after inserting the diode, the (*b*) circuits appear to be producing direct current and that the effects correspond to about half of what you would expect from a 6 V supply. It appears that half of the a.c. supply has been stopped. Use the oscilloscope to check this.

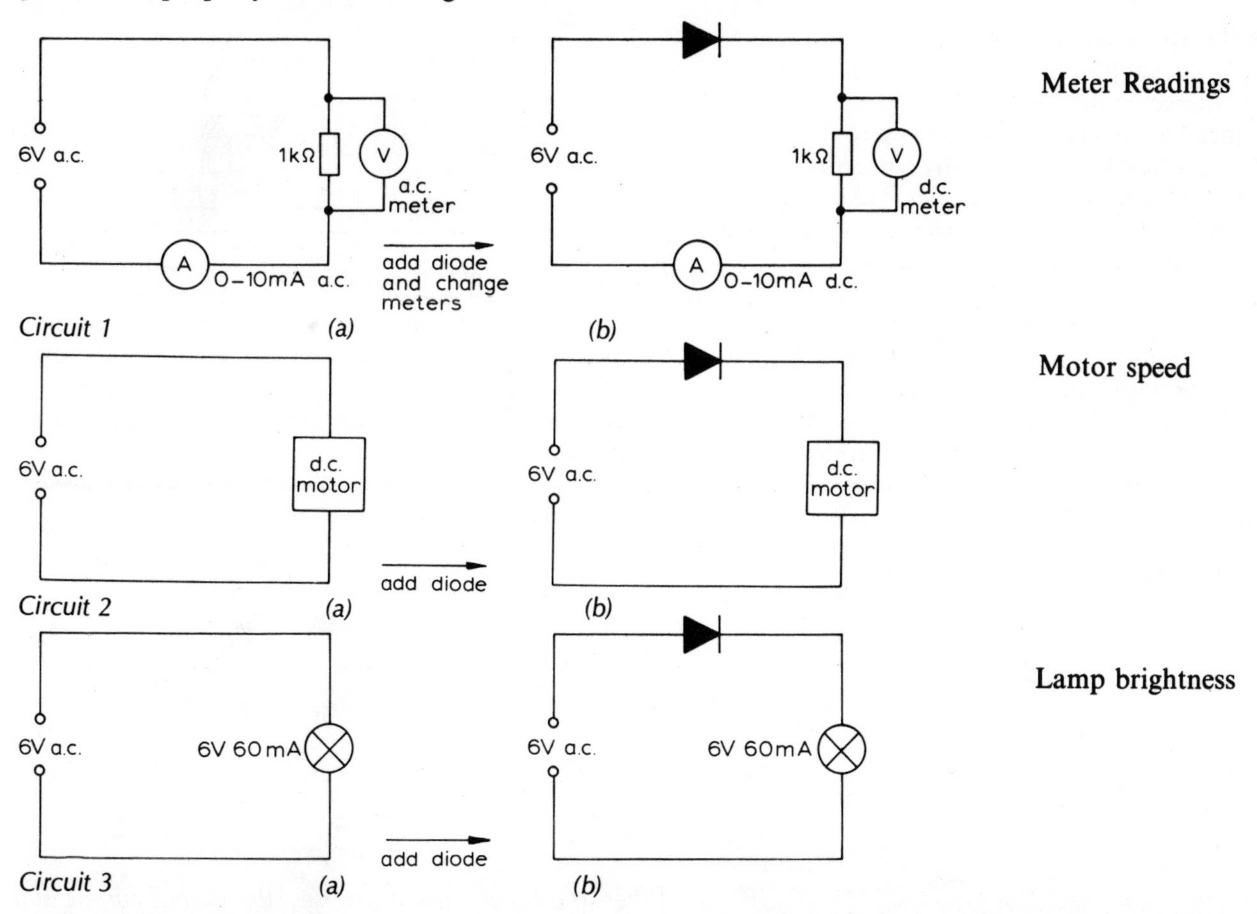

Rectification Display

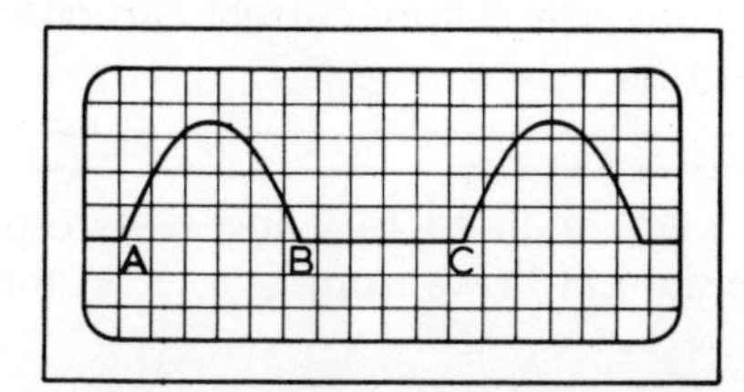

Replace the voltmeter in circuit 1(*b*) with an oscilloscope.

Between times A and B, the supply is unaffected.

Between times B and C no current flows.

This is now direct current.

The direct voltage now averages half of the original alternating voltage because of the half cycle with zero voltage.

This is known as *half-wave rectification.*

Full-Wave Rectification

The circuit here is a very ingenious method of achieving rectification and this arrangement of diodes is called a diode bridge. It is such a common circuit that it is often sold with all four diodes encapsulated in one component, connected as shown.

Instructions

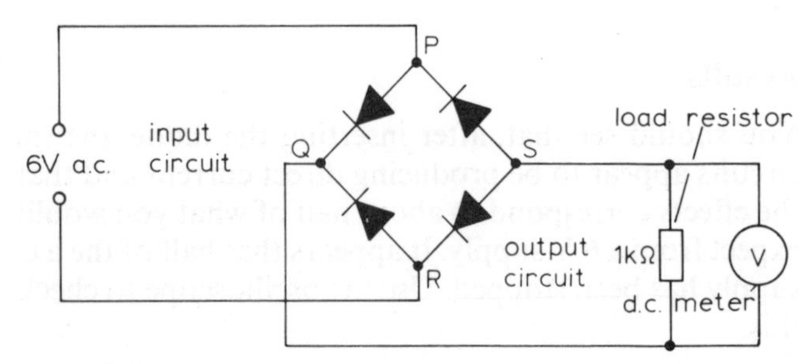

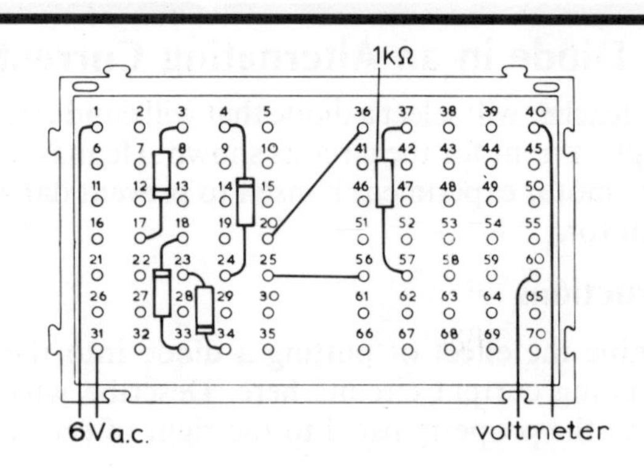

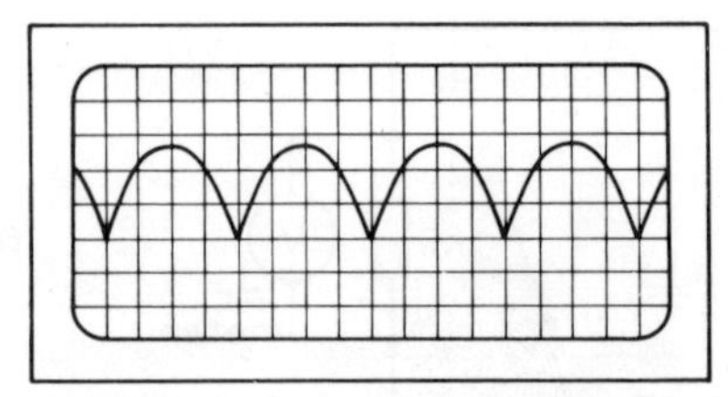

Oscilloscope display of rectification

Assemble the above circuit and connect the voltmeter and the 6 V a.c. supply. It is most important that the diodes are connected the right way round, so take care. If you do make a mistake and connect up to the power supply, then it is too late to make any corrections since you will have 'blown' two of the diodes. Take a reading with a d.c. voltmeter across the load resistor and then replace the voltmeter with an oscilloscope.

Voltmeter reading =

Notice that now, in the output circuit, the current flows in one direction only. This is *full-wave rectification.* No part of the alternating current input is missing, it has merely all been 'steered' in the same direction round the output circuit. Notice, too, that you still have alternating current flowing in the input circuit. If you do not believe that this is so, then connect your oscilloscope across points PR.

So, how can you input alternating current and output direct current?

Explanation

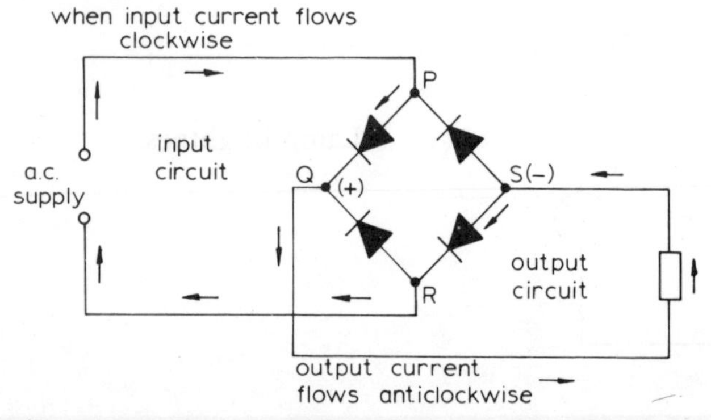

When the input current flows clockwise, the output current flows anticlockwise.

The direction of the current has been marked on the diagrams here. Remember that a diode can conduct current in one direction only. Follow the arrows and check that the current can only go in one direction in the output circuit. (Notice that the current must flow from the a.c. supply, through the bridge, through the load resistor, back through the bridge and back to the a.c. supply.)

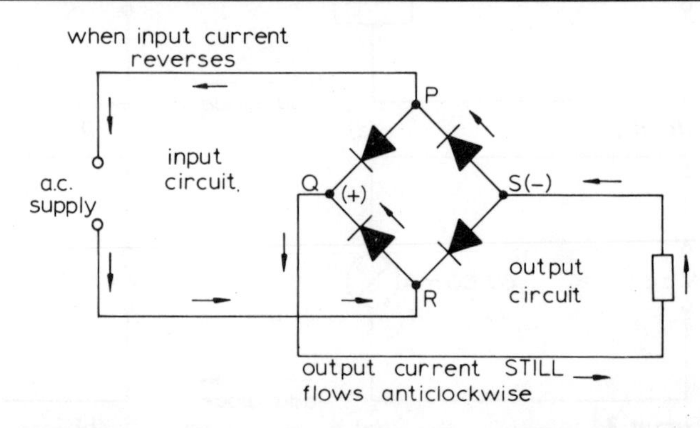

When the input current reverses, the output current still flows anticlockwise.

Thus, although the input current reverses in direction regularly, the output current can only flow in one direction.

Q and S are marked (+) and (–) respectively because current flows from Q to S, in the output circuit.

Rectification—Smoothing

On page 94, you saw how to obtain fluctuating direct current from alternating current. Most electronic equipment needs a steady d.c. supply as shown below. Let us see if we can smooth the output from your last circuit.

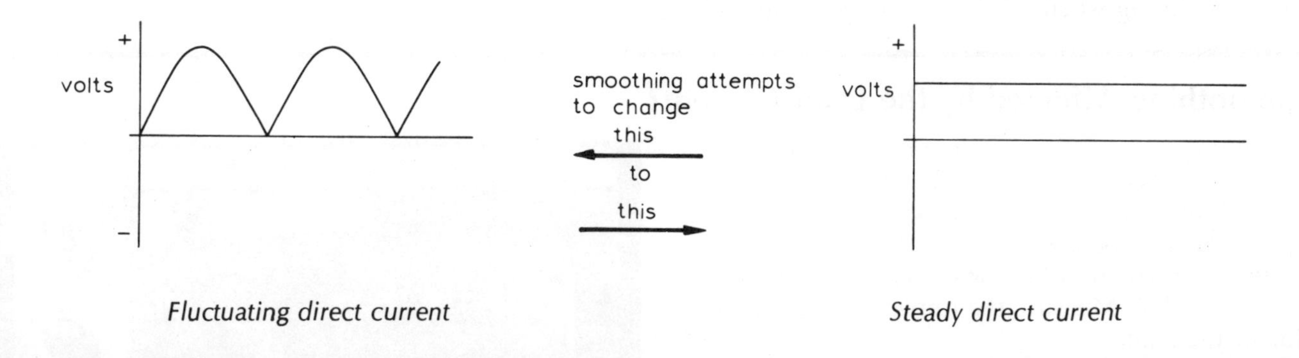

Fluctuating direct current

Steady direct current

Using a Capacitor for Smoothing

Instructions

Put a capacitor in parallel with your load resistor in your last circuit and connect your oscilloscope as shown. Notice that your electrolytic capacitor must be connected with the correct polarity, the bottom rail being positive. Examine the oscilloscope display of your output circuit.

Clearly, the output is a lot closer to the steady direct current we are looking for and only a small ripple remains. Let us now look more closely at this ripple.

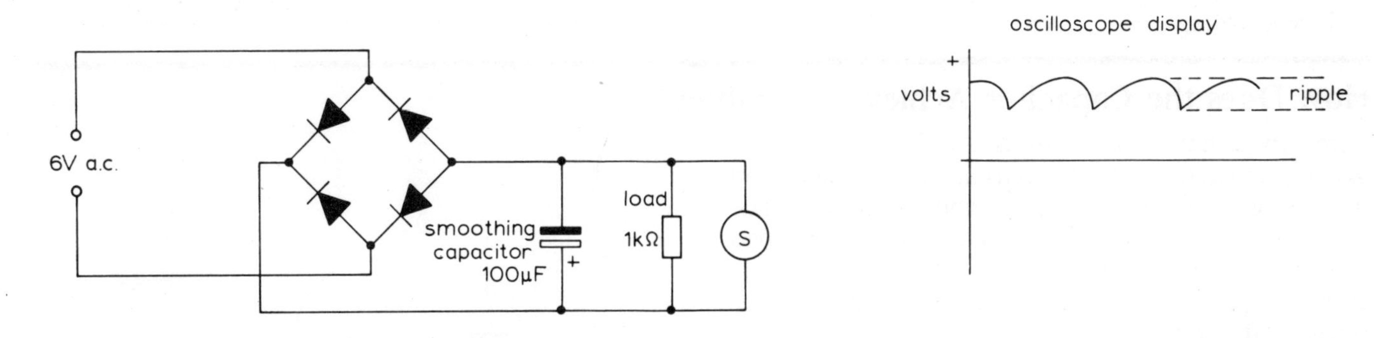

Effect of Capacitor Size on Smoothing

In the table below you will see the four values of capacitor you are to use for smoothing. Find the size of ripple (in volts) left in the output by each capacitor in turn.

1. Set the a.c./d.c. switch on your oscilloscope to alternating current. This will keep the trace in the centre of the screen for you.
2. Put the test capacitor in the circuit, in place of the capacitor already there.
3. Change the volts/cm switch until the ripple occupies as much of the screen as possible.
4. Measure the ripple voltage.
5. Repeat for the other values of capacitance.

Capacitance (μF)	Volts/cm switch	Peak-to-peak height of ripple (cm)	Peak-to-peak voltage of ripple (V)
1			
10			
100			
1000			

Your results will speak for themselves:

> The greater the value of capacitance, the more effective is the smoothing.

Rectification—Drawing Load Current

Up to now, the current which these rectification circuits have had to provide has been small. For instance, with an a.c. input of 6 V and a load resistor of 1 kΩ the load current would be 6 mA (ignoring the voltage drop across the diodes). Clearly, we need a supply which will give us a larger current than this. So you are now to examine the effect on the smoothing when you draw a larger load current.

Is Smoothing Affected by the Load Current?

Use a 9 V a.c. input; a smoothing capacitor of 47 μF; and load resistors of 470 Ω, 1 kΩ, 10 kΩ and 100 kΩ. Start with the load resistor of 100 kΩ and measure the ripple voltage, as before. If this voltage is too small to measure, then you may be able to say that it is 'smaller than —— volts'. Measure the ripple voltage for each value of the load.

1. Load resistor = 470 Ω
   Load current = ___ mA
   Ripple voltage = ___ V
2. Load resistor = 1 kΩ
   Load current = ___ mA
   Ripple voltage = ___ V
3. Load resistor = 10 kΩ
   Load current = ___ mA
   Ripple voltage = ___ V
4. Load resistor = 100 kΩ
   Load current = ___ mA
   Ripple voltage = ___ V

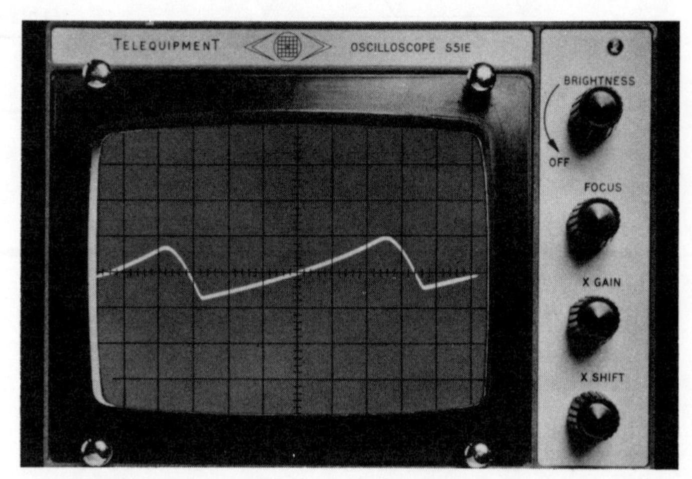

As the load current increases, the smoothing deteriorates.

How Does the Capacitor Achieve Smoothing?

Remember that a capacitor is an electrical bucket. When charged up, it stores electricity until needed. You may be guessing now when it will be needed. The important consideration is the current through the load resistor. The graph shows the input voltage and the circuit diagrams show what is happening at the three times A, B and C.

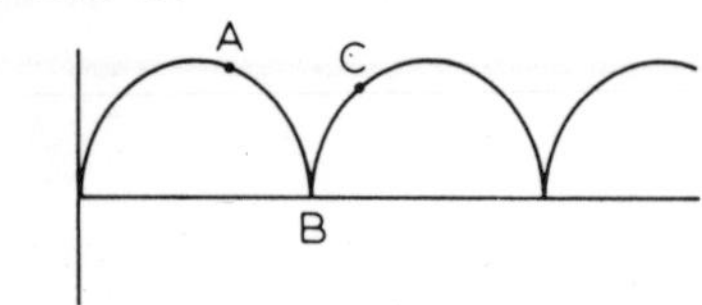

Input voltage supply

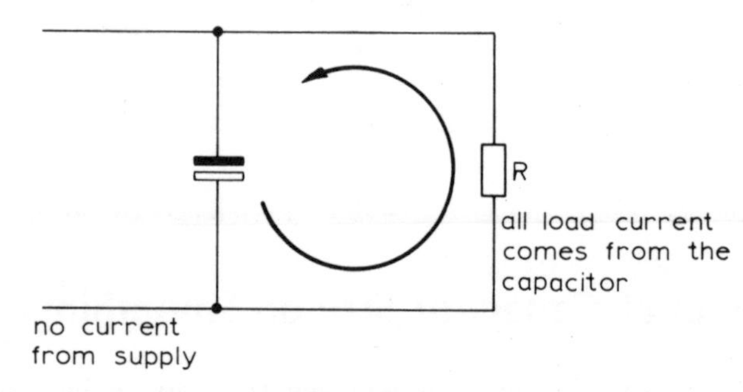

Time B. The input voltage at zero

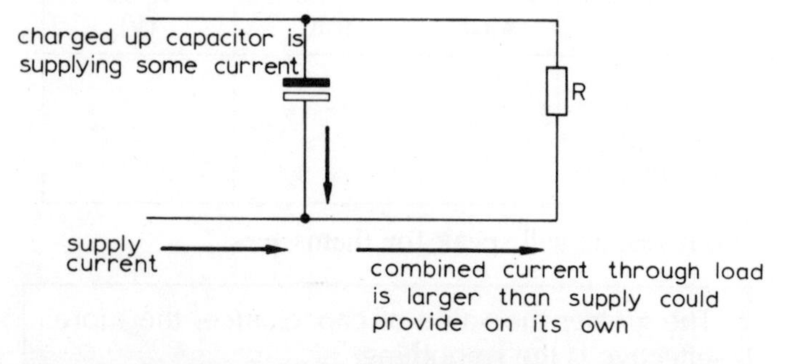

Time A. The input voltage beginning to fall

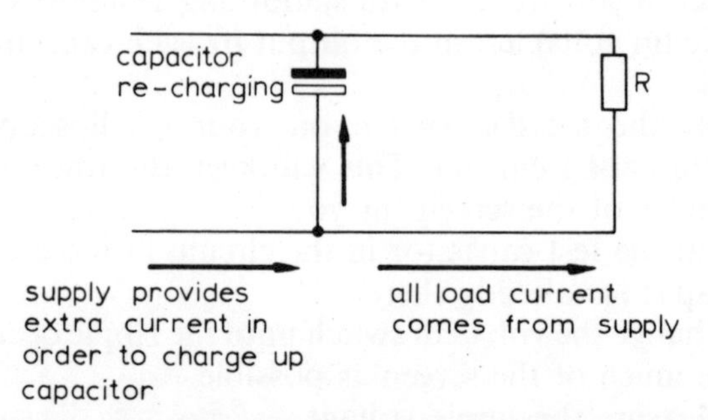

Time C. The input voltage near maximum

10 V Power Supply

This circuit contains all the elements discussed so far and, in addition, it has one refinement—a zener diode. This component is discussed fully on page 144. Extra space has been left for the circuit to be improved by the addition of a transistor, at a later date.

This circuit must not be connected to the mains until it has been tested by your teacher.

Components

1. The four diodes are contained in a single diode bridge.
2. The zener diode is a 10 V, 1·5 W type.
3. The fuse is a 100 mA.
4. 4 mm terminals are used for the output.
5. The transformer is an MT150 type giving two outputs at 12 V and 150 mA.

Construction

1. Follow all the instructions given on page 13.
2. The line cable goes to the fuse holder and then to the switch, both on the front panel.
3. The extra pair of 12 V a.c. outputs from the transformer is taken to the front panel to give an additional output.

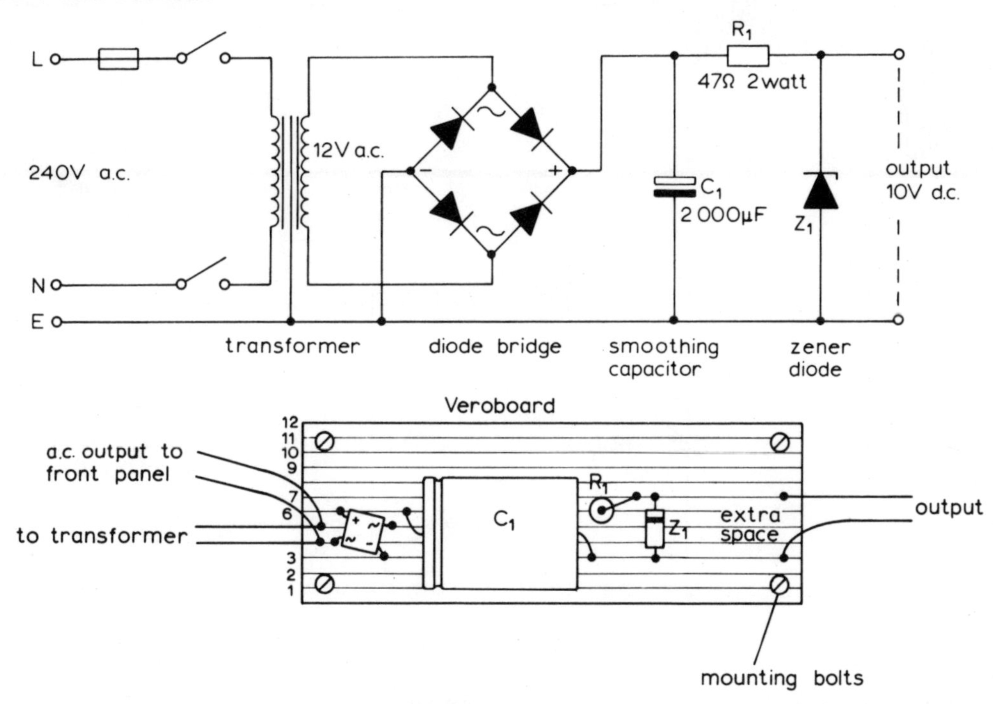

Performance

Stability

As with all voltage supplies, we would like the voltage to be unaffected when current is drawn from it. Find how much the output voltage varies as you increase the load current. Connect the load resistor and a voltmeter across the output.

Load resistor	10 kΩ	1 kΩ	100 Ω
Output voltage			

Current Limitation

This circuit has a useful property—it can never give more than about 250 mA current even if you short circuit the output. Use an ammeter to find the maximum current which the circuit can provide. This short circuit current is, in fact, more than the MT 150 transformer should take, but, for short periods, this protection is very useful. Alternatively, you may include a 100 mA fuse on the front panel.

Note: This circuit has one drawback to it. When no current is being taken from it, a large current flows through the zener diode and it gets rather hot. To counteract this you can:

1. Add on the transistor, shown on page 153.
2. Put a heat sink on the diode.
3. Increase the value of R_1 and be content with a lower output.

The Simplest Radio

Having studied the capacitor, inductor and diode, we can now look at the simplest radio, which depends on these three for its operation. This, so called, crystal radio has the following stages.

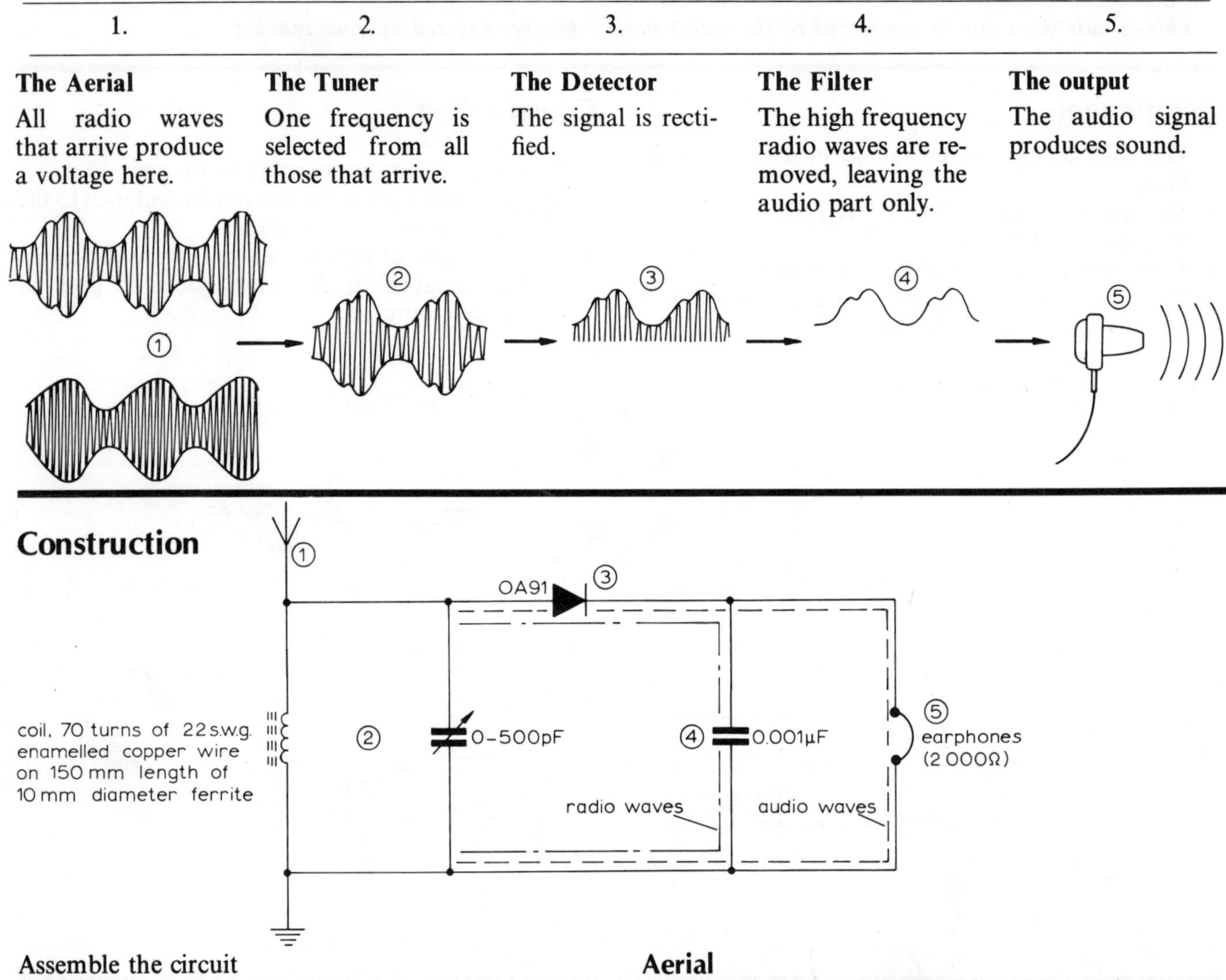

Assemble the circuit on S-DeC.

Earth

A good connection to the metal pipes of either the gas or water supplies will work as an earth. You should check that none of the piping is in plastic, which is, of course, an insulator.

Coil

Wind the coil neatly and firmly and secure both ends with Sellotape. Remember to remove the insulation varnish where you make connections.

Tuning Capacitor

Fit a plastic knob on the shaft of the variable capacitor or you will hear interference whenever you touch the capacitor. Use the mounting board of the S-DeC to hold this capacitor.

Aerial

A long piece of insulated wire, strung at roof level, will serve as aerial.

Operation

Vary the tuning capacitor and find as many stations as you can. The number you receive will depend on which area of the country you live in and the quality of your earth and aerial.

Notes: It may not be obvious why the circuit needs a diode. Examine the waveform at 2. You will see that radio frequency waves give a waveform which carries two audio signals—of equal positive and negative values. The cone of any loudspeaker would feel an equal pull in both directions and so it would not move at all. This is another way of saying that the two audio waves average zero. The diode blocks one of these waveforms so you can hear the other one.

Practical Radio

This circuit has many advantages over the simple crystal radio. In particular, it is portable, since the coil serves as the aerial and it will also receive more stations. The circuit contains an amplifier to give a louder output.

Construction

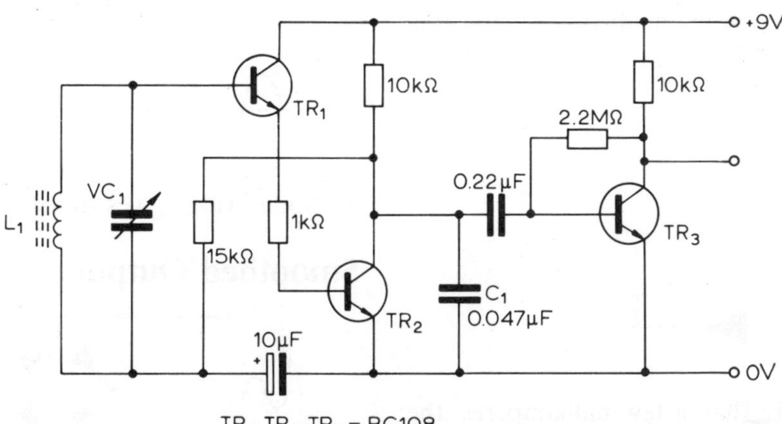

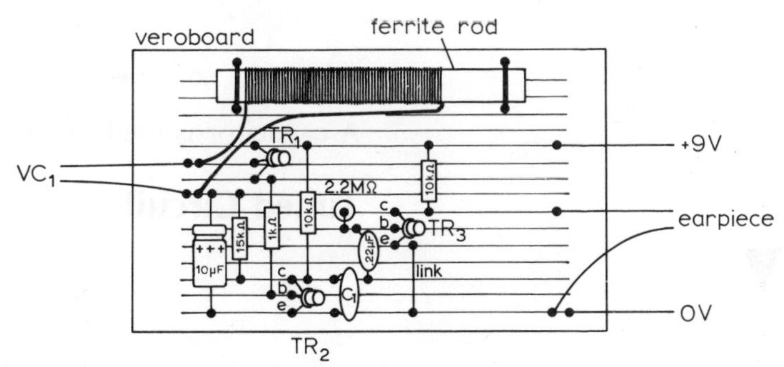

The components are assembled on Veroboard. They are fairly well spaced out, for ease of assembly, so, if you wish to make a more compact radio, you may re-design the Veroboard layout. Follow the usual rules for Veroboard work. Secure the ferrite rod (which is quite heavy) to the board using wire or clips, but make sure that you do not bridge any copper tracks. Also ensure that the battery cannot come loose inside your case. A jack plug may be used so that the earpiece may be removed when the radio is not in use.

The Circuit

The Aerial

The radio waves generate a voltage in the coil in this radio, rather than in the straight wire aerial of the crystal radio.

Tuning

The tuning section, L_1 and VC_1, is identical to that of the simple radio. It selects one of the radio frequencies and rejects the others.

Detection

In this case, the job of the diode is performed by transistors TR_1 and TR_2. These two block one half of the signal and allow the other half to pass.

Filtering

The radio waves are removed via C_1 in the same way as in the crystal radio. The audio waves continue to the output.

The Diode—Summary

Current Conduction

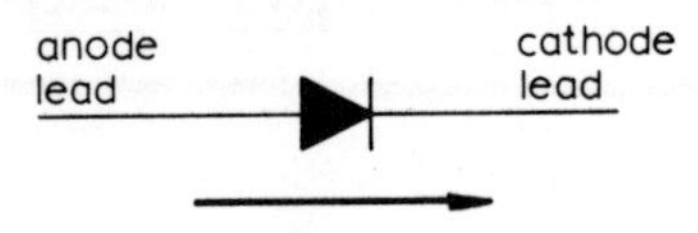

The diode conducts current in one direction only. It is then said to be 'forward biased'.

Voltage Drop

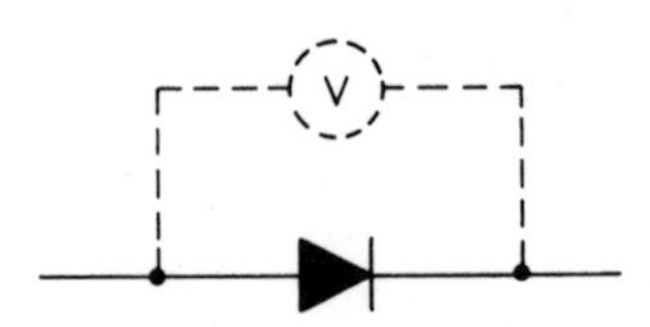

With a current of more than a few milliamperes, the voltage drop is constant: silicon ~ 0.6 V; germanium ~ 0.2 V

Diode Protection

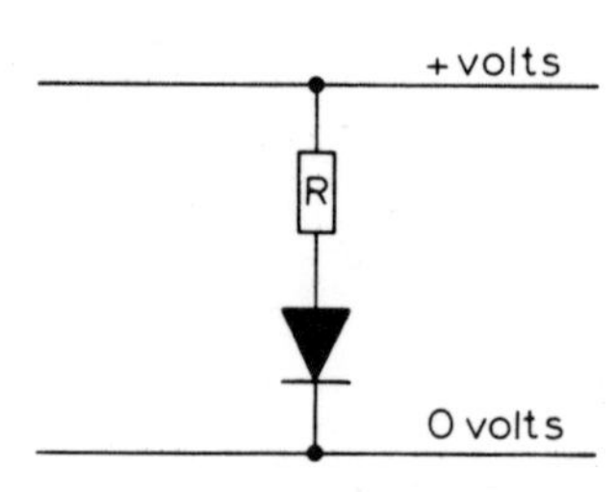

There is a maximum current permitted for every diode, so protection by a resistor is sometimes necessary.

Half-Wave Rectification

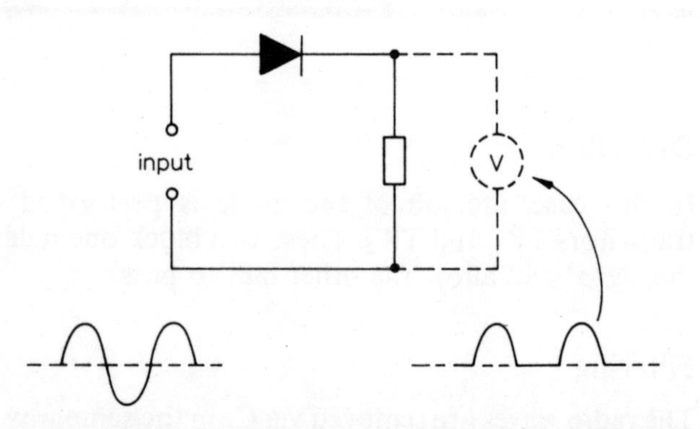

A diode achieves half-wave rectification. Alternating current is converted to direct current.

Full-Wave Rectification

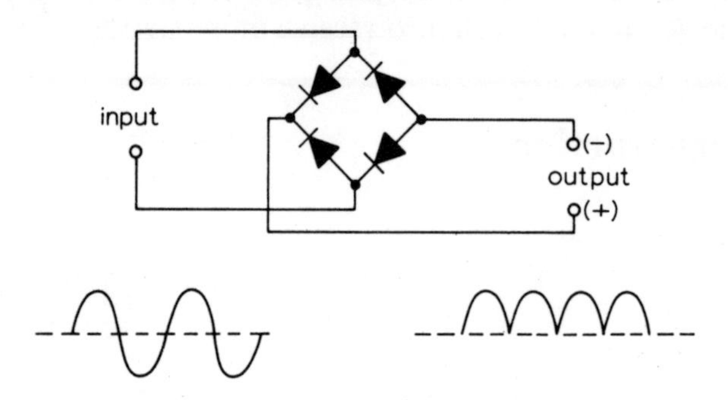

A diode bridge gives full-wave rectification.

Smoothed Output

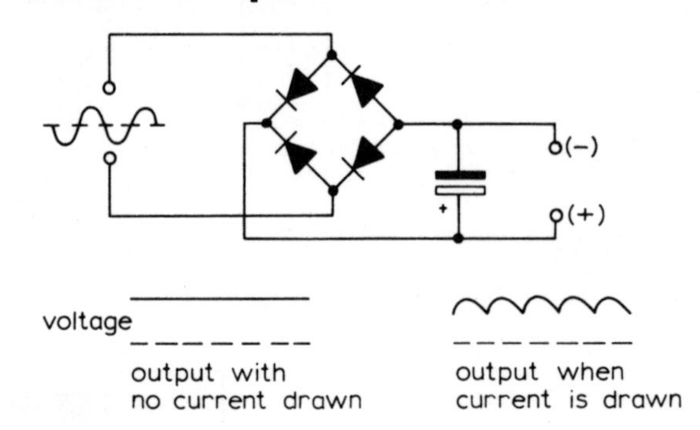

A capacitor is used to give a smoothed output

Tuned Circuit

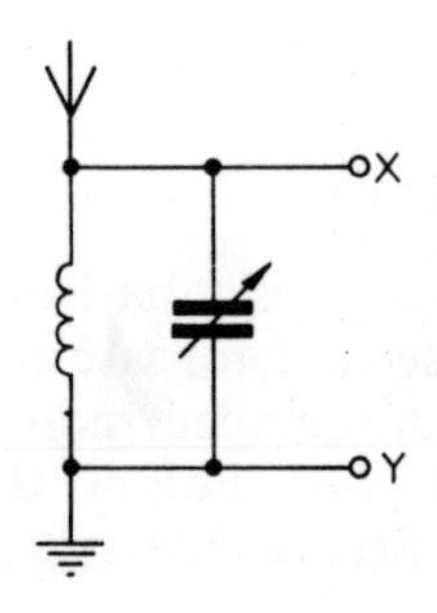

This tuned circuit selects one radio frequency, because one frequency gives a higher voltage across XY than others.

Detection

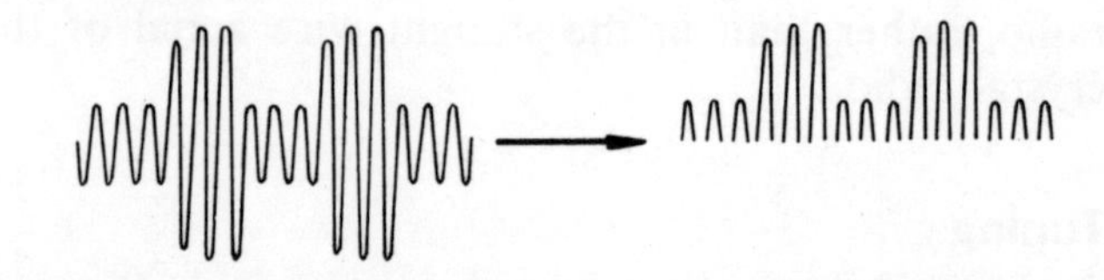

When a diode rectifies a radio signal, the diode is said to 'detect' the radio signal.

7 THE TRANSISTOR

Introduction

The transistor is the second, but most important, semiconductor device that we shall study. Like the diode, it is made from either silicon or germanium and it now takes the place of the valve which is found in older equipment. The transistor is also the father of a family of semiconductors which include the uni-junction transistor, the silicon controlled rectifier and the Triac, all of which you will meet in a later chapter.

Construction

You will remember that the diode (junction type) is composed of a piece of silicon or germanium with two distinct regions in it, called n-type and p-type material. The transistor is also a single piece of silicon or germanium with *three* distinct regions of n-type or p-type material in it. These two kinds of material are produced by the process of doping in which an impurity is allowed to diffuse into the crystal—two different impurities being used to produce the two different kinds of material. The immense range of different transistors now available is a result of differences in the amount of doping and the sizes of the three different regions. There are only two different arrangements of the three regions in a transistor: these produce the npn and pnp type of transistor.

There is no difference in the performance of the npn and pnp type of transistor in a circuit, except that all the voltage polarities must be the opposite way round for the two types. However, silicon npn types are the most common type of transistor, partly because they are easier to manufacture and partly because they are more robust and less sensitive to temperature variations. You will see that we shall concentrate on this type in the following pages. If you understand the explanations of how the npn transistor works, you will quickly adjust when you happen to need to use a pnp type.

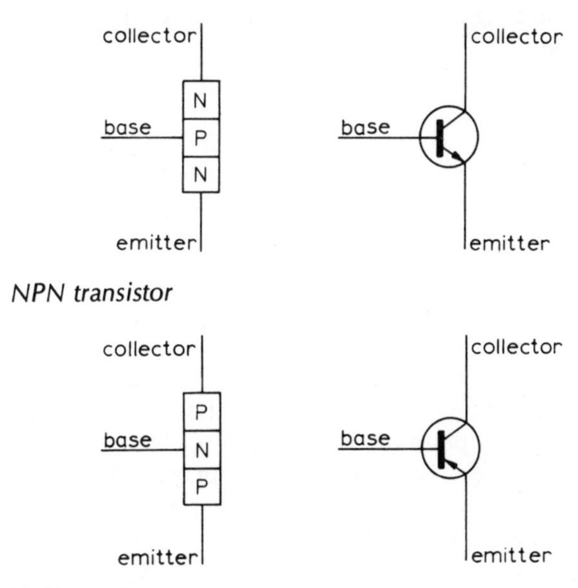

Identification

The three different regions in a transistor are connected to three external leads, as shown in the diagrams and these are called collector, base and emitter. A transistor cannot be used in a circuit until its three leads have been identified and incorrect connection to a circuit will usually lead to damage.

Every transistor is identified by a type number, which is usually found stamped on the case. Once you have this number, you can then consult the manufacturer's literature and this will answer four questions for you:

1. Is it a silicon or germanium type?
2. Is it an npn or pnp type?
3. Which are the collector, emitter and base leads?
4. What is its circuit specification?

Types

When you use a circuit diagram you will usually find diagrams identifying the transistor leads for you. These diagrams show you the view of the transistor *from below* with the leads marked e, b and c. The diagrams for the transistors you will use on this course are shown opposite.

Until we know how a transistor works in a circuit there is little point in looking at its specifications but, by the time you have completed this chapter, you will be able to understand the information given by manufacturers about their transistors.

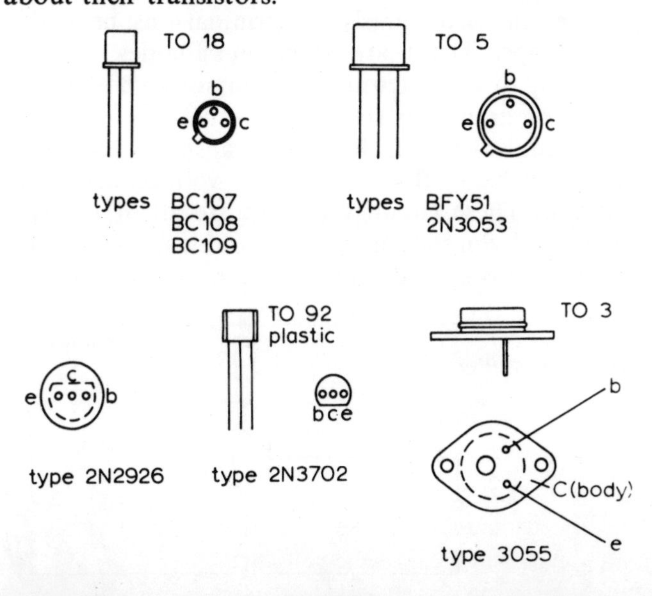

Transistor Circuit Diagrams

You will notice that, now you are studying the transistor, your circuits become more complex. Some of the shorthand used in circuit diagrams can, at first, make them look more complicated rather than simplifying them. The circuit diagram 1 shows a very common way of representing a simple transistor circuit and yet this can be very baffling at first sight.

Although Circuit 1 does not appear to have a complete current path, Circuit 2 shows what the abbreviations mean. It shows that there are in fact, three current paths, the current from the input voltage, the current through the output (or load) and the current through the transistor.

Voltage Levels

Where you see, for instance, 9 V and 0 V marked on a circuit diagram, it means that a 9 V battery is connected across the two points.

Where you see V_{in}, it means the circuit is completed by connecting an input voltage across the two points shown. This supply will also have its own resistance.

Where you see V_{out}, it means that the circuit produces a voltage there and that it is available for driving a load. You must therefore connect a device across these two terminals and a current will then flow through this device.

Once you understand what Circuit 1 stands for, you will probably find it easier to use this shorter type of diagram rather than the full one of Circuit 2.

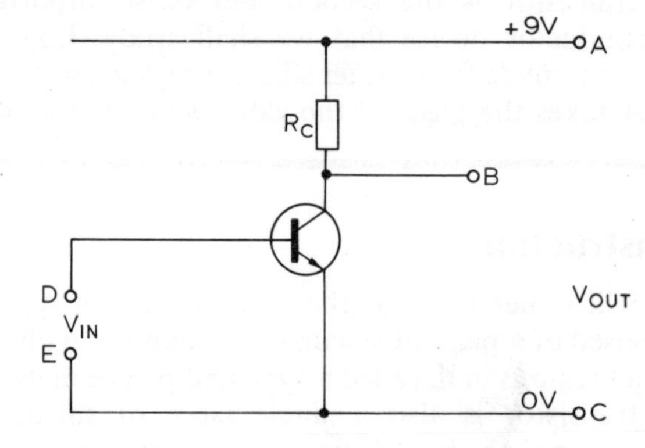

Circuit 1

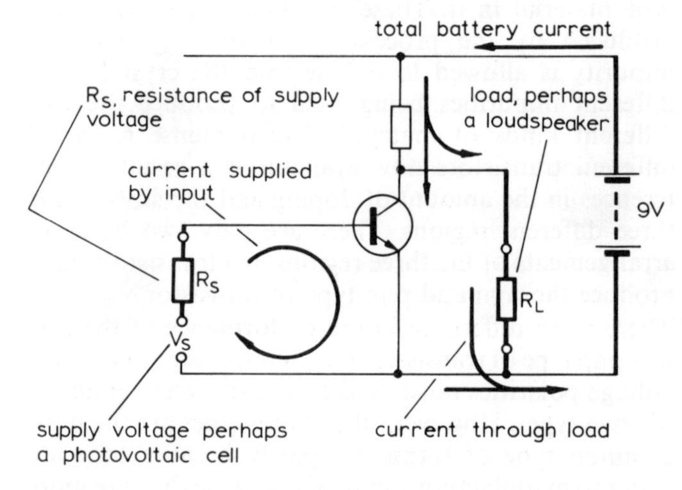

Circuit 2

Modes of Operation

There are two current loops when a transistor is used – input current and output current. Since there are three terminals, obviously one terminal must be used in both loops. There are three possibilities, labelled according to which terminal is common to both loops.

The diagrams all show npn transistors but there is an equivalent set for pnp transistors, the only difference being that the battery voltage has to be reversed. These circuit diagrams are all of the type shown and you should have no trouble drawing in the current paths should you need to.

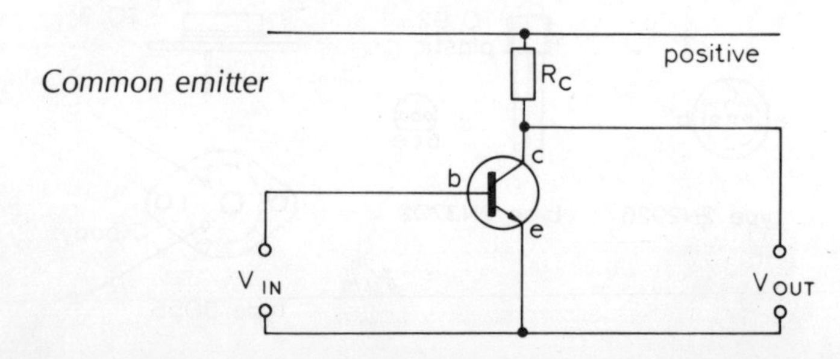

Common emitter

The common emitter is by far the most usual arrangement and we shall concentrate on this one. Learn this arrangement and notice, also, the abbreviations on the diagram: R_c is the collector resistor; R_e is the emitter resistor, and you will also shortly use R_b for the base resistor.

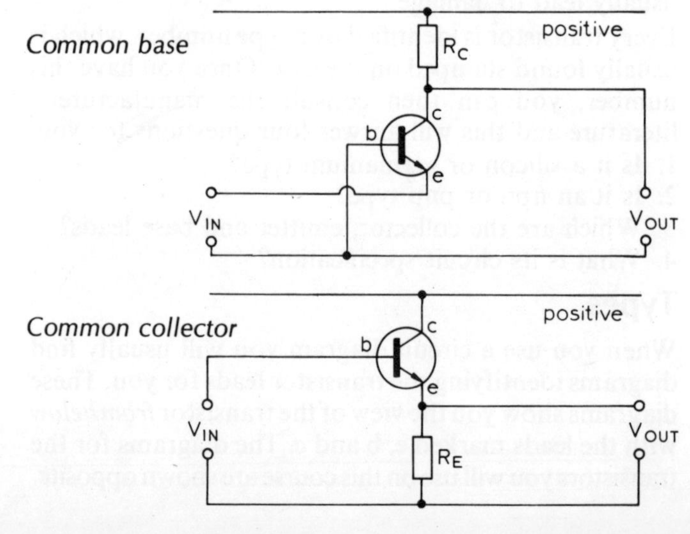

Common base

Common collector

Current Flow in a Transistor

Just as there was only one current path for a diode, so a transistor passes current in one direction only. You must learn the current paths for the npn and pnp transistors, as shown here.

The arrows show the direction of conventional current flow through the transistor. You will see that the three currents are labelled according to the terminals of the transistor:

I_b is the base current;
I_c is the collector current;
I_e is the emitter current.

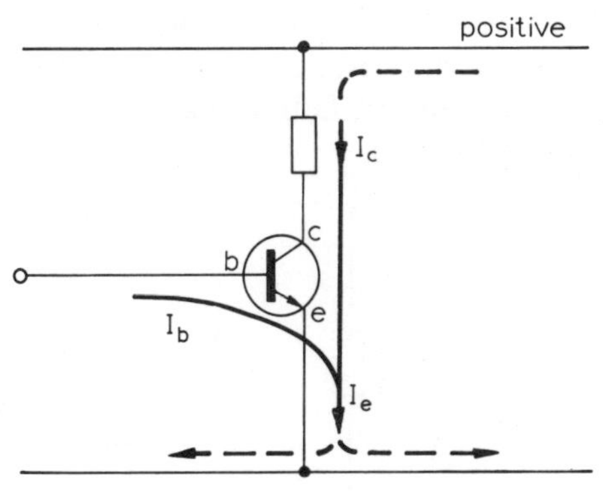

NPN transistor—common emitter mode

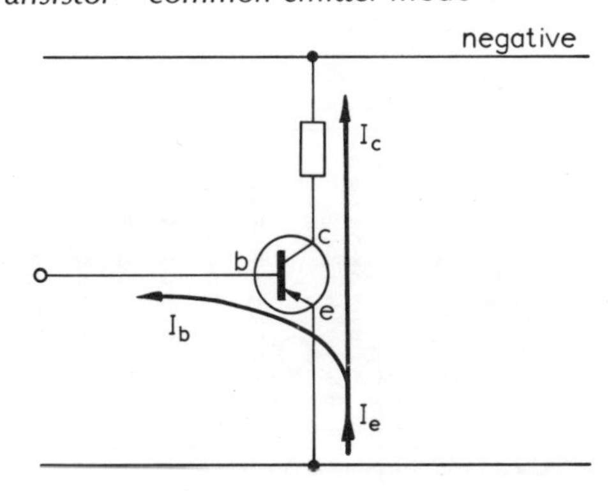

PNP transistor—common emitter mode

NPN Transistor

The current which arrives at the transistor is $I_b + I_c$ and the current which leaves is I_e. Since no current is lost,

$$I_e = I_b + I_c$$

PNP Transistor

The current which arrives at the transistor is I_e and the current which leaves is $I_b + I_c$. Since no current is lost,

$$I_e = I_b + I_c$$

Biasing a Transistor—Giving it the Correct Voltages

The process of supplying a transistor with the correct voltages for its operation is known as biasing. To produce the current flow as above, the transistor has to have battery polarities as shown in the circuit diagrams below. (For your experiments you will use a 9 V supply which should have leads of the correct colour to help your assembly work.)

Instructions
Set up the two circuits shown, ensuring correct connection of the batteries and correct identification of the transistor leads. Using your milliammeters, find the current flowing at the three places shown in the circuit. You should recognise these as I_b, I_c and I_e.
These circuits show you *how* to connect up a transistor but not *why*. Learn these simple circuits and the way the components are connected, in particular the battery voltages.

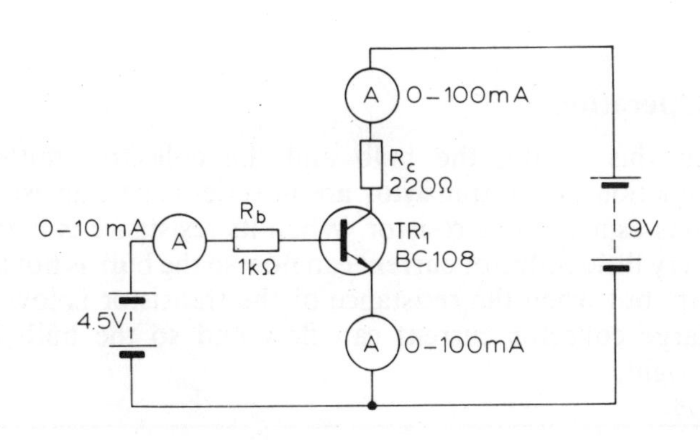

NPN transistor (BC108)

I_e = —— I_b = —— I_c = ——

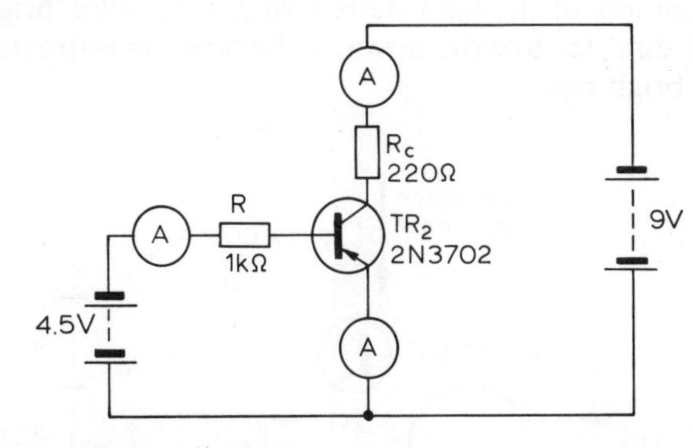

PNP transistor (2N3702)

I_e = —— I_b = —— I_c ——

Why We Use a Transistor

Effect of Changing Base Current

If you have enough resistance meters you may do this experiment yourself, or, failing that, your teacher can demonstrate it for you.

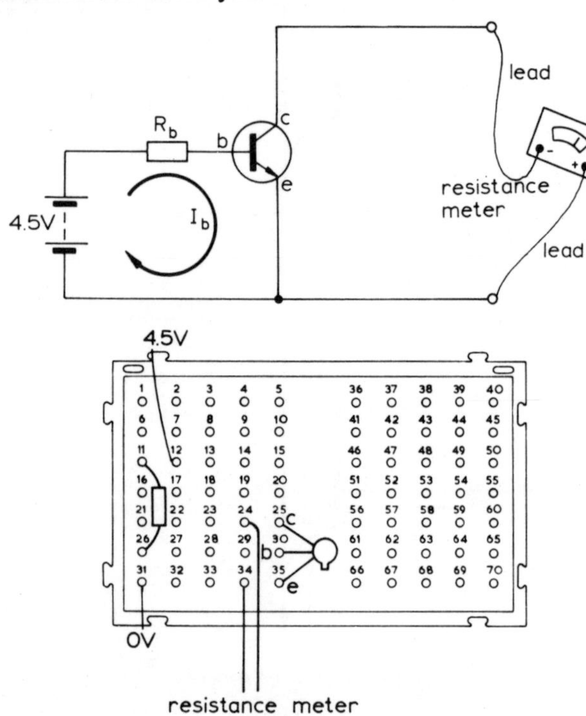

Instructions

Assemble the circuit shown, using $R_b = 3{\cdot}9\ \text{M}\Omega$ and then find the resistance of the transistor between the collector and emitter terminals. Change R_b for the next value in the table and repeat the resistance reading. Work down the table and record your results.

R_b (kΩ)	Resistance of collector/emitter junction
3900	
1000	
470	
100	
47	
10	
1	

Results

Obviously, as you decrease the size of R_b, then the base current increases.

> As the base current increases the resistance between collector and emitter terminals decreases.

The transistor acts as a variable resistor, controlled by base current.

The Simplest Transistor Circuit

You are now going to assemble the simplest transistor circuit—again it is the common emitter mode and we shall use an npn transistor.

Instructions

Use the circuit shown below using each value of R_b in turn. The different values of R_b will give you different base currents and you can record how this affects the brightness of the light bulb. Using words like 'bright' and 'dim', try to write down, in the table, an estimate of the brightness.

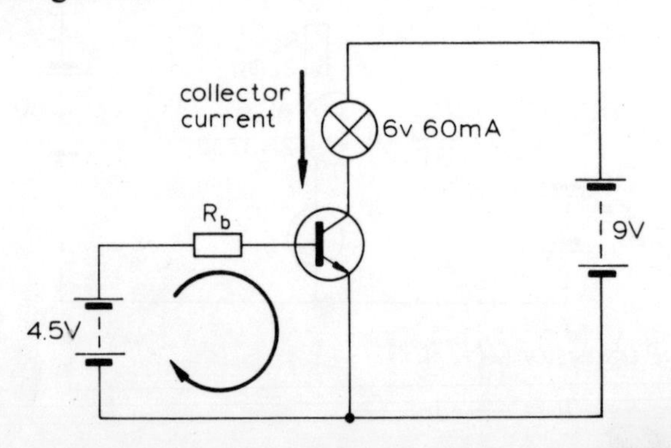

R_b (kΩ)	Brightness of bulb
1000	
470	
100	
47	
10	
1	

Operation

In this circuit, the bulb and the collector/emitter junction of the transistor are in series. The transistor acts as a variable resistor. When its resistance is high, very little collector current can flow so the bulb is not lit up, but when the resistance of the transistor is low a large collector current can flow and so the bulb is bright.

> As the resistance of the collector/emitter junction changes, the collector current changes.

The Base Current

The circuits here use two values for R_b from your last experiment and we shall use them as examples. We come across two problems in talking about transistors:

First, no two transistors are exactly alike so the figures we quote for, say, a BC108 in a circuit would probably be different if that transistor were replaced by another BC108. This means that many figures given must be treated as approximate, unless they were actually measured in an experiment, and your experimental results will be different from anyone else's.

Secondly, the BC108 can take a base current anywhere between 0 and roughly 10 mA and a collector current between 0 and roughly 100 mA, without damage. So you must bear in mind that when we talk of a 'large' base current we mean, say, 10 mA whilst a 'large' collector current may be 100 mA (for the BC108, of course).

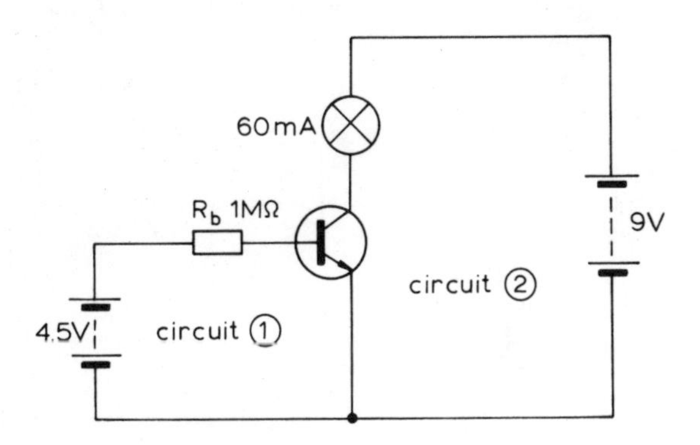

Circuit A

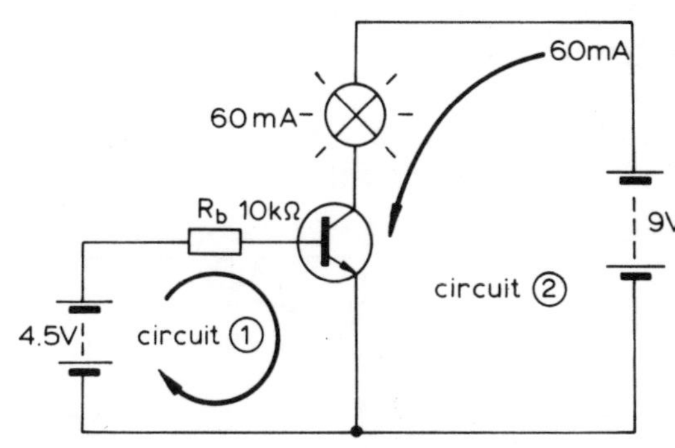

Circuit B

What should be clear from Circuits A and B is that in each case the base current (in circuit 1) controls the current through the light bulb (collector current, circuit 2):

Circuit 1 controls circuit 2

Your response to this may well be: 'So what?'. Wouldn't it be easier to throw away the transistor, one of the batteries and the base resistor and just use the bulb with a variable resistor and one battery? The usefulness of the transistor becomes clear when you look at the sizes of the currents involved.

Sizes of Currents

Look at the currents flowing in circuit B above.

The Collector Current

The current flowing through the bulb, since it is brightly lit, will be about 60 mA, but you can check this with a meter.

The Base Current

Again, we could find this out with a milliammeter but you will often find it quicker to use this approximate calculation. We assume that there is no resistance through the base/emitter junction so that all the 4·5 V is driving current through the 10 kΩ resistor.

In circuit 1 $R_b = 10\,k\Omega$ $V = 4{\cdot}5$ volts

$$I_b = \frac{V}{R_b} \quad I_b = \frac{4{\cdot}5}{10\,000} = 0{\cdot}00045\ \text{A} \quad \text{or } 0{\cdot}45\ \text{mA}$$

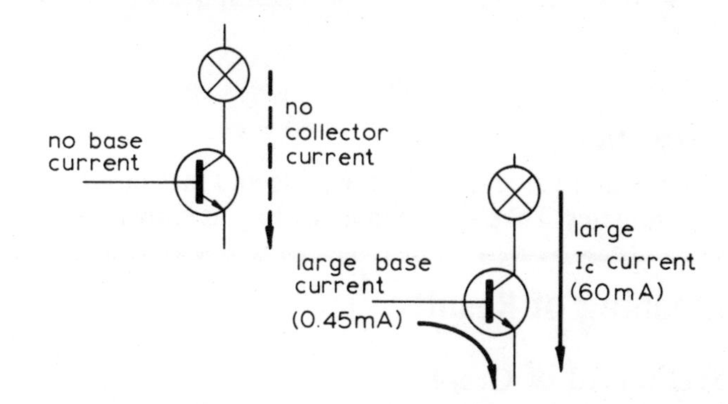

So you can see that a base current of 0.45 mA will allow a much larger collector current (about 60 mA) to flow. The base current controls the collector current. Without that base current the collector current will not flow. Reduce the base current and the collector current goes down.

Usefulness

Now think back to the introduction in the book to transducers. Here are devices which produce or handle very small currents and here we have a way in which we can put them in a circuit and they will switch on a light bulb. Put their small currents into the base of a transistor and on comes the bulb, just as above. This is not a very sophisticated use for a transistor, but it's early days yet.

Base and Collector Currents

Base Current Controlling Collector Current

Use this circuit to give a series of readings of base current and collector current.

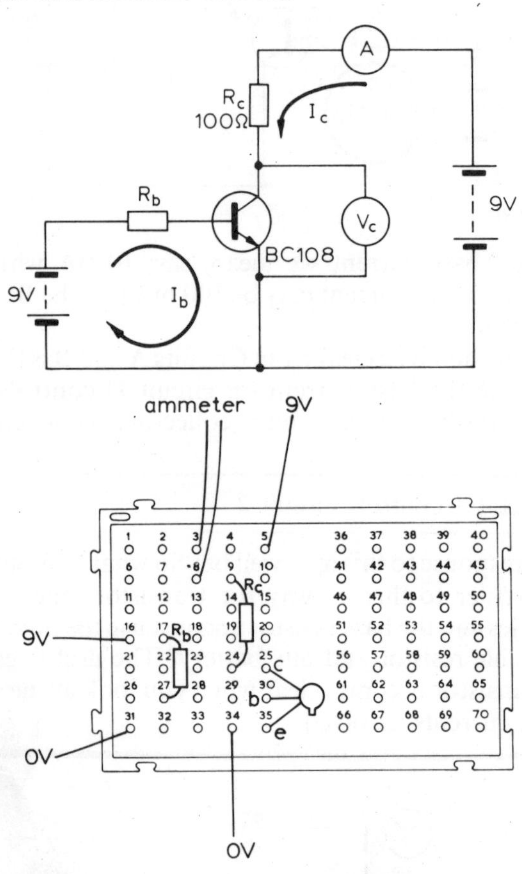

Instructions

Use the first value of R_b in the table and measure I_c on the ammeter. Take the voltage reading as shown in the diagram. Now repeat for all other values of R_b in the table and record the results. Using the method shown on the previous page, calculate the base current that must be flowing for every value of R_b and record these also in the table. Then plot a graph of I_c and I_b and compare it with the one shown below. Transistor performance can vary considerably so don't expect your figures to be the same as those shown, but the graph shape should be similar.

R_b (kΩ)	I_b (by calculation)	I_c	V_{ce}
2200			
1000			
470			
220			
100			
47			
33			
22			
10			
1			

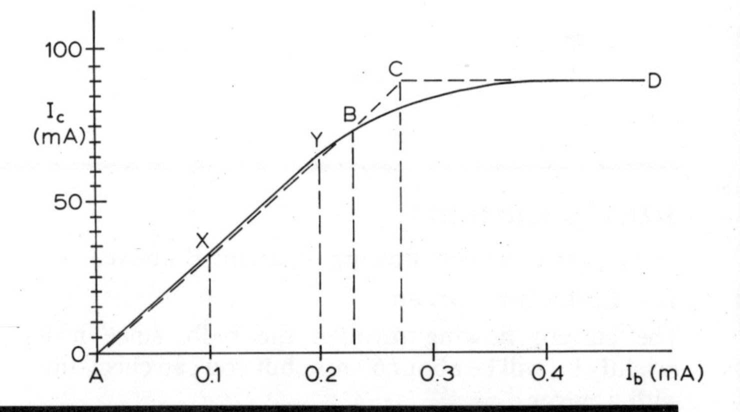

Meaning of Results

Section AB of Graph

You already know that a base current will switch on a much larger collector current but the graph shows that it does not switch on a collector current of any old size. The graph shows that I_b and I_c are proportional.

Double the base current and you double the collector current. As an example, look at the readings for points X and Y on the graph above.

Point X	*Point Y*
$I_b = 0{\cdot}1$ mA	$I_b = 0{\cdot}2$ mA
$I_c = 33$ mA	$I_c = 66$ mA

This behaviour is very important and will explain how the transistor can work as an amplifier. From now on, to make the later work easier, we shall assume that the graph has the simpler shape shown dotted on the display above. You will see that using a graph with just two straight line sections can simplify calculations. Keep your results for V_{ce} for use later.

Section CD of Graph

Here you can see that the collector current reaches its maximum. At this figure the transistor's resistance has fallen to its minimum. Increasing the base current any more doesn't change the transistor's resistance so the current doesn't increase. We say that the transistor is saturated.

> At saturation, the transistor's resistance is at its lowest.
>
> At saturation, the collector current is at its highest

So What is Amplification?

You have seen that the most important factor in the operation of the transistor is the difference in size of the base and collector currents. The base current switches on a much larger collector current. We say there has been a current gain and it is calculated thus:

$$\text{Direct current gain, } h = \frac{I_c}{I_b}$$

For the two points on the graph, mentioned previously, you should be able to see that h works out as below:

$$h \text{ at point X} = \frac{33}{0{\cdot}1} = 330$$

$$h \text{ at point Y} = \frac{66}{0{\cdot}2} = 330$$

This process is also called amplification—making a large signal from a small signal. We say that the small signal has been amplified.

However, do not be confused by the process of amplification. The small signal has not itself been increased in size. Amplification merely means that a larger copy of the signal has been made—in our circuit, a copy which is 330 times larger.

Also note that this larger current which flows in the collector circuit comes from the battery. The transistor does not make the large current itself, it only controls the current from the battery (by changing its own resistance, of course).

Amplification of the current, from, say, microphone:

Means a larger copy of the microphone current is made to flow in another circuit

Does not mean that the microphone is made to produce more current.

An Amplification Exercise

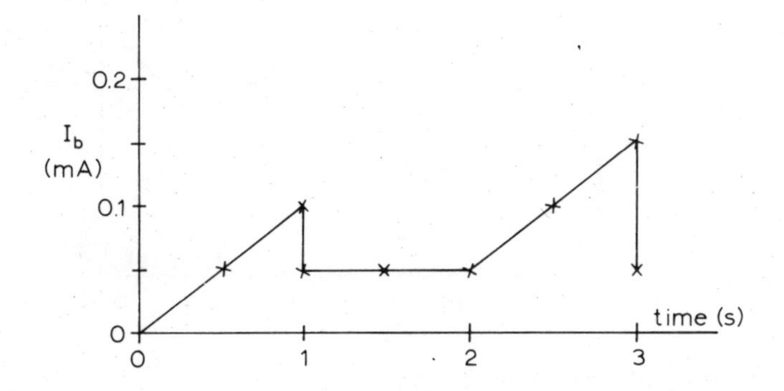

Graph 1 Base current (input)

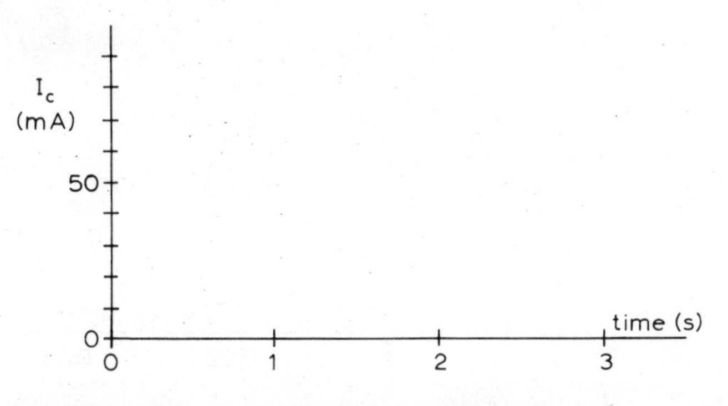

Graph 2 Collector current (output)

Instructions

Graph 1 shows 3 seconds output from an imaginary transducer. Let us find out what would happen if this current were fed into the base of the transistor in your last circuit. Use either the graph on the previous page or your own graph, and for every value of I_b crossed on the graph, above, find the value of I_c and then plot these values of I_c on Graph 2. (You must use only the section AB of your graph. See page 106)

What you are doing is finding the collector current that would be 'switched on' when the base current from graph 1 is put into your transistor circuit.

Results

For currents in the linear part of the I_c/I_b graph, a transistor will pass a collector current that is an exact magnified copy of the base current it receives.

So this is how the transistor manages to amplify. It passes a collector current which is always the same multiple of the base current input. So the waveform in the output is the same shape as the input, but a larger size.

Amplification—Exercises

Exercise 1

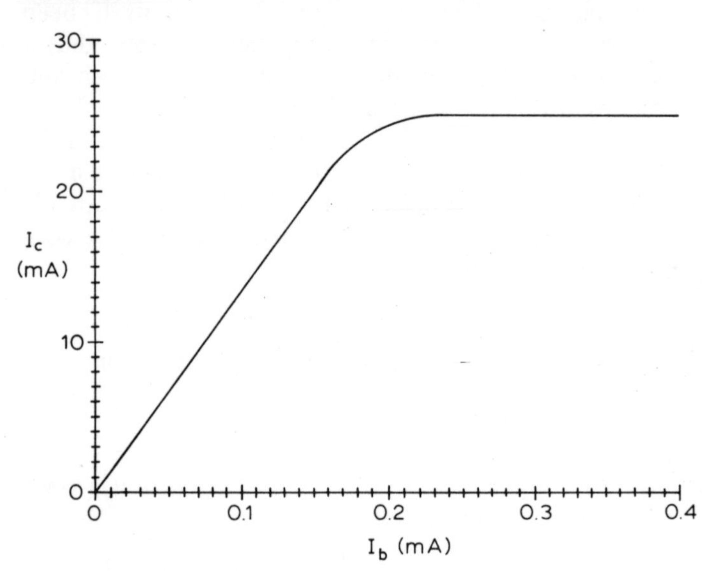

Graph 1 Collector current against base current

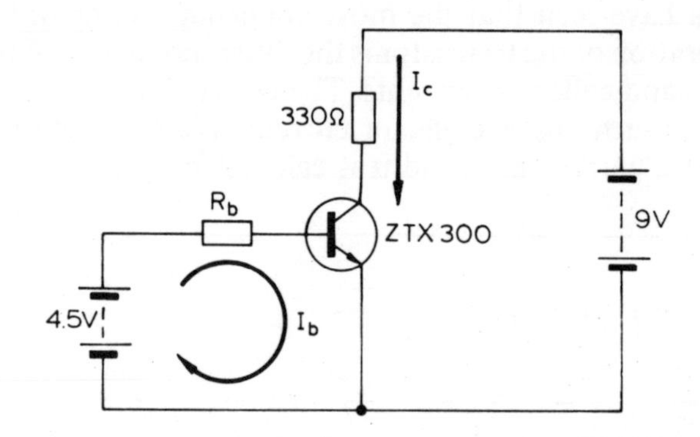

The graph shows readings taken during an experiment which was of the same kind as your last one, but with a different transistor. It shows the collector current (I_c) that flowed when a small base current (I_b) was supplied. Answer the following questions:

1. What is the value of I_c when $I_b = 0{\cdot}1$ mA?
2. What is h, the current gain, for these figures?
3. What is the value of I_c when $I_b = 0{\cdot}15$ mA?
4. What is the current gain for these figures?
5. What is the collector current at saturation?
6. What is the smallest base current that will produce saturation?

Exercise 2

The circuit is now changed slightly so that the base current is now supplied by a signal generator, but the collector/emitter circuit is unchanged.

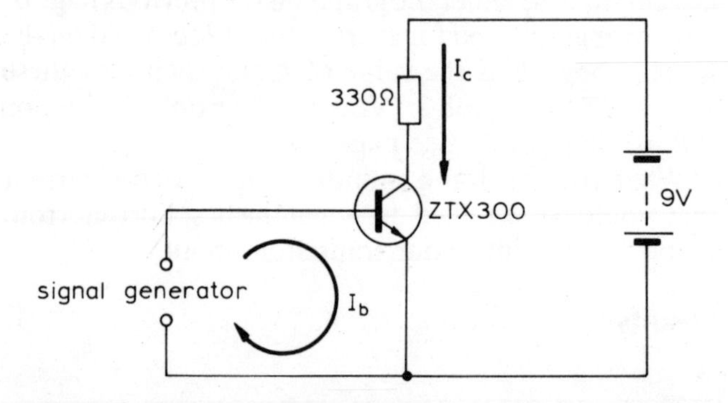

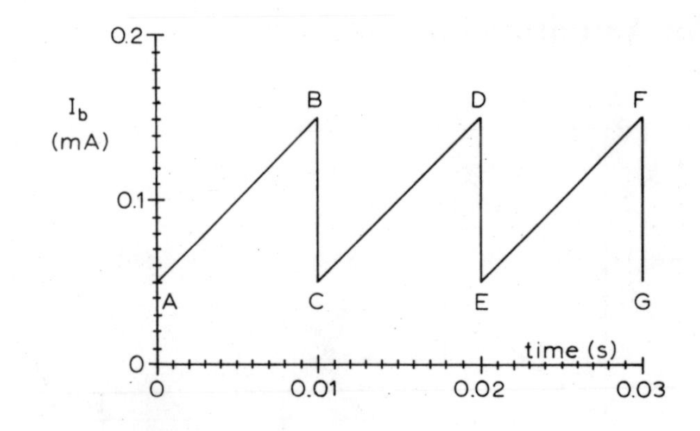

Graph 2 Signal generator input

Graph 2 shows the base current which the signal generator puts into the circuit. Find the collector current which results.

Graph 1 gives you the values of collector current you need. Find the values of collector current that would be switched on by the values of base current at points A, B, . . . , G. Use these values of collector current to plot Graph 3, taking care to plot each value of collector current at the correct time on the graph. You now have a graph which shows the amplification that has taken place.

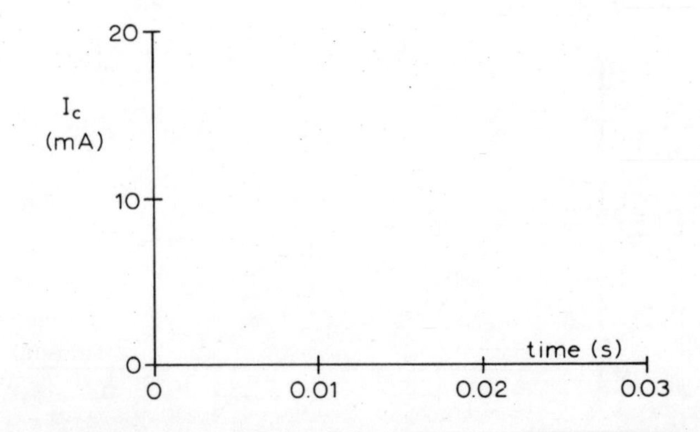

Graph 3 Collector current

Simplifying Transistor Circuits

We can now simplify the transistor circuit by using only one battery. The base bias is now provided by the same battery as the bulb. Circuit 1 has two current paths which are shown in circuits 2 and 3.

Use R_b values of 1 MΩ, 100 kΩ, and 1 kΩ and observe the collector current increasing up to its saturated value, by watching the change in brightness of the bulb.

This circuit achieves exactly what your previous circuits did, but at the saving of one battery.

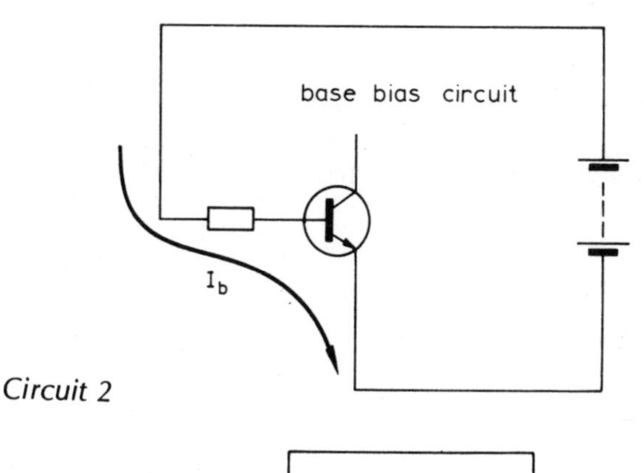

Circuit 2

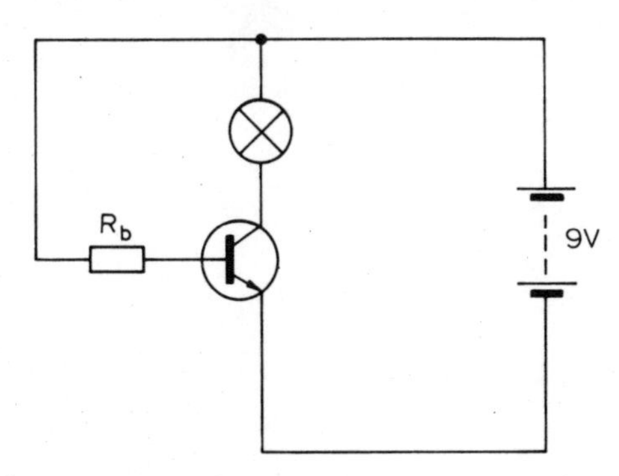

Circuit 1

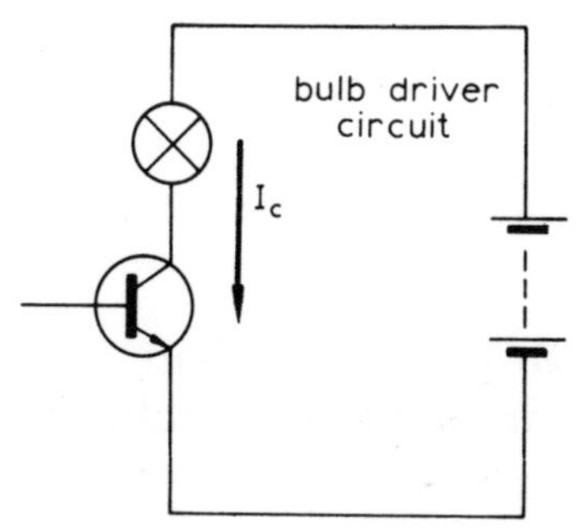

Circuit 3

Using a Voltage Divider

The next circuits for you to try use a potential divider to provide the base bias. Remember that all methods of biasing an npn transistor must make the base positive with respect to the emitter—this means that the base must be at a higher positive voltage than the emitter. Check that this is so in these two circuits. *Note*: You are to be asked to find the voltage at point A in the diagrams here. Remember that this means you are to find the voltage of A above ground (0 V) so you will connect your voltmeter across AB.

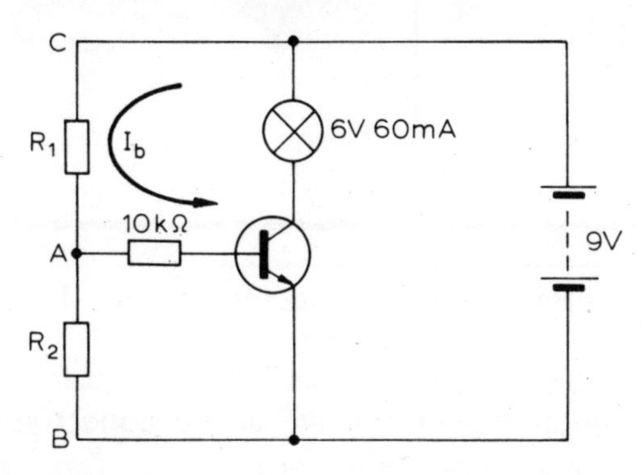

Instructions

Keep R_1 at 1 kΩ in these circuits and vary R_2 or VR_1 so that the voltage at point *A* varies. In this way you will be controlling the base current by controlling the base voltage.

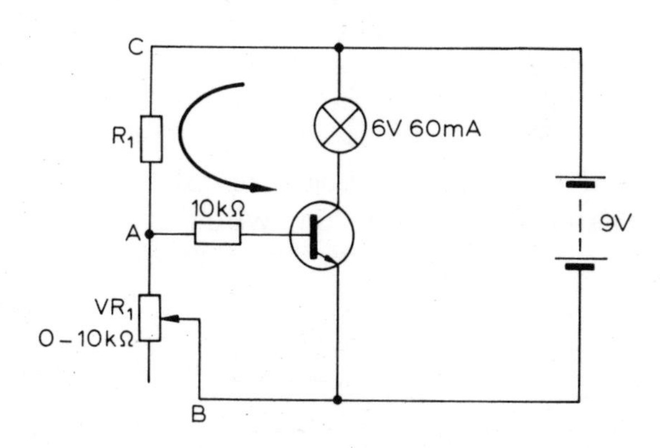

VR_1 gives you the same range of values as the individual resistors in the table. Sweep VR_1 from one extreme (0 Ω) to the other (10 kΩ) and you will see the brightness of the bulb vary as the base bias varies.

R_2 (Ω)	Voltage at A	Effect on bulb
100		
270		
470		
1000		
10 000		

All your transistor circuits will operate from one battery in the remainder of the book.

Looking at Collector Voltages

Remember that a transistor acts like a variable resistor, controlled by its base current. As its resistance varies, so does the voltage at the junction between collector and collector resistor. The three diagrams here show the operating conditions of the transistor, in the common emitter mode.
(Remember, also, that when we talk of a 'large' base current we are meaning, perhaps less than 1 mA.) The

1. Transistor used as 'off switch'.

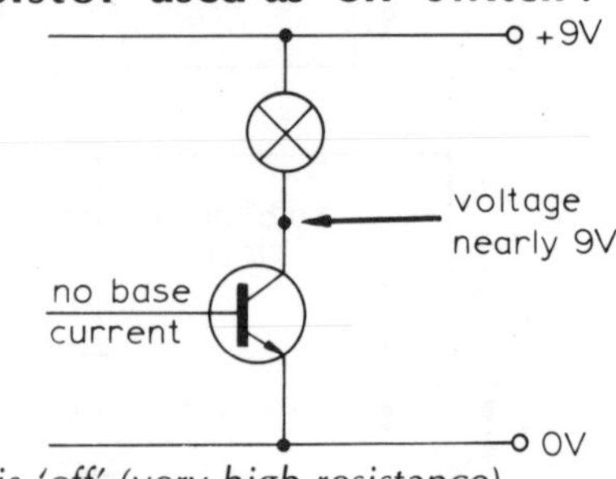

Transistor is 'off' (very high resistance)

2. Transistor used as amplifier.

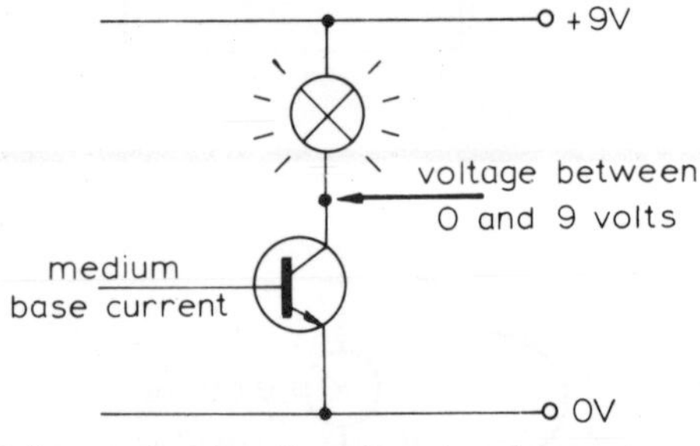

Transistor is conducting ('medium' resistance)

3. Transistor used as 'on switch'

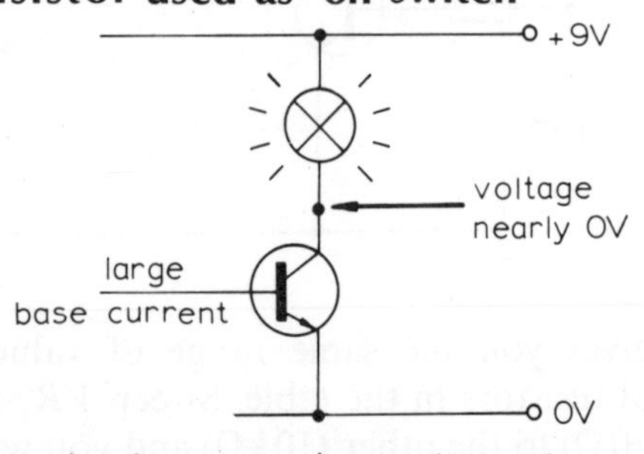

Transistor is 'hard on' (very low resistance)

voltage measurements, mentioned above, should be memorised. If you do not recall taking these measurements (page 106) then it would be good practice to take them again.

If you think of the transistor as a variable resistor, you will see that the collector voltages are explained quite simply by the voltage divider effect. Look at sample figures, given below, for a BC108 transistor.

Typical Figures

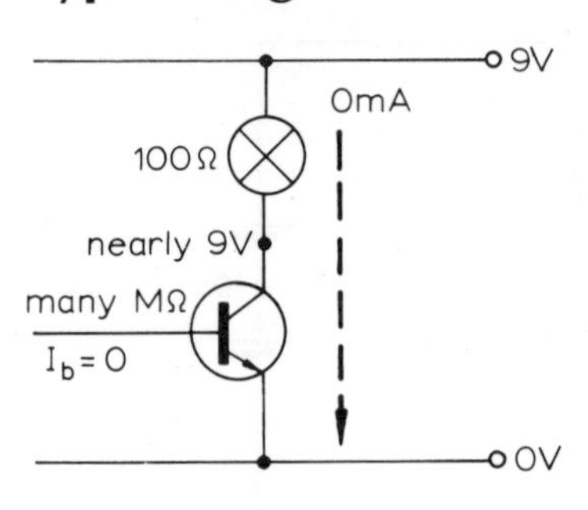

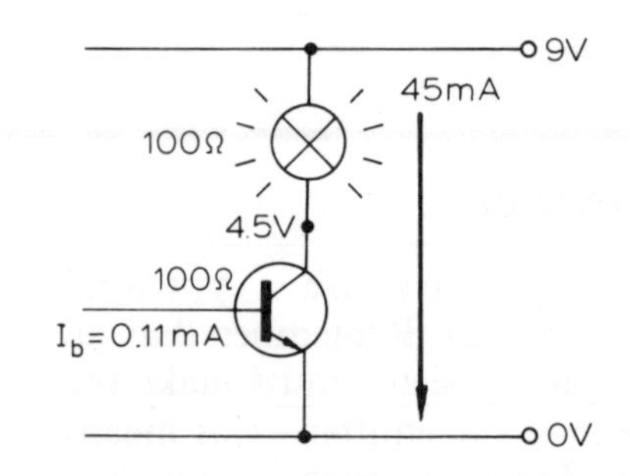

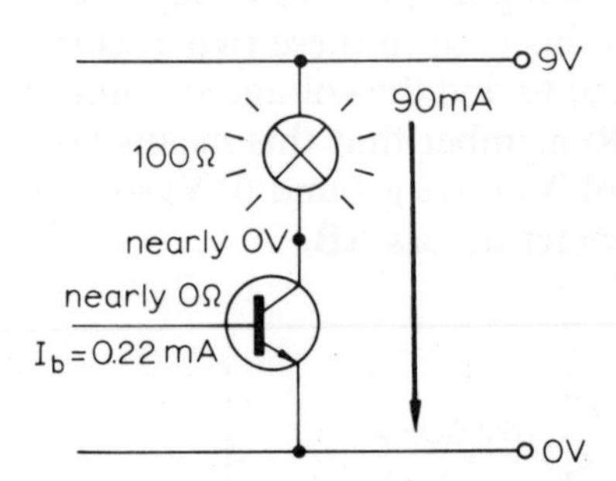

The Saturated Condition

The third example above represents what is also called the saturated condition. This occurs when the resistance of the transistor is virtually 0—obviously, its resistance cannot fall lower than this. In this case, the load resistor (100 Ω) and the battery voltage (9 V) fix the collector current (by Ohm's Law, 90 mA). The collector current can never be larger than this with a 100 Ω load and so is said to be saturated. Increasing the base current no longer increases the collector current.

At saturation, the collector current is at a maximum for that circuit.

At saturation, the voltage at the collector/load junction falls nearly to zero.

At saturation, increasing the base current does not increase the collector current.

Now examine your next project to see these rules in operation.

The Quiz Indicator

The next two projects are designed to help with quiz games. They enable the players in the quiz game to tell which team was first to call.

The circuits have a push switch and bulb for each team, arranged so that, when one team has pressed its switch and lit up its bulb, it is then impossible for the other team to switch on its bulb. In this way, there are no arguments over which team was the first with the answer! You can also modify the circuits so that they test the speed of reactions of competitors.

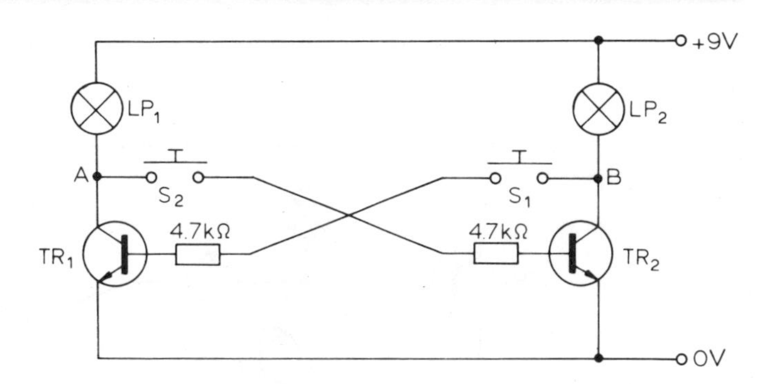

Construction

The diagram shows assembly on tag-board, with tags left empty for the modifications given on the next page. A case would need two cut-outs for the bulbs and two for the switches.

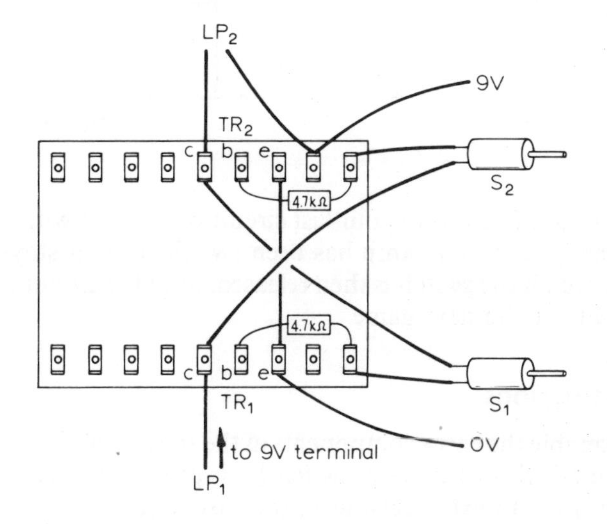

Testing

Press S_1. Lamp 1 should light up and lamp 2 now cannot be lit. Release S_1.

Press S_2. Lamp 2 should light up and lamp 1 now cannot be lit.

Your circuit is now working correctly, if it performs these two tests.

The Circuit

This circuit provides an excellent opportunuty for studying the transistor in its saturated and off conditions. Draw out the following part circuits:

1. Base current for TR_1.
2. Collector current for TR_1.
3. Base current for TR_2.
4. Collector current for TR_2.

Examine the circuit closely and answer the following questions:

1. Which switch controls the base current to TR_1?
2. Which lamp carries the collector current for TR_1?
3. Which switch controls the base current to TR_2?
4. Which lamp carries the collector current for TR_2?

Voltages

Measure the following voltages:

Switches open

1. Voltage at point A (TR_1 collector)
2. Voltage at point B (TR_2 collector)

S_1 closed

3. Voltage at point A
4. Voltage at point B

S_1 and S_2 closed

5. Voltage at point A
6. Voltage at point B

Using these readings, you are now in a position to understand how the circuit works and, in particular, why LP_2 fails to come on when S_2 is pressed. The steps involved are:

Close switch S_1

1. Base current flows into TR_1.
2. TR_1 switches on and LP_1 lights up.
3. Point A drops to nearly 0 V.

Close switch S_2

1. Base of TR_2 is now connected to point A, which is at 0 V.
2. No base current can flow into TR_2.
3. TR_2 stays cut off and LP_2 stays off.

Quiz Indicator with Memory

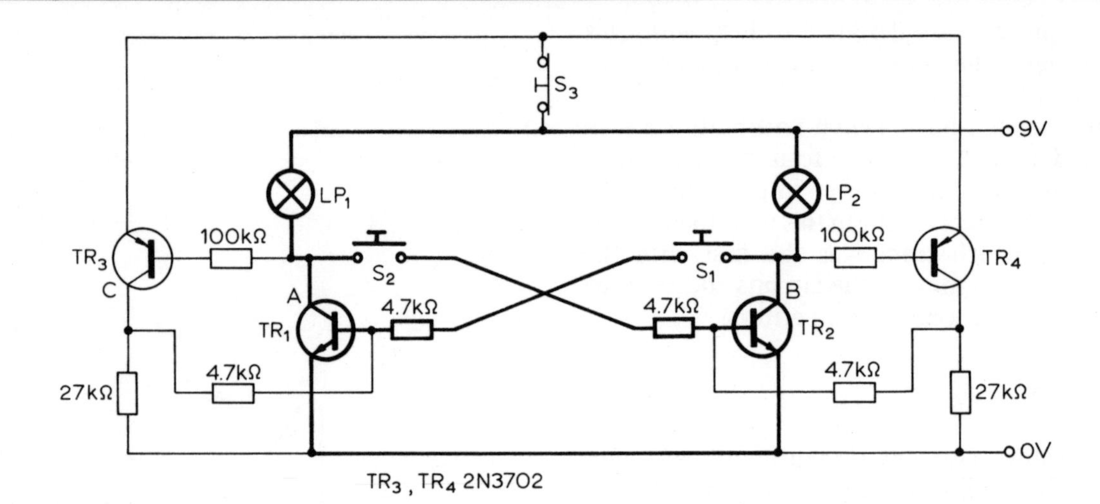

This modification to your last circuit provides it with a 'memory'. Once a lamp has been switched on, it stays on, even if the switch is then released, until the circuit is re-set for the next game.

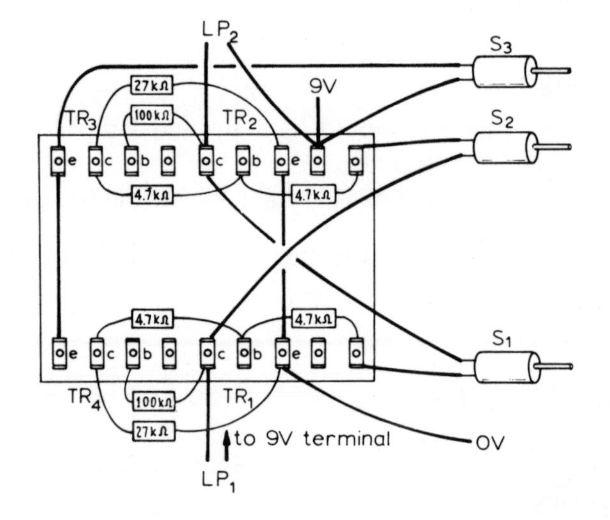

Instructions

Assemble the extra components in the spaces left on the board as shown. A re-set switch, S_3, is provided and this will need an extra cut-out in the case used.

Testing

Press S_1. Lamp 1 should light and stay on when switch released.
Lamp 2 cannot now be lit.

Press S_3. Lamp is extinguished, and circuit is re-set.

Press S_2. Check the operation of lamp 2 in the same way.

The Circuit

You should recognise the central part of the circuit from the previous page. TR_3 and TR_4 form the 'memory' of the new circuit.

Close switch S_1 and read the following voltages:
TR_1 collector (point A)
TR_2 collector (point B)
TR_3 collector (point C)

Question

If switch S_1 is now released, from where can TR_1 draw its base current to keep TR_1 on? Draw out this base current path.

Note: You have used pnp transistors for the first time in your circuits. You should be able to identify the base and load resistors for each. Notice that the bias voltages are the opposite polarities to those used for npn transistors. The collector is negative with respect to the emitter. The base is negative with respect to the emitter.

Controlling the Transistor

Using an L.D.R. to Control a Transistor

You are going to use a transducer, the light dependent resistor (l.d.r.) as part of a voltage divider, to control the transistor. Remember that the l.d.r. has:
In darkness, a very high resistance (1 MΩ); in bright light, a low resistance (200 Ω). These figures refer to an l.d.r. manufactured by Mullard, type ORP 12.

Instructions

Construct Circuit 1 and find out what happens when the l.d.r. is put in bright light and complete darkness. Use a voltmeter to find the voltages asked for. The voltage at A controls the base current and so controls the current the bulb.

Repeat the tests with the l.d.r. and R_1 interchanged, as shown in Circuit 2.

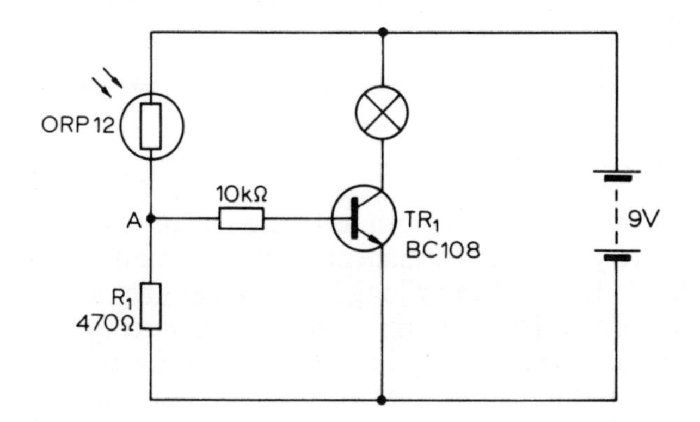

Circuit 1

	In sunlight	In darkness
Voltage at A Effect on bulb		

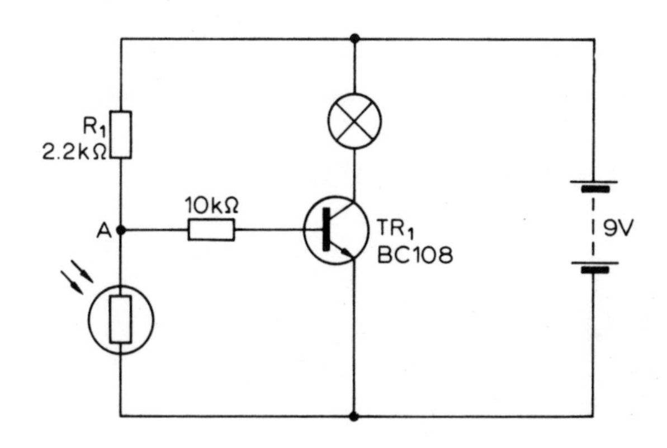

Circuit 2

	In sunlight	In darkness
Voltage at A Effect on bulb		

More Control

The switching of the transistor is controlled by the voltage at A. Using a variable resistor, VR_1, means that you can adjust your circuit and choose the light level at which your circuit switches on and off.

Place the unit close to one of the room lights. Try to adjust VR_1 so that your own bulb switches on when the room light is switched on and goes off again when the room light goes off. Perhaps in this unit you can see the makings of a burglar alarm—the burglar breaks the beam of light falling on the l.d.r. and so switches off your warning bulb. You may find it more useful to have a circuit which switches on a light bulb when the burglar enters, in which case you could modify Circuit 2, above.

Switching

Here you see the l.d.r. controlling the light bulb but we cannot say that the light bulb is being switched, in the normal meaning of the word. Besides being fully on or off, the light bulb can also be set anywhere in between these two. You will see shortly circuits which have been designed to operate in only the 'on' or 'off' conditions and not in between.

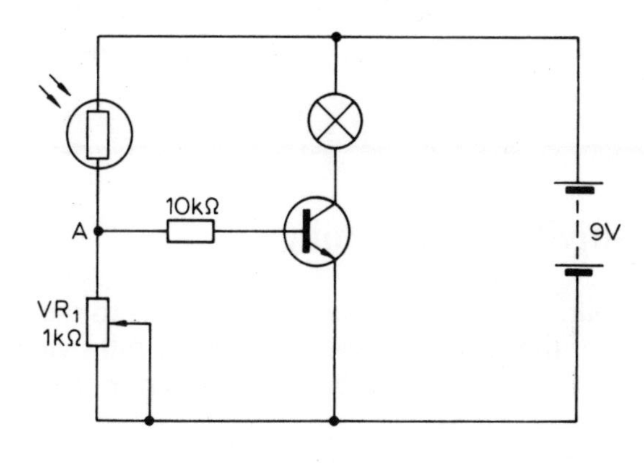

Circuit 3

More Sensitive Control

You have seen the l.d.r. controlling the transistor, but only to produce a very slow switching action. You are now going to examine the effect of using two transistors to achieve a faster switching action. The two circuits to be used could actually be controlled by an l.d.r. (although you would have to change the value of VR_1) but, for the sake of experience, you are to use a thermistor.

A typical thermistor, which would be suitable for these experiments, has the following characteristics: resistance at 25° C = 1 kΩ; resistance at 125° C = 50 Ω

A thermistor is a resistor whose value varies with temperature.

You see that this thermistor, like most others, has a resistance which decreases as the temperature rises. This is put to use in the two circuits shown here to control the transistors.

Constructing Two Alarms

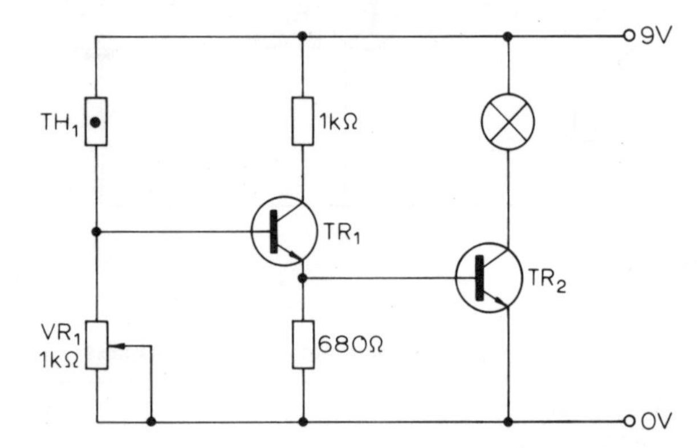

Fire alarm

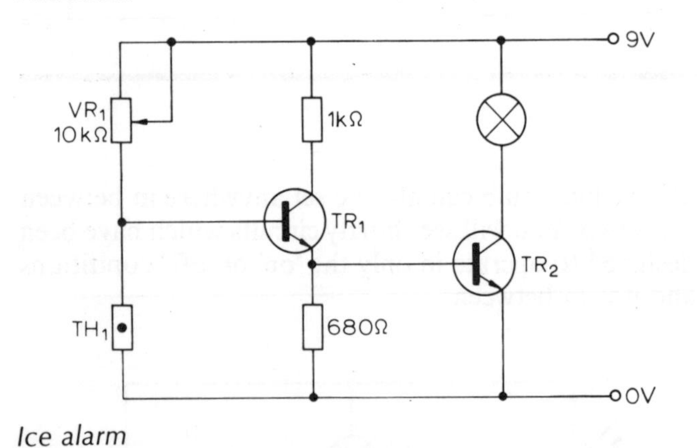

Ice alarm

Instructions

Assemble either, or both, of these circuits on S-DeC before you make permanent versions. The thermistor should be attached by long leads so that you can use it as a probe. The warning of the change in temperature given by these two circuits is provided by the light bulb.

When the temperature gets too high (fire alarm) or too low (ice alarm) the bulb comes on. If you replace the bulb by a relay then the circuit could actually do something useful such as switching on a heater, rather than being just a warning device.

Testing

For testing and setting up these circuits you will need a liquid at a known temperature. Half fill a beaker of water and heat it, using a bunsen burner and tripod and monitor its temperature with a thermometer. Place the thermistor in the water but take care that you cannot drag the beaker over with the trailing leads. Decide on the temperature at which you would like your fire alarm to operate and adjust VR_1 so that the bulb comes on at this temperature.

The procedure for setting up the ice alarm is the same, except that the thermistor is put in a beaker containing a small amount of water surrounded by an ice bath.

Sensitivity of switching

The point of adding the second transistor to the basic circuit of the previous page is to achieve full switching (fully off to fully on) of the circuit when in use.

Instructions

Place your thermistor in the water bath again and change the temperature very slowly whilst observing the bulb or recording the voltage level at the collector of TR_2. Find what temperature change is needed to switch the bulb from off to on. This tells you how sensitive your circuit is to temperature change.

Temperature switching range = °C

The Multivibrator Family

A special family of electronic circuits called multivibrators has the following characteristics:

1. It is a family of three circuits called the *astable*, *monostable* and *bistable*.
2. Each circuit contains two transistors.
3. Each transistor is either in the fully on or off state.
4. The two transistors in each circuit are always in the opposite state.
5. When the transistors change from on to off for any reason (or vice-versa) they do so very rapidly.

You will study each circuit in turn but here you can see how each earned its name: the three prefixes a-, mono- and bi- stand for O, 1 and 2, respectively.

To understand how these three circuits operate, examine the effect of a capacitor in a transistor circuit. Remember that, whilst a capacitor is charging or discharging, it *appears* to pass an electric current.

Astable
Each transistor has no stable state. This means that the transistors switch from on to off at regular intervals without any outside signal.

Monostable
Each transistor has one stable state. This means that if any outside signal should switch the transistors to their unstable state, they will automatically switch themselves back again, after a certain time delay.

Bistable
Each transistor has two stable states. This means that each transistor is happy to stay either on or off and only changes when switched by an outside signal.

Effect of Capacitor on Transistor Operation

Instructions

Assemble the circuit shown but do not connect the capacitor to point B yet. Connect a voltmeter across AC and now complete the circuit. Observe the voltmeter readings and the effect on the bulb and then disconnect the circuit and discharge the capacitor. Now repeat the experiment with the voltmeter across DE.

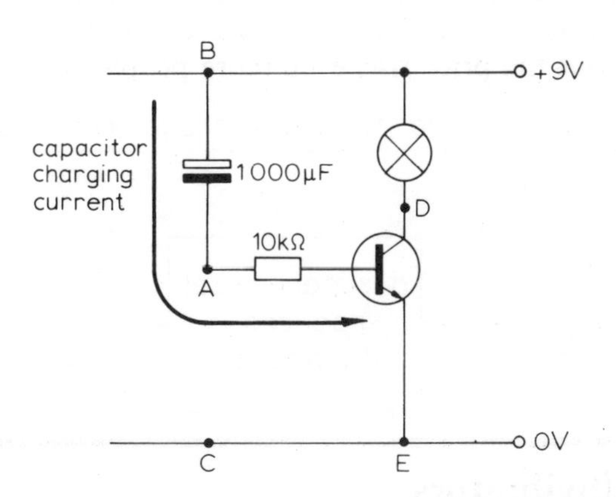

Results

The experiment shows quite clearly how the charging of a capacitor can control the state of a transistor. The voltage at A gradually falls to zero and the transistor switches off whilst the voltage at D gradually rises to the supply voltage as the transistor goes off. In particular, this simple, one transistor, circuit shows:

Capacitor charging current can switch on a transistor.
A fully charged capacitor can block the base current to a transistor.
The switching from on to off occurs very slowly.

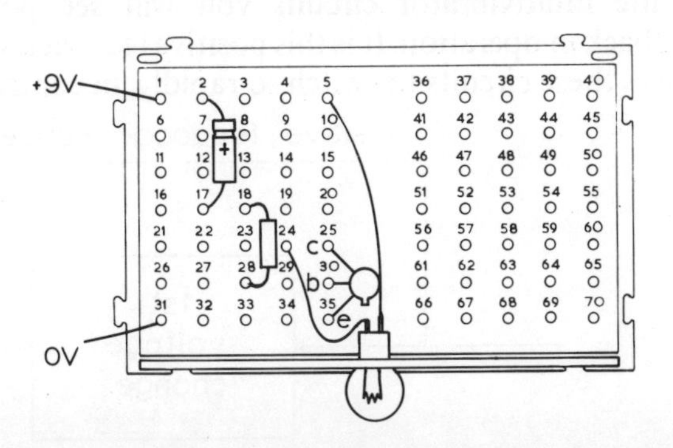

Feedback in Multivibrators

You are shortly to assemble the first of the multivibrator circuits which all depend on the feedback caused by the the addition of the second transistor. The nature of feedback is shown in the diagrams, there being two types, positive feedback and negative feedback.

Here you see the process of feedback. A change in the state of a circuit causes a change at a second point in the circuit and a fraction of this second change is fed back to the original. The sort of change we are talking about may be, say, an increase in voltage. However, feedback is not an effect which is only found in electronic circuits, since we can apply the term to everyday situations.

Positive feedback will increase the size and speed of the first change

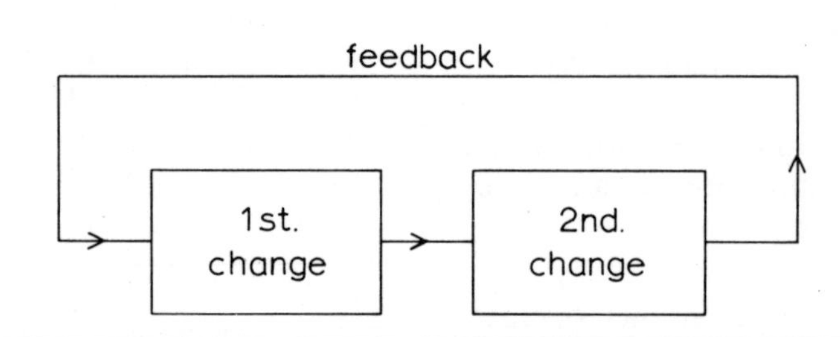

Negative feedback will decrease the size and speed of the first change

Negative and Positive Feedback

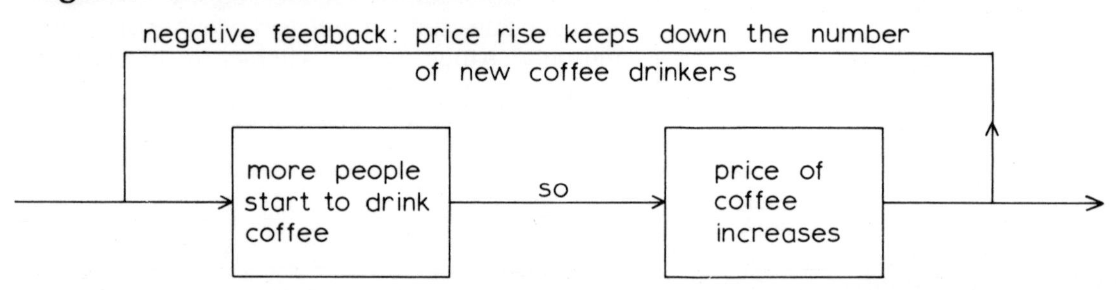

Here you see the effect of negative feedback. The second change decreases the size of the first change. Without the price rise, even more people would have started to drink coffee. Now look at an example of positive feedback:

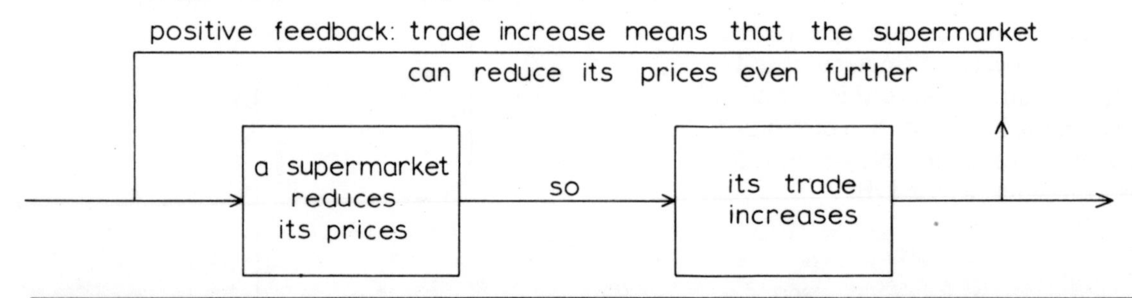

Multivibrators

In the multivibrator circuits you will see positive feedback in operation. It is this positive feedback which causes these circuits to switch so rapidly, in contrast to the very slow switching of the circuit on the last page which had no feedback.

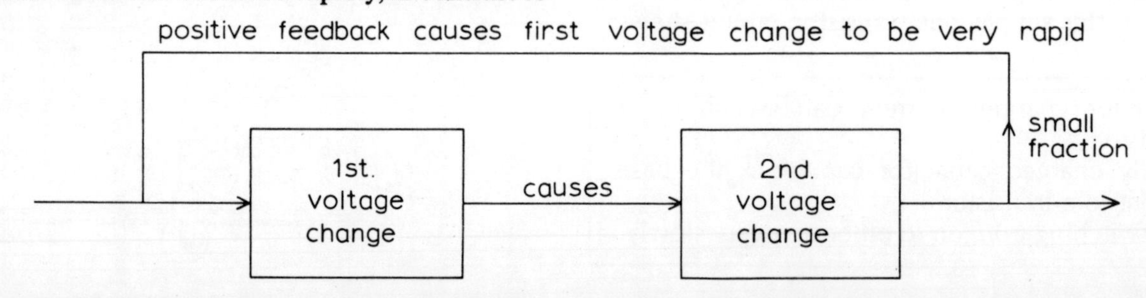

The Astable Multivibrator

Characteristics

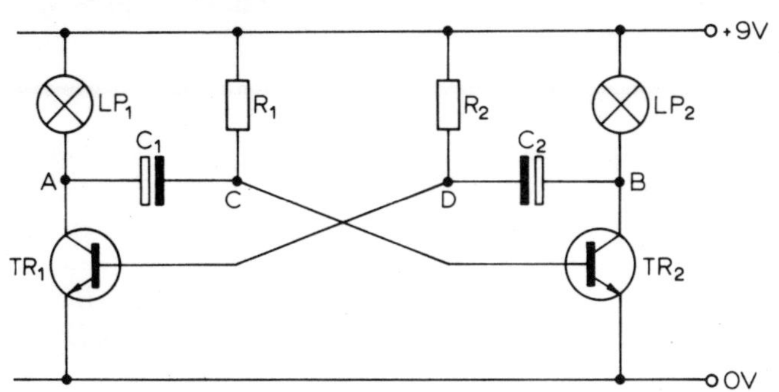

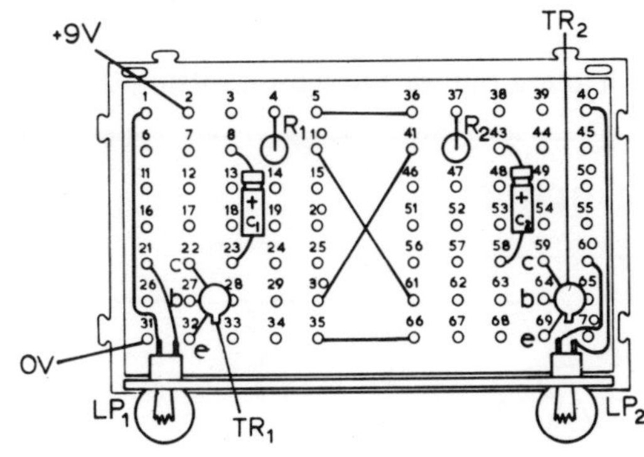

Instructions

Assemble the circuit, as shown, and find the voltages at points A and B as the circuit switches. Satisfy yourself that the transistors switch from the off state to the on state very rapidly, in contrast to the one transistor circuit. Record the number of flashes per minute as you change the component values, as shown in the table. C_1 with R_1 and C_2 with R_2 each form a charging circuit and, as you know from the chapter on capacitors, as the component values increase so does the time for the capacitor to charge. This increase in the time constant causes an increase in the length of each flash period.

C_1, C_2	200 μF	200 μF	200 μF	1000 μF	33 μF
R_1, R_2	10 kΩ	22 kΩ	4·7 kΩ	10 kΩ	10 kΩ
Rate of flashing					

The Circuit

Draw out the four part circuits which show:

1. Base current for TR_1.
2. Collector current for TR_1.
3. Base current for TR_2.
4. Collector current for TR_2.

Notice that the collector of each transistor is connected to the base of the other transistor, via a capacitor. It is this connection which provides the feedback, as you will see below. Each transistor switches on when the voltage at its base (either the point C or D) rises above 0·6 V. The voltages at these points vary as the capacitors charge up and it is the capacitor charging which causes the continuous switching.

Feedback

Consider the circuit when TR_1 is on and TR_2 is off. This means that point A will be near 0 V and point B will be near 9 V. There are several steps involved in describing the feedback but no step is new to you so follow the steps carefully. We join the circuit with capacitor C_1 beginning to charge up:

The effect in square 6 adds to the original change so the voltage at C rises even more rapidly and switching becomes rapid.

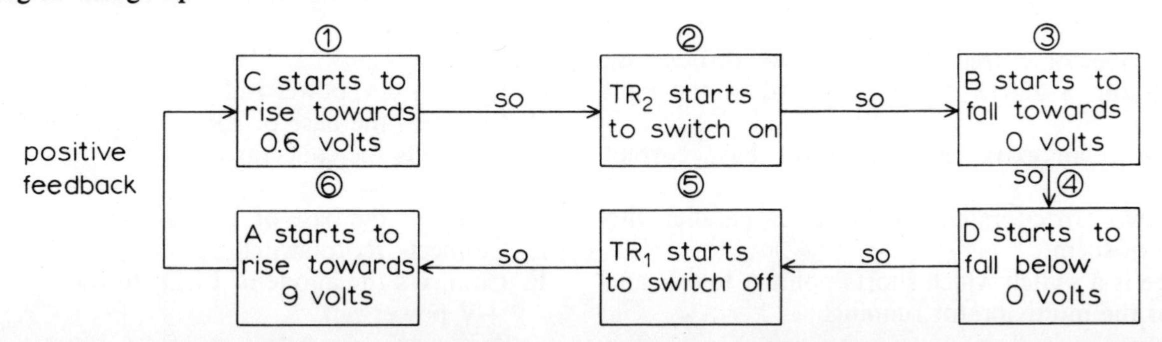

Electronic Heads and Tails

This electronic toy will 'call' heads or tails when switch S_1 is pressed. It does this in a completely random fashion and can be used in many games. It is a multivibrator which switches very rapidly, and then stops when you press the switch leaving one of the light emitting diodes (l.e.d.s) on.

Construction

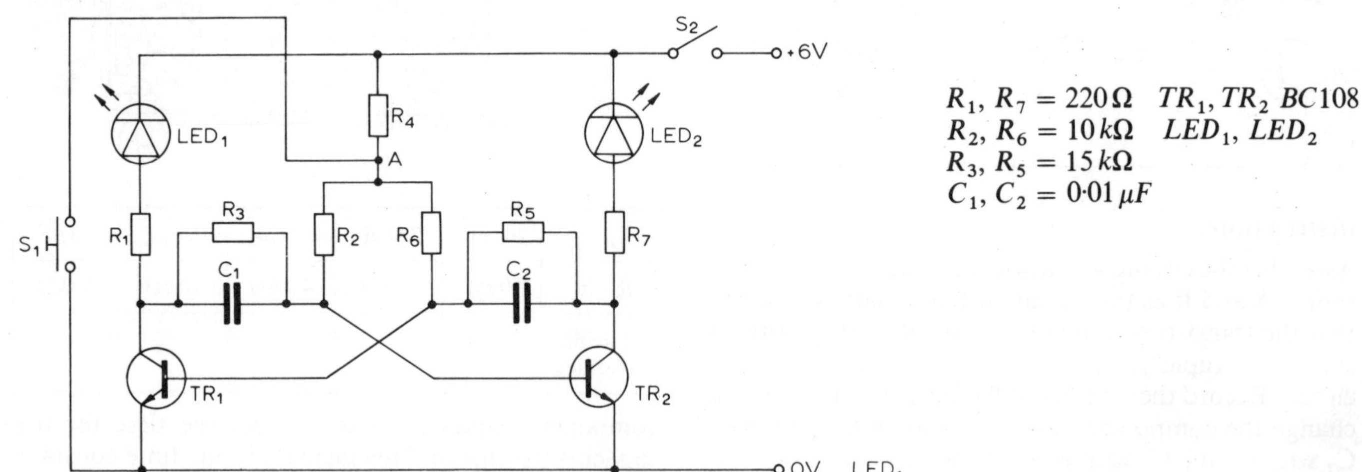

$R_1, R_7 = 220\,\Omega$ TR_1, TR_2 $BC108$
$R_2, R_6 = 10\,k\Omega$ LED_1, LED_2
$R_3, R_5 = 15\,k\Omega$
$C_1, C_2 = 0{\cdot}01\,\mu F$

As usual with p.c.b. work, check that your capacitors are the right size for the spacings shown on the diagram. Follow the instructions for p.c.b. assembly. Note that R_7 may have to be mounted vertically, because of the small space allowed for it. Your case should have cutouts on the front panel for:

S_1, which is the switch which 'calls' heads or tails.
S_2, which is the on/off switch to save your battery.
LED_1, LED_2 } which light up the heads or tails signs.

Use transfer stencils to mark one of the LEDs as heads and the other as tails. Make sure that you identify the leads of the LEDs correctly and assemble them the correct way round in the circuit.

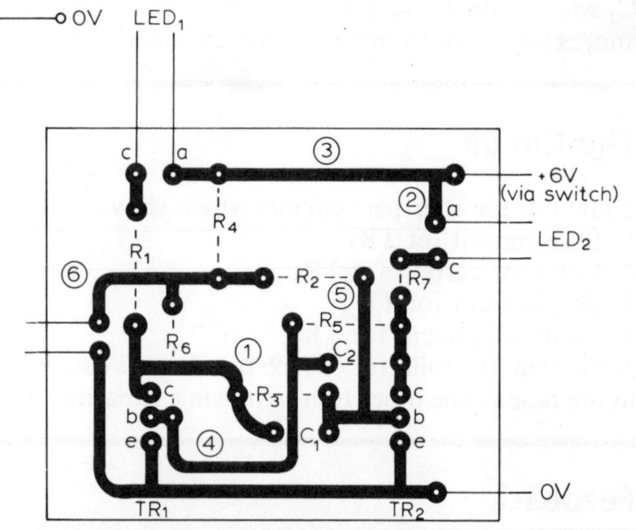

The Circuit

You should recognise the basic structure of the multivibrator from the previous page. However, this circuit has some extra components. Identify the following and give the component numbers from the circuit diagram:

1. The collector current for TR_1 flows through an LED and a 220 Ω resistor, instead of the light bulb.
2. There is an extra resistor, in the base current circuit.
3. The two capacitors have a resistor in parallel with each of them.
4. There is a switch which shorts point A to 0 V and stops the multivibrator running.

Circuit Exercises

It is a good exercise in circuitry to check the circuit diagram against the p.c.b., diagram. You will find that some of the conductors on the p.c.b. are numbered. Use these numbers to describe which conductor does each of the following:

	Conductor Number
A. Is the +6 V power rail	———
B. Connects the base of TR_1 to capacitor C_2	———
C. Connects the collector of TR_1 to capacitor C_1	———
D. Connects the base of TR_1 to C_2.	———
E. Connects R_2 to switch S_1.	———
F. Connects the anode of LED_2 to the +V power rail.	

A Door 'Bell'

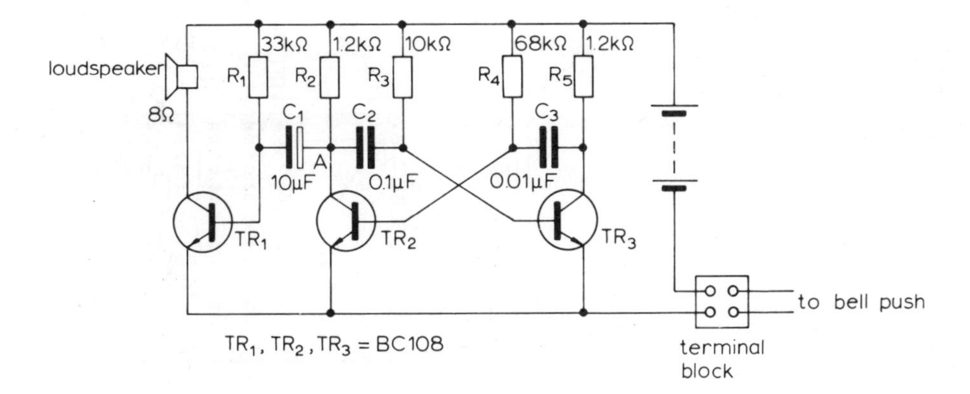

This circuit is the electronic equivalent of a door bell. Closing the push-button will result in a note being emitted from the loudspeaker.

Construction

The circuit is assembled on 4 mm Veroboard. Follow the usual rules of construction, ensuring that you cut the board to size, drill and make the breaks in the copper strip and clean the copper before you start soldering. The circuit board, battery and loudspeaker are housed in a suitable case which is positioned where the sound can be heard. The length of wire to the bell-push must be known before this can be connected but use of a terminal block means that this wire can be fitted at home.

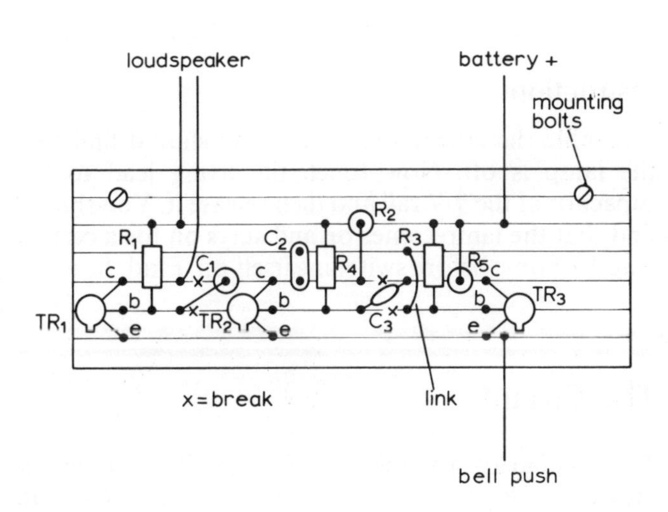

The Circuit

You should recognise two part circuits—an astable multivibrator and a common emitter amplifier. Capacitor C_1 connects the two part circuits together, passing the a.c. signal only and no direct current.

The Astable Multivibrator

Use a d.c. voltmeter and measure the voltage between:

TR_2 collector and 0 V (bottom rail) ———
TR_3 collector and earth ———

Remember that when a transistor in an astable is:

At cut off, the voltage at the collector is near 9 V.
Hard-on, the voltage at the collector is near 0 V.

Thus, when a transistor oscillates between these two extremes, the voltage at its collector averages about mid-way between these two figures, say 4 V.

The Amplifier

Connect an oscilloscope across the loudspeaker terminals and:

Measure the frequency of the output ———
Measure the voltage of the output ———
Draw the waveform of the output

Connect the oscilloscope from capacitor C_1 (point A) to the bottom rail

Measure the voltage (which is the input to the amplifier) ———
Draw the waveform.

Remember that transistor TR_1 draws its base current via R_1. Examine the waveform output and decide whether its shape has been amplified exactly or not. If not, you may change the base bias by altering the value of R_1. Report on your results.

The Monostable Multivibrator

Investigating the Monostable Multivibrator

Remember that this circuit has one stable state, to which it always returns. In the circuit shown, the stable state is transistor TR_1 (and the lamp 1) off and transistor TR_2 on.

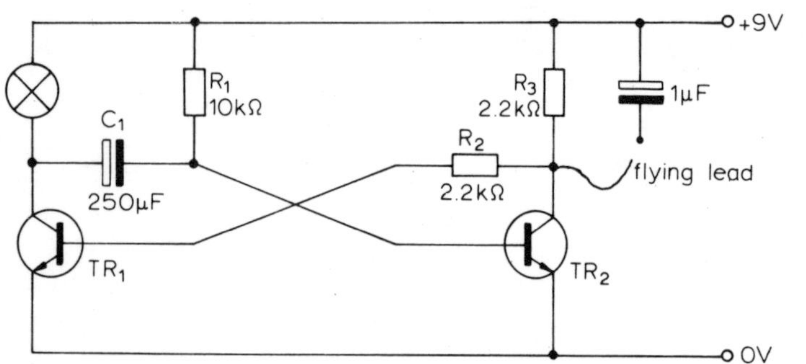

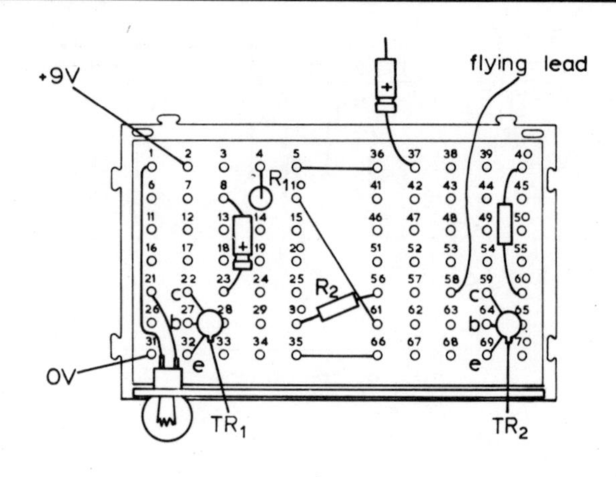

Instructions

Assemble the circuit shown and you should find that the lamp is off. Now touch the flying lead to the capacitor at the 9 V rail and then release it. You should find that the lamp comes on and stays on for a certain length of time before switching itself off. Find the effect on the time of illumination of different component values, as in the tables below.

$R_1 = 10\ k\Omega$ C_1	Time of illumination
250 μF	
500 μF	
1000 μF	

$C_1 = 250\ \mu F$ R_1	Time of illumination
22 kΩ	
47 kΩ	
100 kΩ	

The Circuit

The similarity to the astable circuit is obvious. However, TR_1 does not have a capacitor in its base lead and you will remember that it was the increasing voltage as a capacitor charged up which enabled the transistors to switch on.

The results above show you, again, that it is the time constant of a *RC* combination which controls the switching times. With large values of R_1 and C_1, the period of illumination becomes quite long.

The switching of TR_2 is controlled by the voltage change in the R_1 C_1 combination.

Feedback

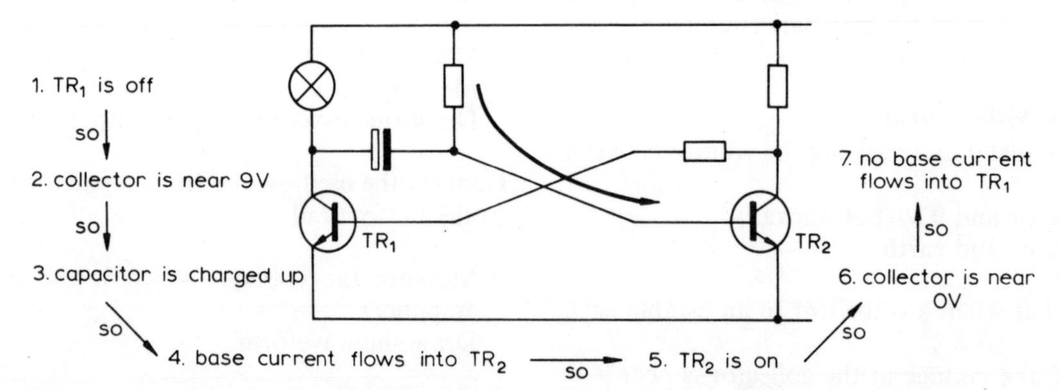

Examine why the feedback in this circuit should tend to keep TR_1 switched off and TR_2 switched on. The logical steps involved are similar to those we found in examining the astable. Consider the situation when TR_1 is off and why feedback reinforces this situation.

So event 1 leads through a chain of events to event 7 and this last one reinforces number 1. So the circuit remains in this state unless you disturb it and this feedback results in the circuit always returning to this state.

The Monostable Timer

You have seen that the monostable can switch on a light bulb, for instance, for different periods of time, depending on the value of components used. We can put this to good effect in making a timer which can control light bulbs or (using a relay) any other device you have.

The variation in the timing period is achieved by the use of a variable resistor which controls the rate at which the capacitor charges up. Two circuits are shown below: the first switches a relay and the second uses a transistor which can take a much larger current than the BC108 transistor which we have normally used.

Construction

The case for each of these timers needs cutouts in its front panel for:

1. On/off switch to conserve the battery when the timer is not in use (S_1).
2. Timing switch, to start the timing period (S_2).
3. Variable resistor, which controls the length of the timing period (VR_1).

A further hole and grommet are needed to take the cables to the device being switched.

Circuit 1

This is a simple adaptation of the circuit from the previous page, with a relay replacing the light bulb. The circuit being switched by the relay contacts is not shown but the relay must be firmly fixed inside the case.

Circuit 1

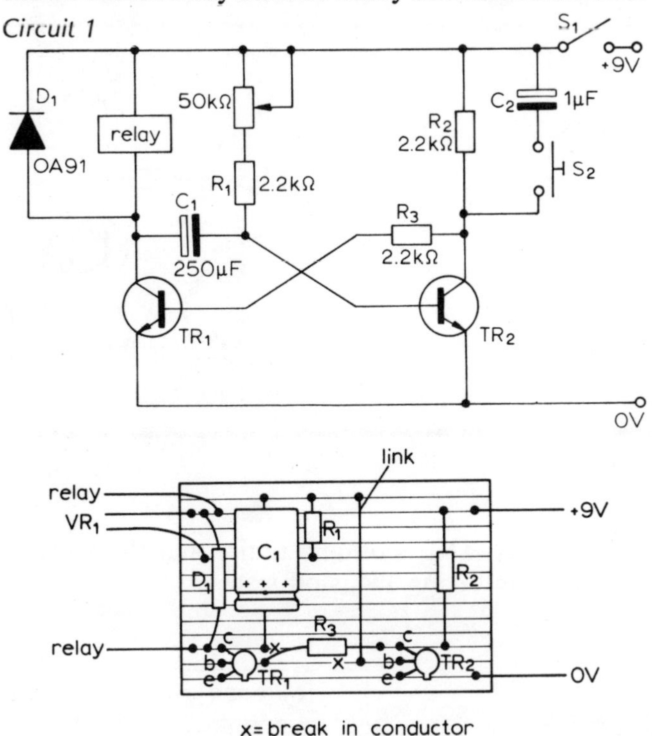

Circuit 2

The BFY51 in this circuit will take a current of about 1 A but should be fitted with a push-on heat sink if you intend to use it near its limit. However, you have to bear in mind that a small capacity battery, such as a PP6, is not designed to supply this size of current and you may wish to consider the use of a power pack to supply 1 A.

Circuit 2

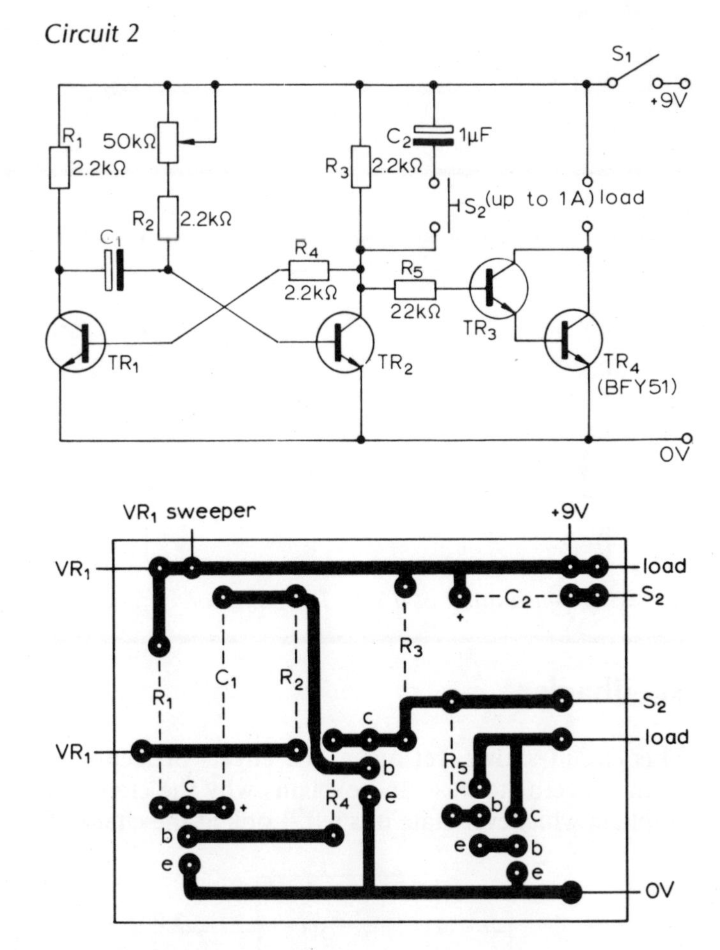

Testing

VR_1 should give timing periods from 1 s to 10 s. Your next task is to calibrate the timer. Find the setting of VR_1 which gives timing periods of 5 s and 10 s and mark the positions on the panel, using transfers. Then mark in the settings for every intermediate second. You now have a timer with which you can dial any timing period you need (within limits) for any device that can be switched by a relay.

The Bistable Multivibrator

Investigating the Bistable Multivibrator

The last of the multivibrator family, the bistable, has two stable states. The two transistors operate like the light switches in your house—each transistor stays either off or on until switched by an outside agent but they never switch themselves automatically as the first two multivibrators did. These two transistors always remain in the state in which they were left.

Instructions

Connect up the circuit as shown and you should find that one of the transistors is off and the other is on. They will remain that way until you do something to switch them over. Use the flying lead and touch it (for a moment) to either the top or bottom rail. You will find that you can switch the circuit over by choosing the correct voltage to input (that is, +9 V or 0 V).

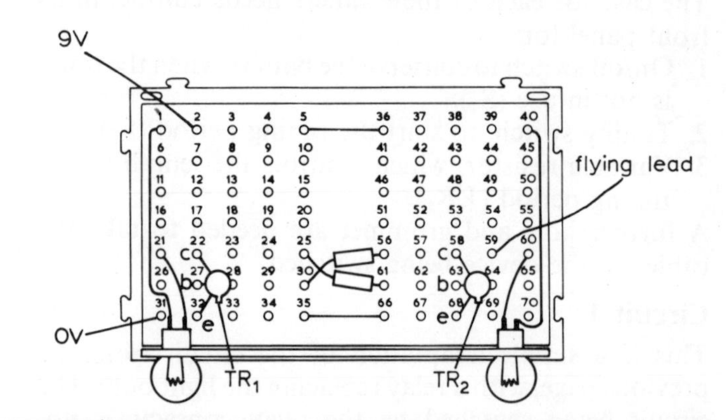

Results

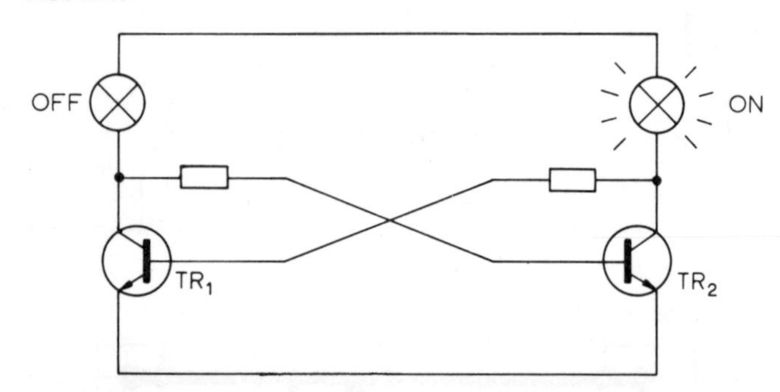

To switch—use input of 9 V to TR_2 collector.

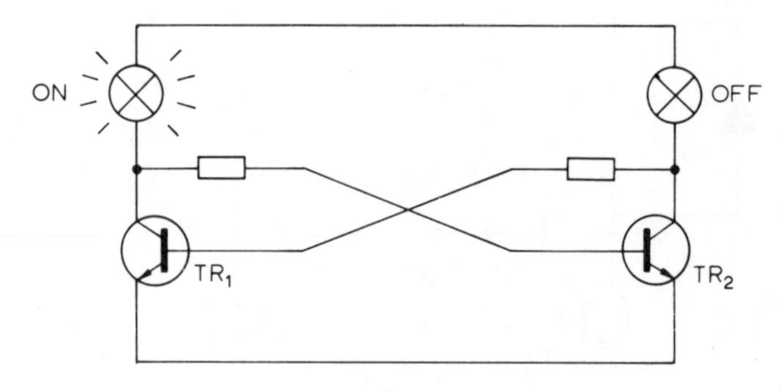

To switch—use input of 0 V to TR_2 collector.

Feedback

This circuit shows, yet again, the effects of feedback from collector to base and explains why the circuit is stable in whichever state it is left. Look at the chain of events if, say, TR_1 is off and notice that these events result in reinforcing TR_1's off state.

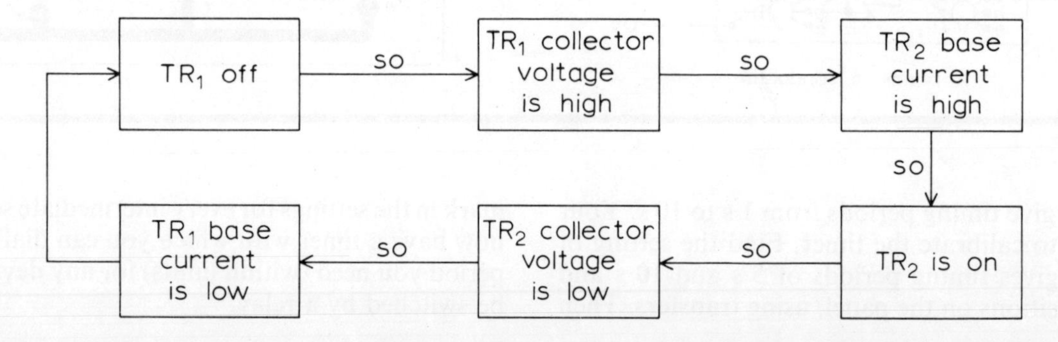

More Useful Bistables

The bistable is the circuit which is at the heart of many of the operations performed by a computer. To make a bistable circuit which will be more useful for your later computer experiments there are modifications to make.

Bistable with Separate Indicator

Assemble the circuit shown. Use S-DeC if you do not wish to keep the circuit for the binary adder in Chapter 12. Carefully label the leads for (+) and (−) supply and output.

TR_3 and the lamp form an on/off indicator, for the bistable, allowing it to switch more quickly.

Operate the bistable, using the flying lead and confirm that it switches by connecting A or B in turn to the earth rail.

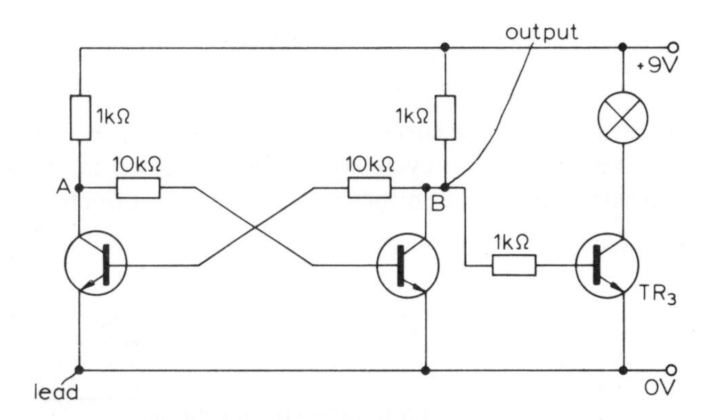

Bistable with separate indicator

Bistable with One Switching Point

Add the resistors, diodes and capacitors as shown. This circuit now requires input voltages at one point only, the lead labelled input. The circuit will switch for changes from high to low voltage applied to the input lead.

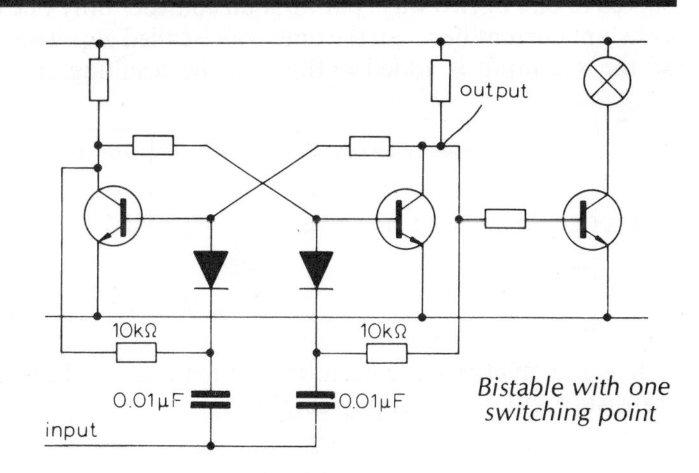

Bistable with one switching point

Addition of a Gate

Add in the extra transistor stage, as shown, and connect the input lead to the collector. This makes a bistable with a very simple switching action.

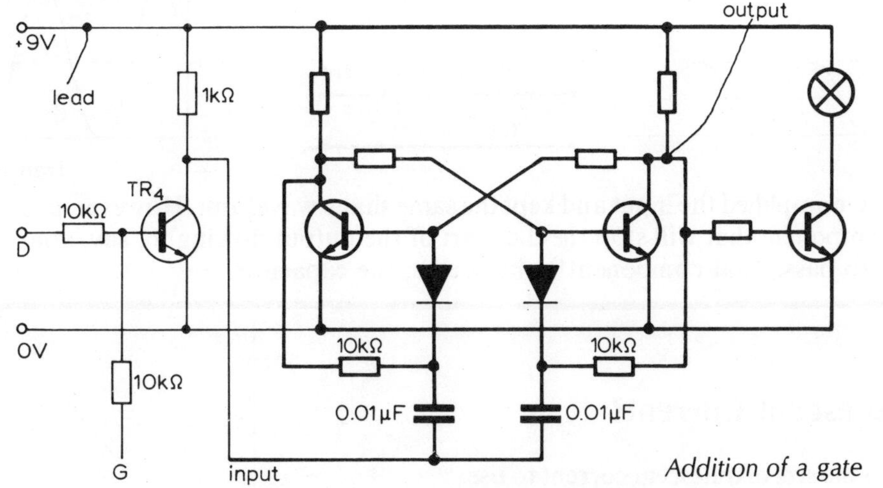

Addition of a gate

Connect G to 0 V

Tap the 9 V lead to point D slowly several times and note that the bistable switches for each tap. You now have a bistable which switches at one input point. This is very important for using a bistable in a computer.

Connect G to 9 V

Now you will notice that it is impossible to switch the bistable. Hence the use of the term gate. In this case, the gate has been closed by taking G to high voltage.

Keep this bistable for the experiments in Chapter 12.

8 AMPLIFICATION OF ALTERNATING CURRENT

So far we have only put direct current into the base of a transistor. We have to deal with alternating current in a special way because:

A transistor conducts load current in one direction only.

The diagram below shows you what would happen if you put alternating current into the base of a transistor. The positive half of the cycle would be amplified in the same way as our previous direct current sources but the transistor will not conduct in the opposite direction, so the negative half would be cut off.

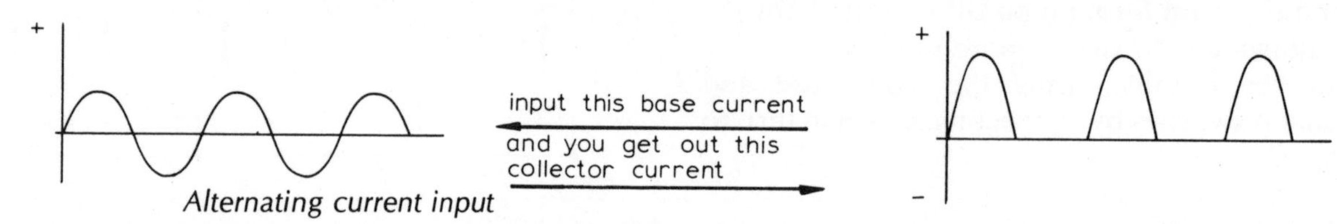

Alternating current input

In considering this problem, we have used the figures obtained for a BC108 transistor (page 106). The problem is solved in one of two ways but we shall consider only one of them here. The base is given a d.c. bias so that a small constant current flows all the time—also called a quiescent current. This current is shown in the first diagram and so the a.c. input is added to this and the resulting current put into the base is shown.

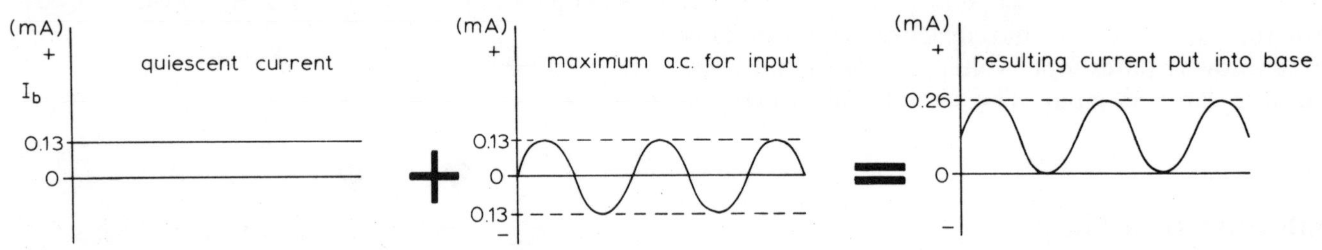

It is this fluctuating current that is fed into the base of the transistor and, since it is now direct current, the transistor will quite happily amplify it.

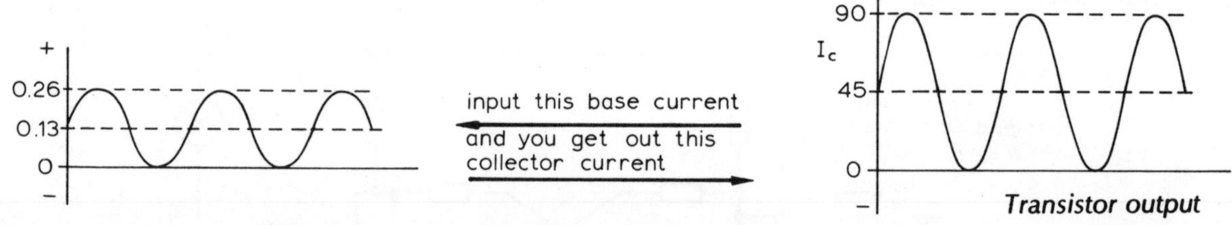

Transistor output

In this way you have amplified the input and kept the same shape waveform. However, it is still in fluctuating d.c. form. We need a component that will stop the d.c. part of the output flowing to, say, your loudspeaker but will allow the a.c. part to pass. That component is, of course, the capacitor.

How Much Quiescent Current?

Here is how you find the size of quiescent current to use. Go to the I_c/I_b graph for the circuit you are using and find the straight line part of the graph AB. Select the point, Y, which is half way along the line and use this value for your base current, I_Q. This produces a collector current of half the maximum and so gives a collector voltage of half the supply voltage.

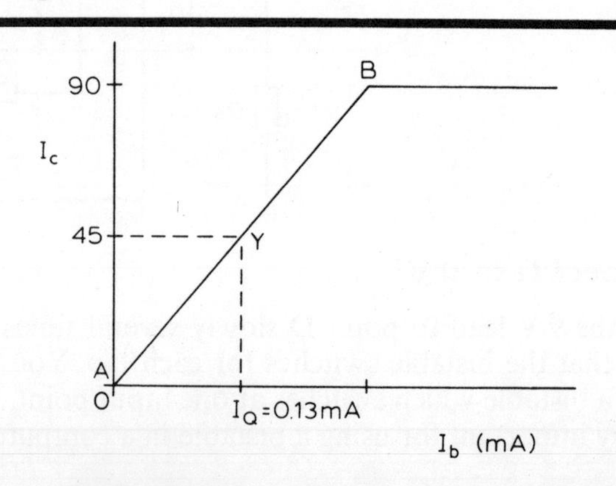

Amplification of Alternating Current—the Practice

The Transistor as an Amplifier

The values used on this page are those used on the last page. Your transistors will almost certainly give different values from those used by the author, so use your value of R_b to set the transistor near its mid-point of operation (collector voltage of 4·5 V).

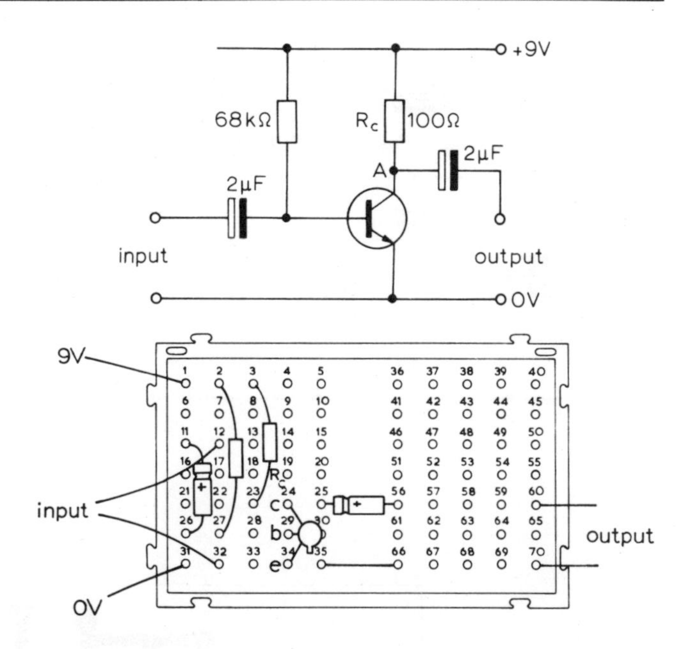

Instructions

Assemble the circuit shown and use a signal generator at the input and connect an oscilloscope across the output. Have the oscilloscope switched to alternating current and then increase the input until your output records nearly 9 V, peak to peak, as shown on the screen. Set the V/cm switch to obtain a large display on your oscilloscope screen and then record this oscilloscope display and the others asked for.
Your oscilloscope will show, quite clearly, the amplification of the signal.

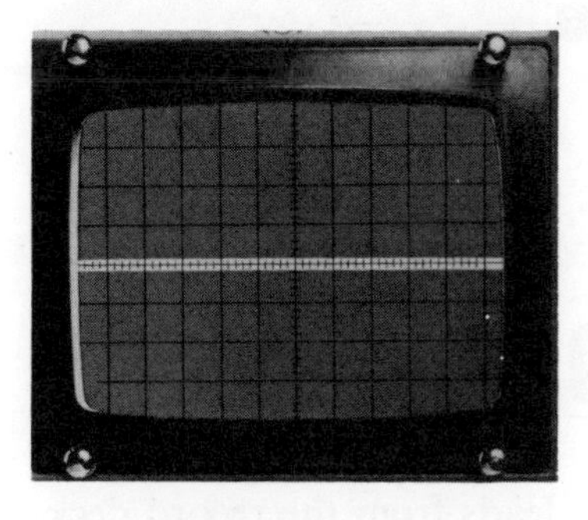

1. Input displayed on same setting as the output shown in 3

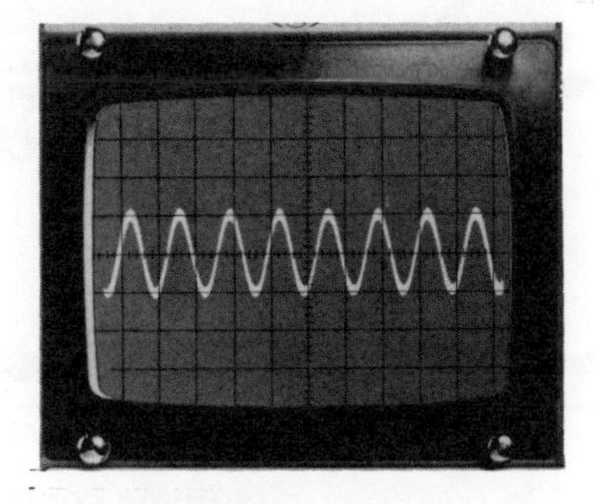

2. Input displayed on more sensitive setting. Peak to peak voltage = —— V

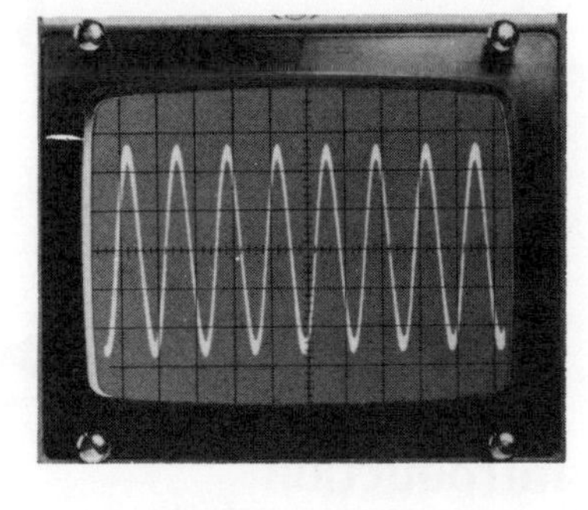

3. Output. Peak to peak voltage = 9 V

Voltages

We have considered amplification chiefly in terms of current, but, since the oscilloscope is a voltmeter, you were in fact looking at voltage amplification.
The table and diagrams show what happens to the collector voltage when the collector current changes.

Point	Current (mA)	Voltage drop across R_c (100 Ω)	Collector voltage at point A
X	45	4·5	4·5
Y	90	9	0
Z	0	0	9

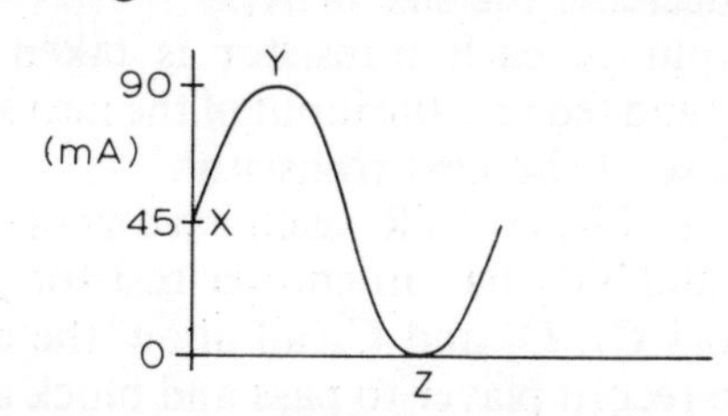

Collector current

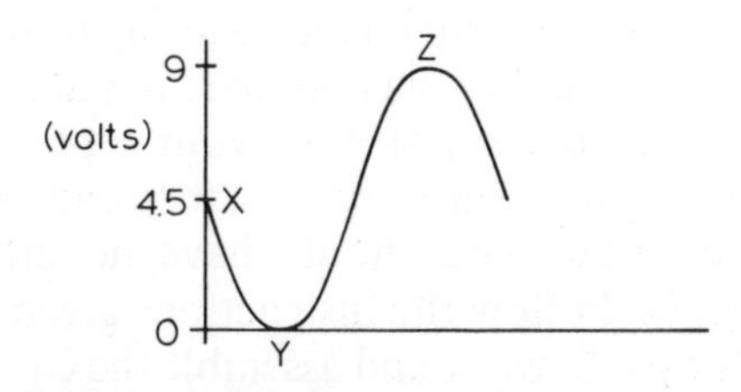

Collector voltage at point A

A Practical Amplifier

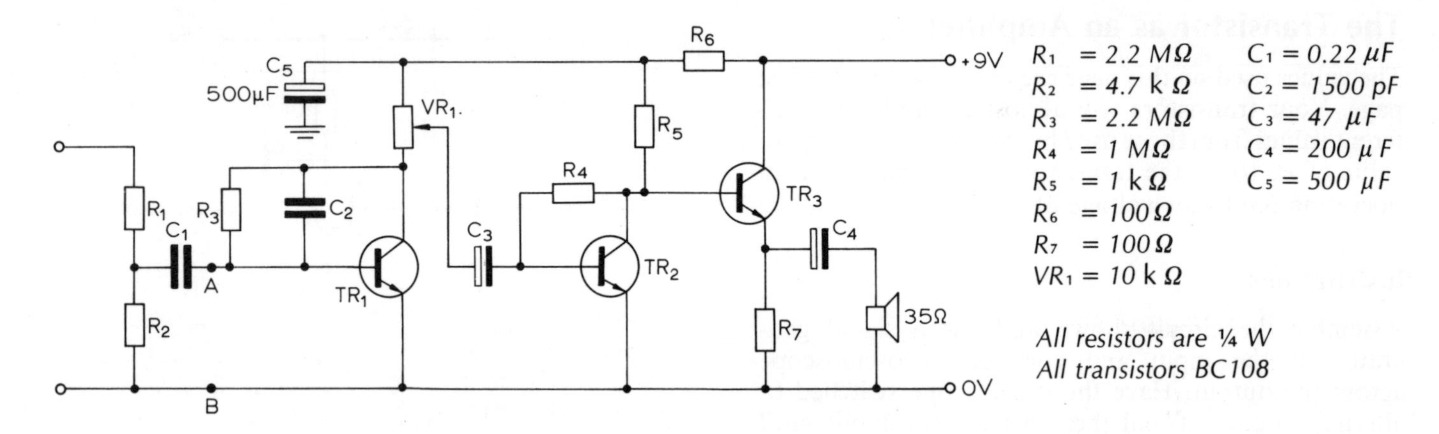

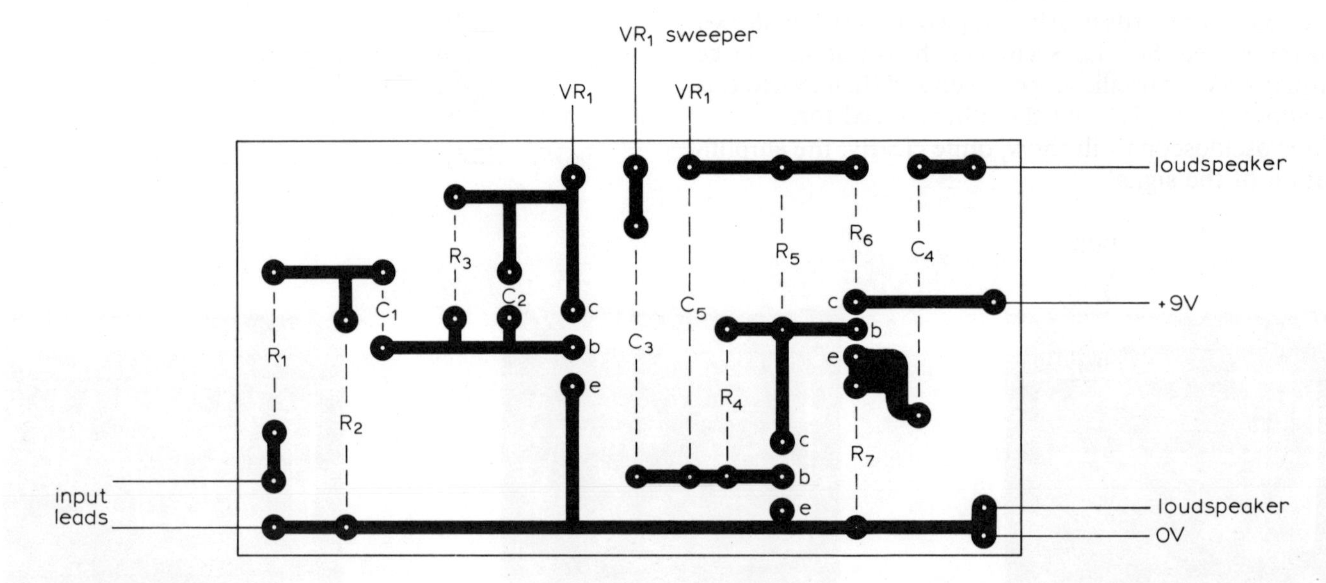

Introduction

This is a simple three transistor amplifier, suitable for a record player with a crystal cartridge. If you wish to use it for an intercom, with a crystal microphone then omit R_1, R_2 and C_1 and connect the microphone to points A and B. You may have to change the value of C_2, by trial and error. The p.c.b. layout is quite spacious, so that it is easy to compare it with the circuit diagram. You may design your own board if you wish to make it smaller.

Construction

Check the capacitors you intend to use against the p.c.b. diagram, to ensure that the hole spacings are correct. If the spacings are not suitable for your capacitors, then adjust the diagram. Since the resistors and transistors are standard sizes, you should have no difficulty in fitting these in. Follow the instructions given earlier in the book for p.c.b. work and assemble the circuit. Your case will need cut-outs for the loudspeaker, VR_1 (volume control) and the leads from the record deck. You will probably wish to insert a switch in the battery leads.

The Circuit

Whilst this circuit is more complicated than the simple one-transistor circuit, you should recognise some of its features:

1. R_1 and R_2 form a potential divider at the input. If you wish to increase the size of your input, then you should decrease the size of R_1.
2. The output of each transistor is taken from its collector and fed into the input of the next stage, that is, the base of the next transistor.
3. Transistors TR_1 and TR_2 each have a collector load resistor and TR_3 has an emitter resistor.
4. Capacitors C_1, C_3 and C_4 all allow the a.c. signal from the record player to pass and block any direct current.

Using Other Transistors

You have seen amplification achieved using the BC108 transistor. Out of the bewildering array of transistors available, two more have been chosen for you on which to practise biasing exercises. The two most important exercises are:

1. Setting the bias at the mid-point.
2. Switching the transistor fully on.

2N2926 Transistor

The 2N2926 transistor is an npn silicon-planar type, usually encapsulated in plastic, and able to take a maximum collector current of 100 mA. Notice that its lead positions are different from those of the BC108. There is, in fact, a family of four 2N2926 transistors, each given a different colour code. You are to use the 2N2926 orange. (The 1 kΩ resistor is not strictly necessary. It is only included to protect the transistor against accidental damage, should you make R_1 too small.)

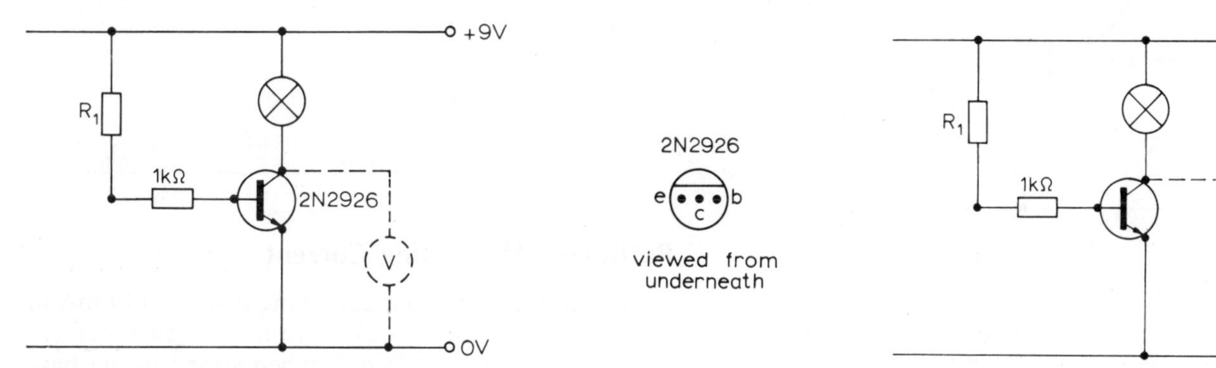

Setting the bias at midpoint

Switching the transistor fully on

Setting the Bias at Midpoint

Start with R_1 at 2·2 MΩ and find the collector voltage. Now decrease R_1 in stages until you find a value which sets the collector voltage at mid-rail, 4·5 V. This means that you have now set up this new transistor as an a.c. amplifier.

Results

Base resistor = ———

Collector voltage = ———

Base current, I_b = ———

Collector current, I_c = ——

Gain, $h = \frac{I_c}{I_b}$

= ———

Switching the Transistor Fully On

Continue reducing the value of R_1 until the transistor is fully switched on. This occurs when the bulb is at its brightest and the voltage across it is a maximum. You have now set up this new transistor to its 'on' position.

Results

Base resistor = ———

Maximum voltage across bulb = ———

Base current, I_b = ———

Collector current, I_c = ——— Gain, h = ———

ZTX300 Transistor

This is a similar type of transistor to the 2N2926 but it can take a higher collector current of 500 mA.

Setting the Bias at Midpoint

Repeat the above experiment with this transistor, to set the collector voltage at mid-rail.

Results

Base resistor = ———

Collector voltage = ———

Base current, I_b = ———

Collector current, I_c = ———

Gain, $h = \frac{I_c}{I_b}$

Switching the Transistor Fully On

Repeat the above experiment so that you just set the transistor fully on.

Results

Base resistor = ———

Maximum voltage across bulb = ———

Base current, I_b = ———

Collector current, I_c = ———

Gain, h = ———

Alternating Current Amplification

Operating the Amplifier

Just a brief look at a transistor radio will show that its amplifier is more complicated than anything studied so far. Why do amplifiers have to be made so complicated to give good results? Look at the problems which occur with transistor circuits on this page and you will see why the very simple amplifier is not good enough.

Reminder: These two diagrams show the operating conditions for the very simplest transistor amplifier. The graph was obtained by the author, and so will vary slightly from the one you obtained, on page 106. You may have to adjust R_b to set the collector voltage at 4·5 V.

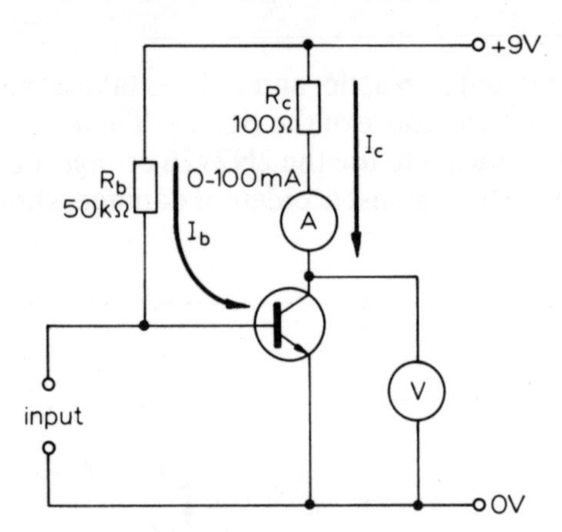

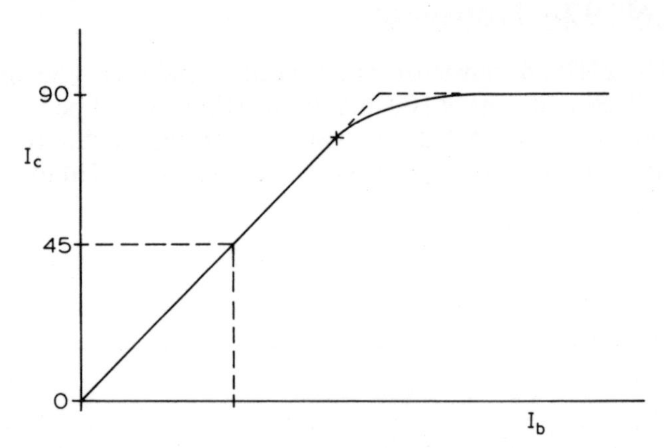

Setting the Bias

1. The base bias current is set at the mid-point of the linear part of the graph; $I = 0{\cdot}14$ mA.
2. This sets the collector current also at the mid-point; $I_b = 45$ mA.
3. This sets the collector voltage at mid-rail (approximately); $V_c = 4{\cdot}5$ V.

This transistor is now set at the ideal state for taking a.c. input, since it can amplify equal positive and negative swings of base current.

Putting in Alternating Current

1. A.C. input to the base can swing from $-0{\cdot}14$ mA to $+0{\cdot}14$ mA. (This would give a total base current swing from 0 to 0·28 mA, when added to the base bias current, above.)
2. This gives a total collector current swing from 0 to 90 mA.
3. This gives a collector voltage swing from approximately 0 to 9 V.

Unfortunately, this ideal state is easily disturbed by two factors which you are now to investigate.

Disturbing the Circuit

Change in Temperature

Assemble the circuit shown above. Gradually heat up the transistor, using a match, and watch the change in current and voltage readings. Record the largest change in I and V. It is better to throw away the transistor after this heat treatment.

Maximum I_c = ———
Minimum I_c = ———
Maximum V_c = ———
Minimum V_c = ———

Change in Transistor

Assemble the above circuit again and take readings with as many different BC108 transistors as you can find. Again find the maximum variation in the current and voltage readings.

Maximum I_c = ———
Minimum I_c = ———
Maximum V_c = ———
Minimum V_c = ———

Results

Your results should show you that this simple circuit is too sensitive and is easily knocked off its operating point. The ideal circuit design is one which is unaffected by changes in temperature and will accept any transistor of the specified type. The next few pages are devoted to studying circuits which are stabilised and the common methods used to achieve stability.

What Happens When the Bias Goes Wrong?

If you look at the I_c/I_b graph on page 128, you will see that, for the circuit we are using, the collector current can vary between approximately 0 and 90 mA and the collector voltage can vary between approximately 0 and 9 V. You are now going to see what happens when the base input tries to take the transistor outside these limits.

The Effect of Changing the Bias Current

Instructions

The circuits shown below are just the simple transistor amplifier circuit again, with a signal generator and oscilloscope connected. Use a frequency of 1 kHz and follow the instructions given with each. Check your oscilloscope display with the one shown.

Base Bias Set at Ideal Mid-Point

Choose R_b to set the collector voltage at about 4.5 V. Increase the input to give the largest possible output, without distortion. Measure the output voltage.

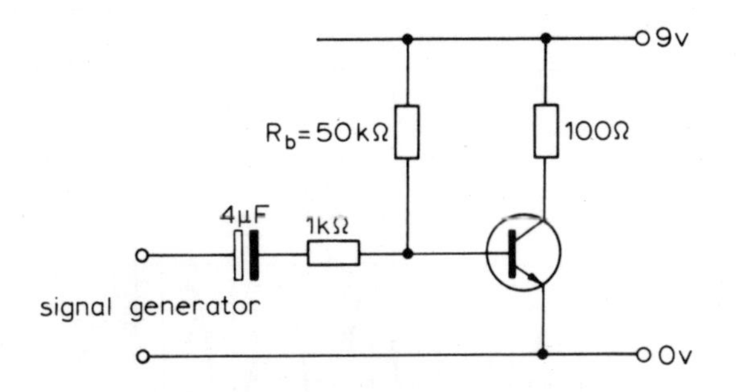

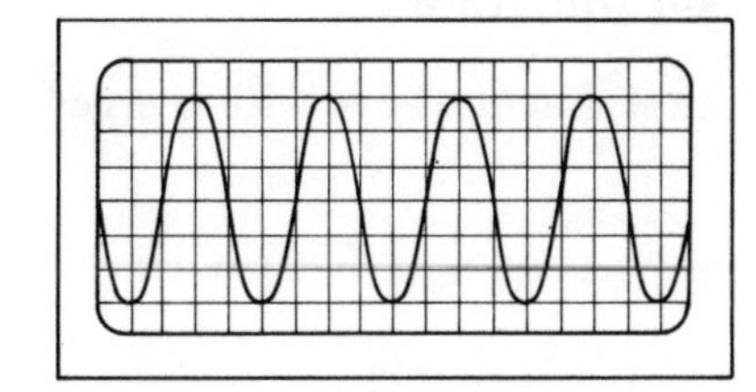

The output shows good amplification up to 9 V.

Base Bias Set at Mid-Point and Input Increased to Maximum

Gradually increase the input from above and note the effect on the size and distortion of the output.

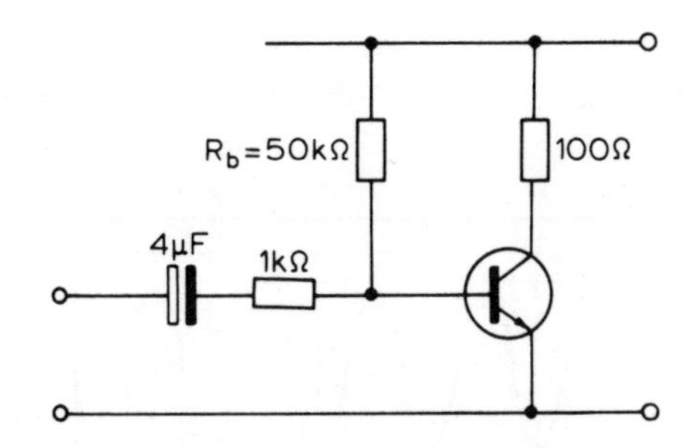

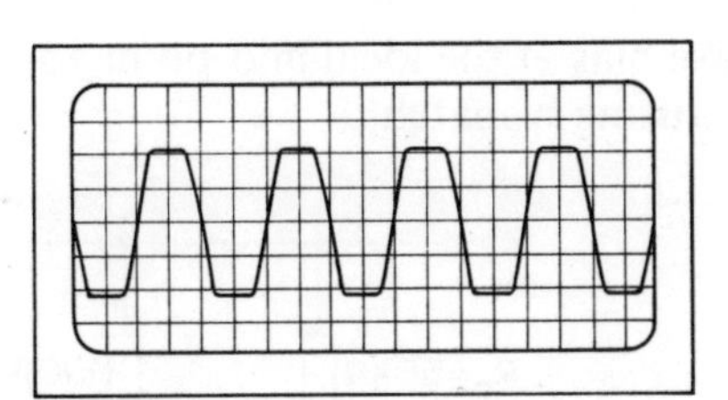

The transistor is suffering over-driving. It cannot amplify the peaks of the input.

Continued on p. 130

Base Bias Set too Low

With a new base bias find the maximum output before distortion sets in and then note the wave shape when it is distorted.

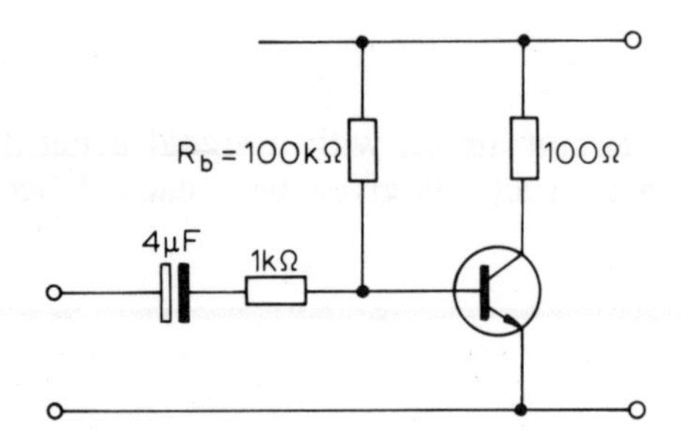

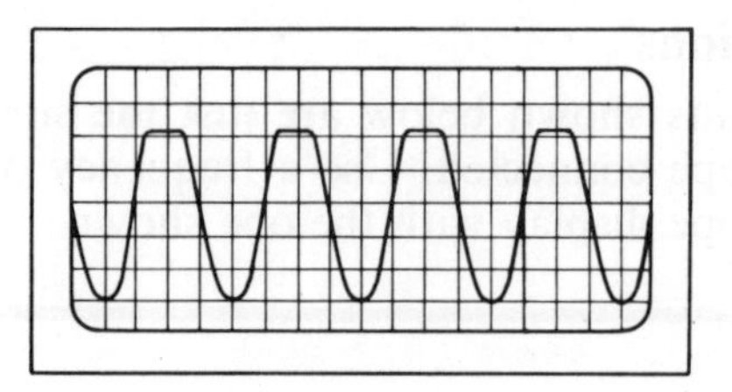

The output is 'clipping'—negative peaks of input go outside the possible range of amplification

Base Bias Set too High

Repeat the experiment above with the new base bias.

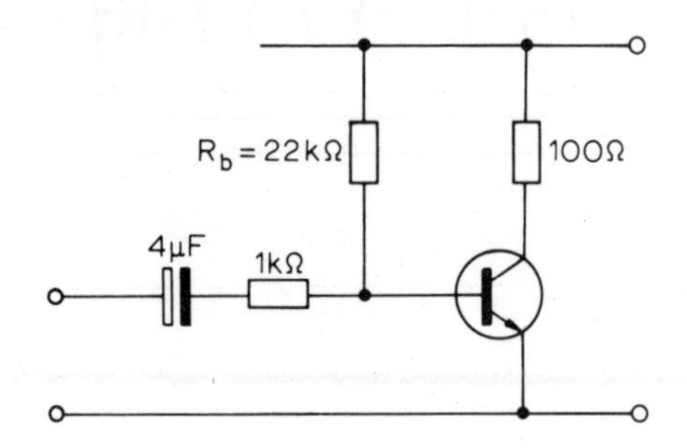

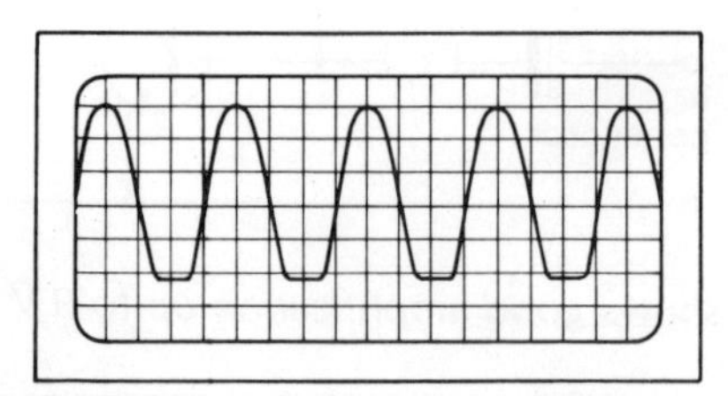

The output is 'bottoming'—positive peaks of input go outside the possible range of amplification

Base Bias Set too High by Increasing Temperature

Set the base bias at the ideal mid-point and set the input to give the largest undistorted output. Now heat up the transistor, using a match.

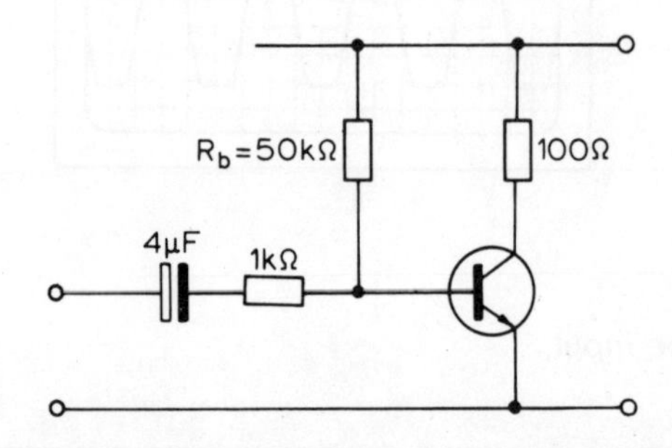

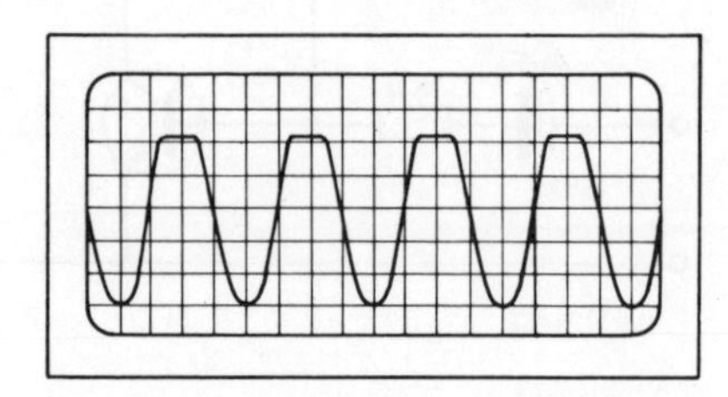

The effect is the same with the base bias set too high.

Theory into Practice

Looking at the Amplified Signal

Let us now put this theory into practice. The circuit here has R_b so chosen to give a base current at the mid-point on the last graph. You can check that your transistor really is set at its mid-point since this should give a collector voltage of approximately 4·5 V.

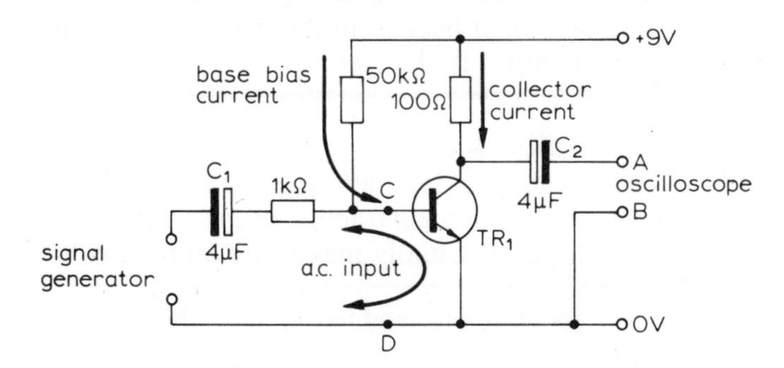

Instructions

Assemble the circuit shown and connect the oscilloscope and your own signal generator (at its minimum output). C_1 is there to ensure that only alternating current is input from the signal generator. This one-transistor amplifier should pass a collector current that is a magnified copy of the input. You can examine this collector current by following the voltage change it produces at the collector, on the oscilloscope.

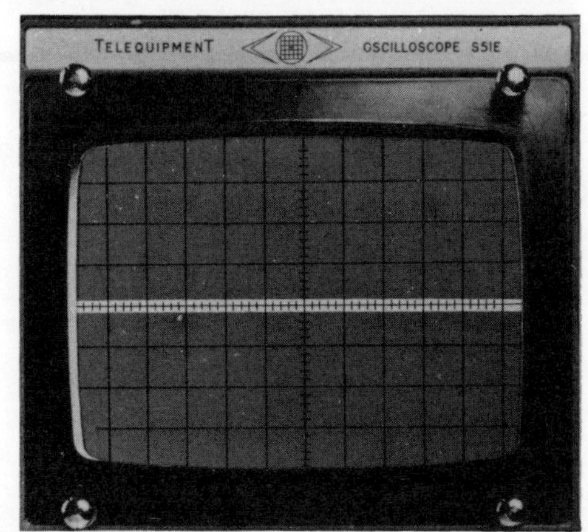

The input voltage taken across CD, on the same oscilloscope setting as used for the output voltage across AB. The input is too small to be seen clearly.

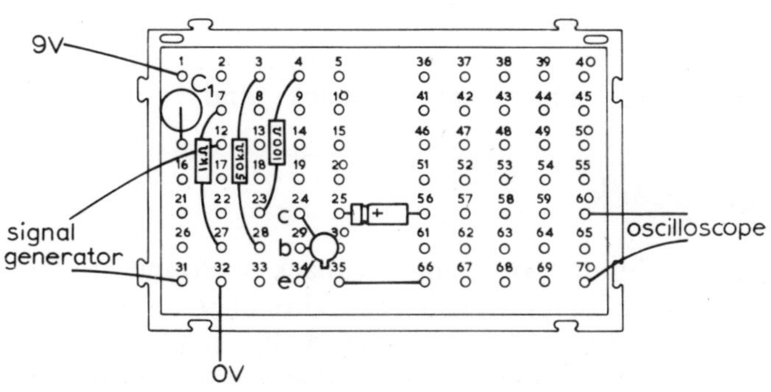

Results

Your oscilloscope displays should be similar to the ones shown here. They show quite clearly the two requirements of a.c. amplification:

1. The output signal is many times larger than the input.
2. The output waveform is the same shape as the input.

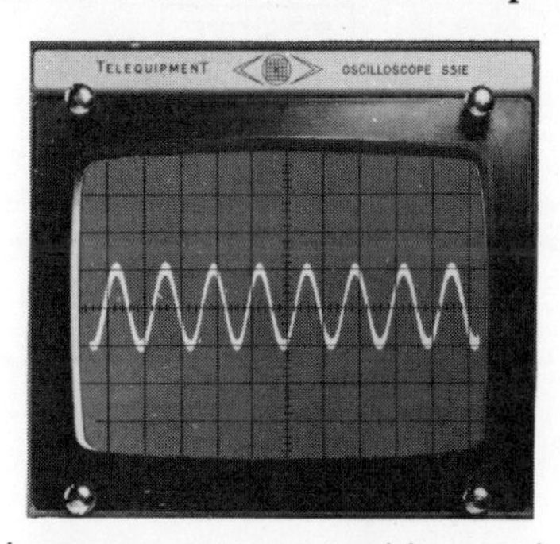

The input voltage on a more sensitive setting. Input voltage = 60 mV

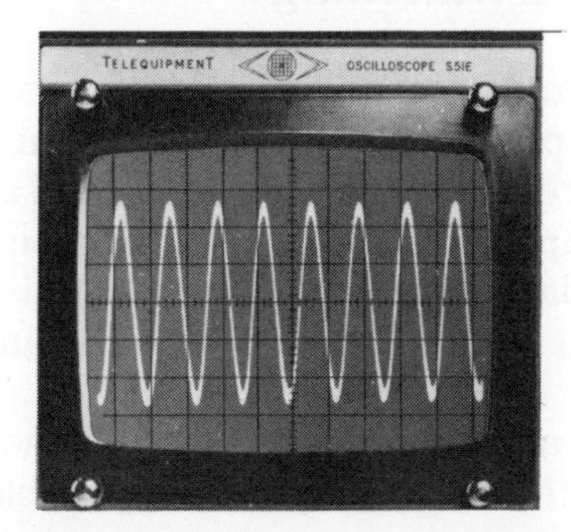

The output voltage taken across AB. This reaches a value of 4 V. The waveform is identical to that of the input voltage taken across CD.

Phase Reversal

The collector resistor and the transistor form a potential divider (see page 40). When the resistance of the transistor falls, two readings change. The current through the transistor increases but the voltage across the transistor falls. This is a normal effect in a potential divider when one of the resistances falls—its share of the voltage also falls. This relationship, an increasing current occurring with a falling voltage is known as phase refersal.

Feedback and Impendance Matching

There are two important terms which you should become familiar with to understand work on amplifiers. They are feedback and impedance matching.

Feedback

The technique of using feedback is found in the majority of amplifiers so its use should be understood. You will find it useful to know the symbol for an amplifier, which is used in many books.

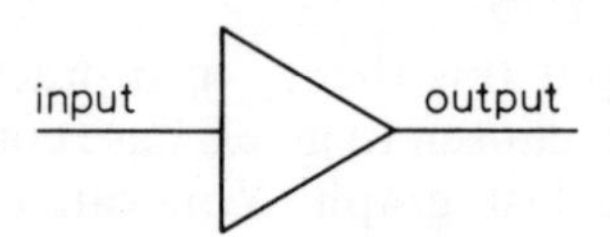

There are two kinds of feedback—positive and negative. They both involve taking a fraction of the output and feeding it back in to the input. In this way, the size of the input is changed.

Positive Feedback

In this case, the output is in phase with the input and so the small fraction that is fed back to the input will increase its size.

Positive feedback is used to increase the size of the input changes.

Negative Feedback

In this case, the output is a half cycle out of phase with the input, so when the small fraction of the output is fed back to the input it actually makes the input smaller.

Negative feedback is used to decrease the size of the input changes.

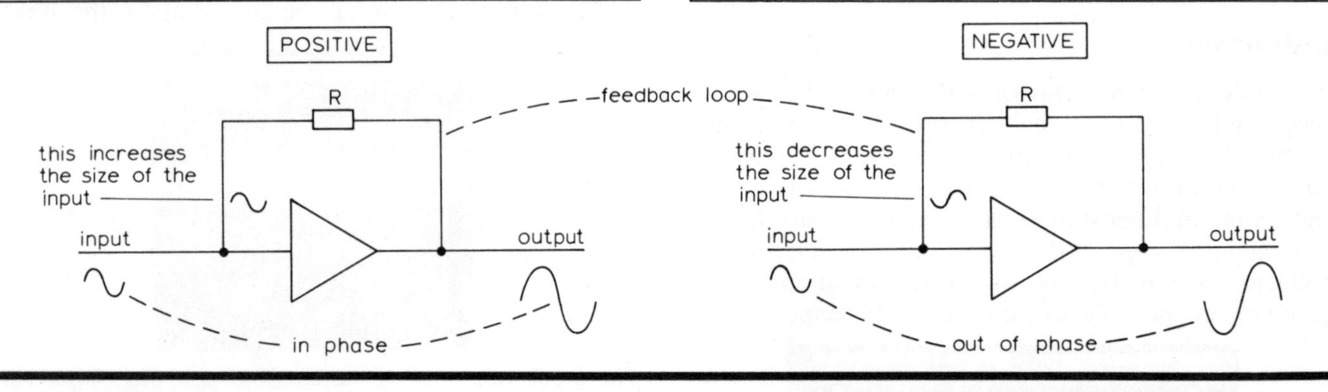

Impedance Matching

Every electronic component and circuit has its own value of impedance. As an example, the two 'black boxes' here represent a microphone (with an output of 1·1 V and impedance of 10 Ω) and an amplifier (with an input impedance represented by the resistor shown). Look now at calculation for an amplifier impedance of 10 Ω.

When the microphone is connected to the amplifier, it is desirable that as much power as possible is 'passed' into the amplifier. The calculation below tells you how much power is passed to the amplifier.

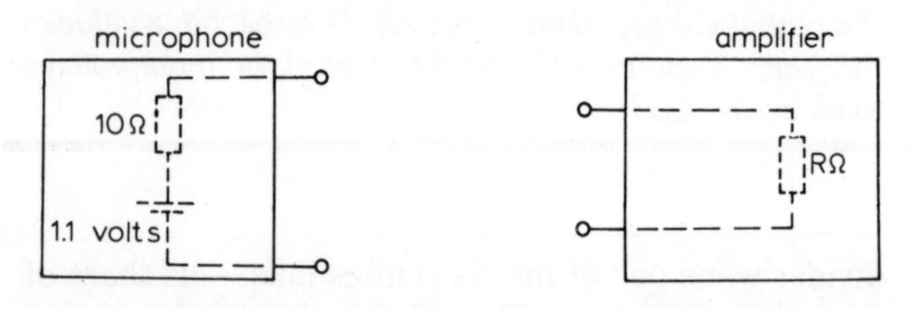

The microphone's output and impedance are represented by a cell and a resistor. The amplifier's input impedance is represented by R Ω

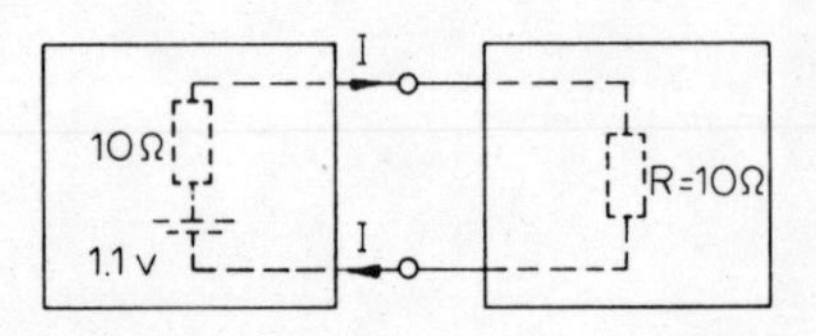

Voltage	1·1 V
Total impedance	20 Ω
Current, I, (Ohm's Law)	0·055 A
Voltage across amplifier	0·55 V
Power in amplifier $= V \times I = 0.55 \times 0.055$	0·03 W

It can be shown generally that:

The maximum transfer of power is achieved when two circuits have identical impedances.

Getting Stable Direct Current Bias

You have seen the need to have a stabilised base bias, so that the collector voltage stays near mid-rail. Variations occur when the gain of the circuit varies for some reason (for example, a change in temperature or transistor).

> If the gain in a transistor circuit increases, then the collector current increases and the collector voltage falls.

If this should happen, the situation is restored by decreasing the base current.

Examine the three methods shown below, for stabilising the bias. For each circuit use two BC108 transistors and find out how well the circuit keeps the same collector voltage, with each transistor.

Using an Emitter Resistor

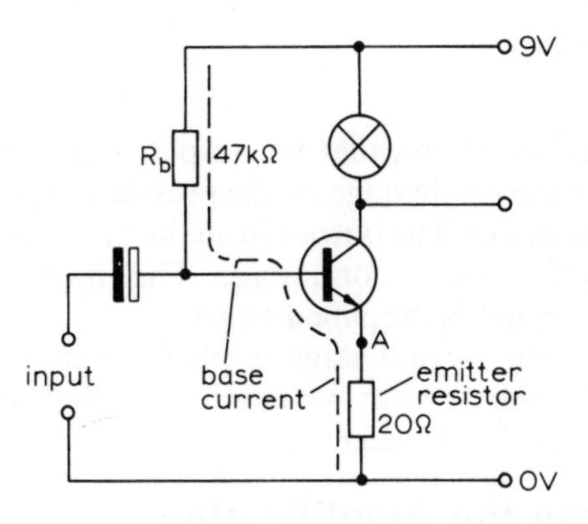

Result
Collector voltage with first transistor ———
Collector voltage with second transistor ———
Explanation of stabilisation
If the collector current should try to increase, the voltage at A would increase, leaving less voltage across R_b. This would decrease the base current and so help to stop the original increase in I_c.
An emitter resistor provides negative feedback.

Using a Collector to Base Resistor

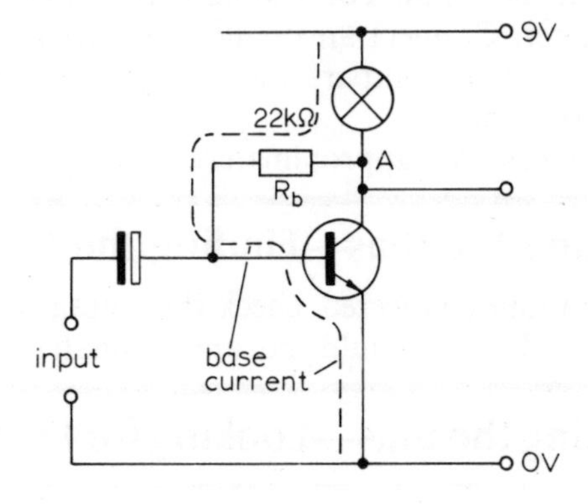

Result
Collector voltage with first transistor ———
Collector voltage with second transistor ———
Explanation of stabilisation
If the collector current should try to increase, the voltage at A would decrease. This means there would be less voltage driving base current through R_b, so the base current falls. This helps to stop the original increase in I_c.
A collector to base resistor provides negative feedback.

Using a Voltage Divider

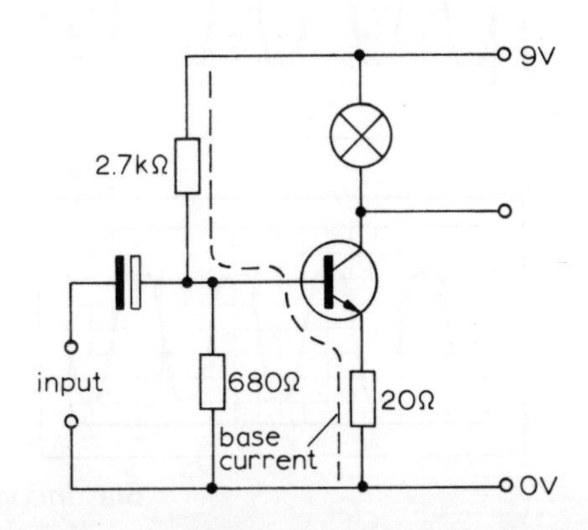

Result
Collector voltage with first transistor ———
Collector voltage with second transistor ———
The voltage divider is considered to be the best method of stabilising the d.c. bias.

Getting More Amplification

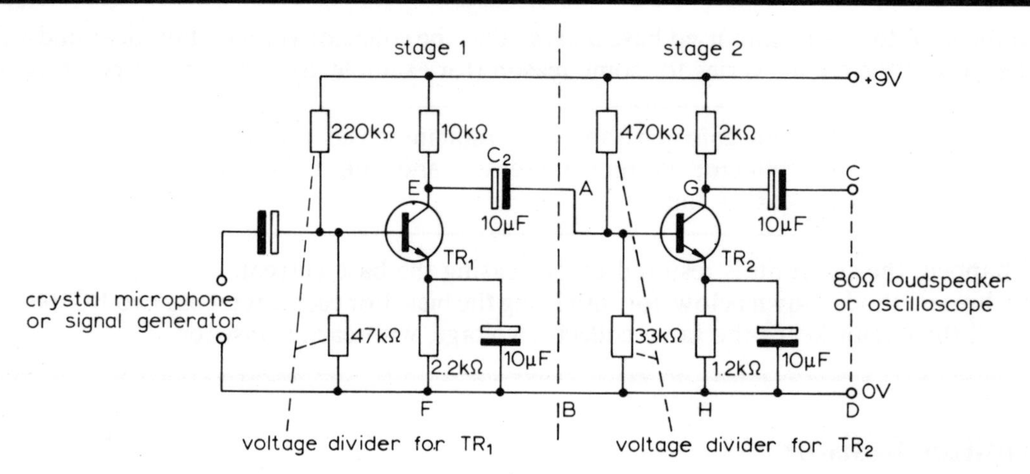

The diagram shows that the amplifier has two stages, each like the single stage on the previous page, but with different values. The output from the first stage is fed to the input of the second stage. This means that the original signal is amplified twice.

Solder the second stage in place on matrix board, changing the resistor values in the first stage and connect the loudspeaker.

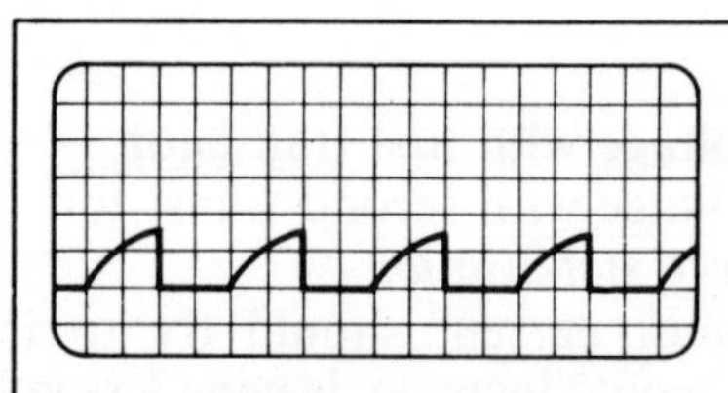

Adding a second stage gives considerably more amplification.

Showing the Amplification

Connect your own signal generator (at minimum output) at the input. You should hear a note from the loudspeaker. Connect an oscilloscope across AB and CD in turn and compare the size of the signal with those shown here.

Output across AB approximately = ___?___ V.

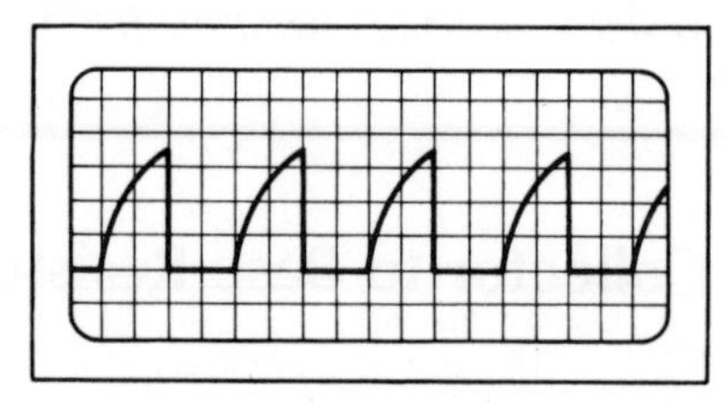

Output across CD approximately = ___?___ V.

Checking the Bias—Finding the Collector Voltage

With no input connected, check the voltage across EF and GH. These should be approximately mid-rail (4·5 V). If they are not, you should try to adjust the biasing.

Checking the Bias—Looking for Distortion

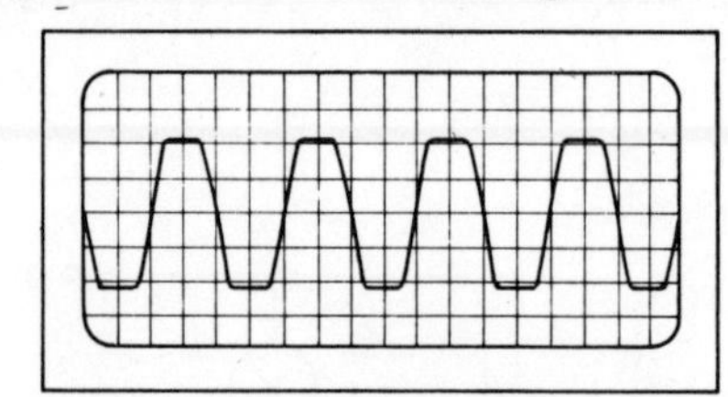

Bias correctly set

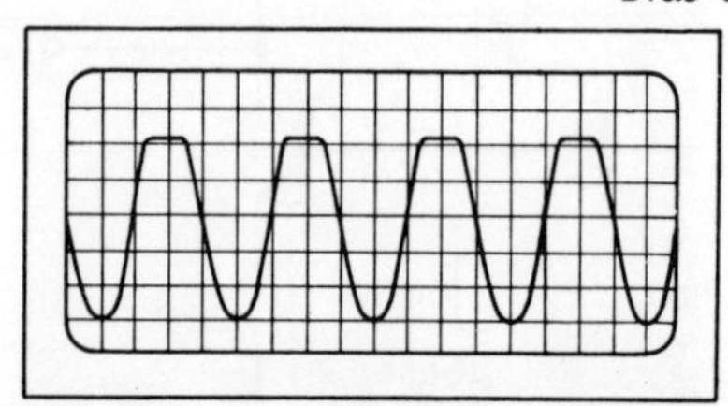

Bias incorrectly set

Connect a sine wave signal generator to the input, starting at 0 V and connect an oscilloscope across CD. Gradually increase the input until the oscilloscope shows distortion of the waves.

Bias correctly set means that overdriving causes equal clipping of peaks. Bias incorrectly set means that clipping occurs on one set of peaks before the other.

Capacitor Coupling

Stage 1 and stage 2 are connected by capacitor C_2. This 'passes' the a.c. signal but blocks any direct current. The oscilloscope shows this effect quite clearly. Connect the oscilloscope across EB then AB, with the oscilloscope switched to d.c. The trace will be in a different position for each one.

Another Two-Stage Amplifier

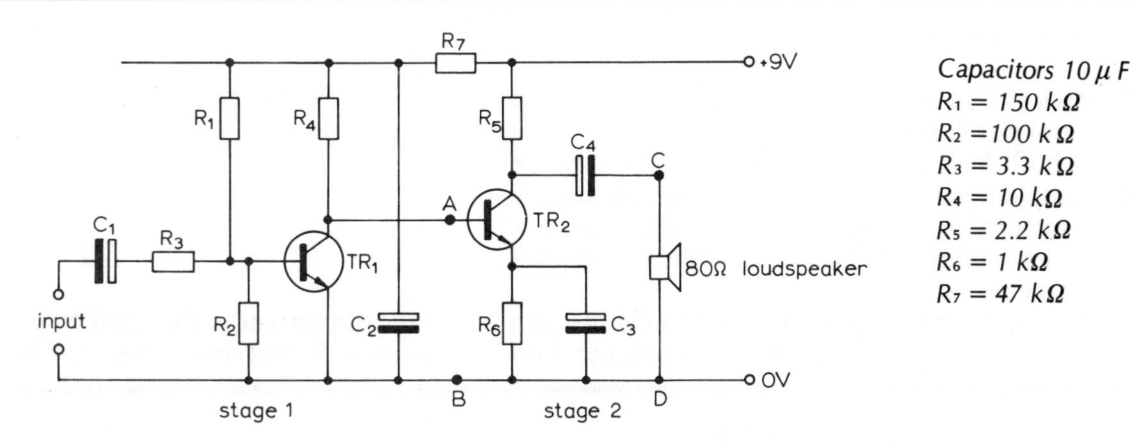

The Circuit

This is a direct coupled amplifier. Look at the circuit and you will see that the two transistors are linked directly—there is no capacitor between them. This means that both alternating and direct current can flow from the first to the second transistor.

Stage 1 Base Bias

R_1 and R_2 form a voltage divider which controls the base current into the first transistor.

Stage 2 Base Bias

The base current for TR_2 flows through R_4, (which is TR_1's collector resistor).

Construction

Instructions

This circuit can easily be assembled on matrix board and pins. A signal generator at the input will let you check the amplification.

Checking the Bias

Take the following measurements:
Collector voltage of TR_1 = ______
Collector voltage of TR_2 = ______
With this type of circuit, it is very difficult to adjust the bias because one stage affects the other so you will not find it easy to correct any problems.

Checking the Amplification

As with the previous amplifier, connect an oscilloscope across the input and AB and CD. The sizes of these three signals will show you the amplification. Draw these three signals to scale in your record of this amplifier.

The Bypass (or Decoupling) Capacitor

This capacitor (labelled C_3) has a very important job to do. It allows the a.c. signal to bypass the emitter resistor, R_6. Connect an oscilloscope across CD and record the size of the signal with and without the capacitor in place. The effect on the amplification is very noticeable.

Output with C_3

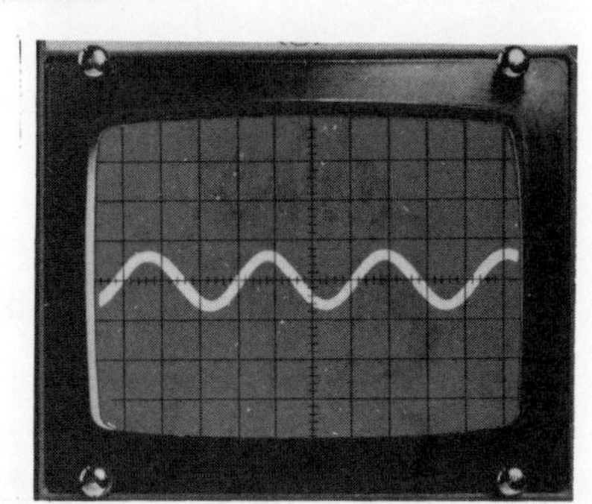

Output without C_3

The Emitter Follower

The Problem of Mis-Matching

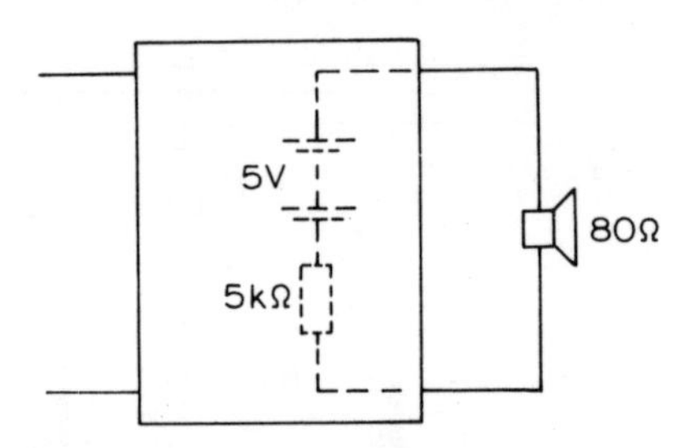

Remember from page 132 that you only get a large transfer of power from one electronic stage to the next if they both have a similar impedance. The diagram represents your last amplifier, which produced, say, 5 V but unfortunately had an output impedance of, say, 5 kΩ which is much larger than that of the loudspeaker it is supplying.

The amplifier has to push current through its own 5 kΩ impedance as well as the 80 Ω loudspeaker. This means the current to the speaker is small and most of the power is 'lost' inside the amplifier.

The Solution to Mis-Matching

The solution lies in the use of an impedance converter. This has a high impedance input and a low impedance output. Now the amplifier and loudspeaker are matched up, and there is a large transfer of power from one stage to the next.

The most common impedance converter is the emitter follower circuit.

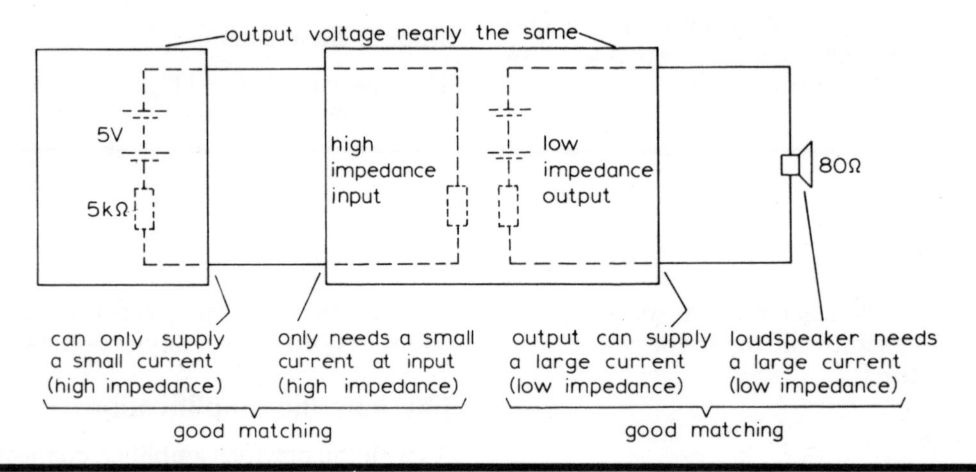

Adding an Emitter Follower

Instructions

Build the emitter follower circuit shown below, which is developed from the circuit on page 135. Use your signal generator and perform the usual amplification checks. You should find that the voltage displays on your oscilloscope of the input to, and output from, the last stage are about the same but the last stage increases the volume from your loudspeaker.

> The emitter follower gives a large output current, because of its own low output impedance.

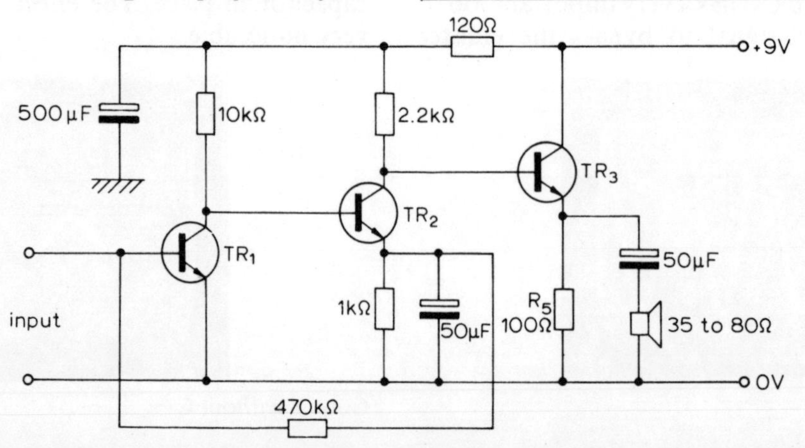

Bootstrapping

Features

Identify the following features in the circuit diagram:
1. TR_1 acts as a common emitter amplifier.
2. TR_2 acts as an emitter follower amplifier. R_3 is its emitter resistor.
3. R_4 provides base bias for TR_1.
4. R_4 provides negative feedback, because the voltage at A is out of phase with the voltage at P.
5. C_1 and C_3 are used to pass the a.c. signal and block direct current.

In addition, you should notice the following new feature:

6. C_2 and the extra collector resistor for TR_1 form a 'bootstrapped' system.

See page 143 for component values.

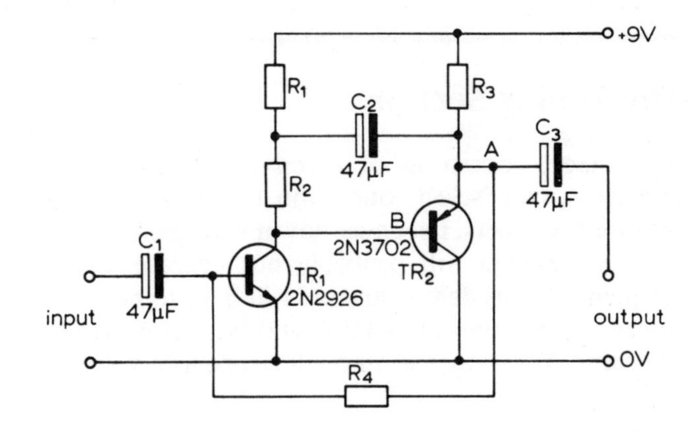

Construction

Assemble the amplifier on S-DeC, matrix board or Veroboard. For inputs, you can use a crystal microphone or signal generator and, to examine the amplification, you can use an 80 Ω loudspeaker and oscilloscope at the output. Check that you are getting satisfactory amplification.

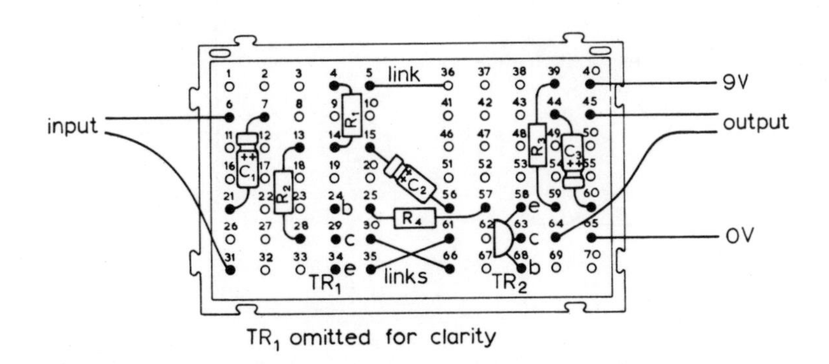

TR_1 omitted for clarity

Advantages

The negative feedback and bootstrapping give the following advantages:
1. *High input impedance*
   This means that the amplifier can be used with high-impedance input sources.
2. *Low output impedance*
   This means that the amplifier can be used with low-impedance loudspeakers, which are, of course, the most common type.
3. *High gain (amplification)*
   Only two transistors are needed to give good amplification.

9 THE RADIO

Introduction

You have already made a small working radio (page 98) but now we will go back to basics and see how the separate stages of a radio work and how they can be improved.

The Tuning Section

This stage, consisting of a coil (inductance) and a capacitor, can 'select' one radio frequency (r.f.) and 'ignore' the others. Let us see how it manages this. Since your voltmeters will probably not operate at radio frequencies, this demonstration will have to be done at lower frequencies (a few kilohertz) but the principle is the same. You may use an oscilloscope as a voltmeter.

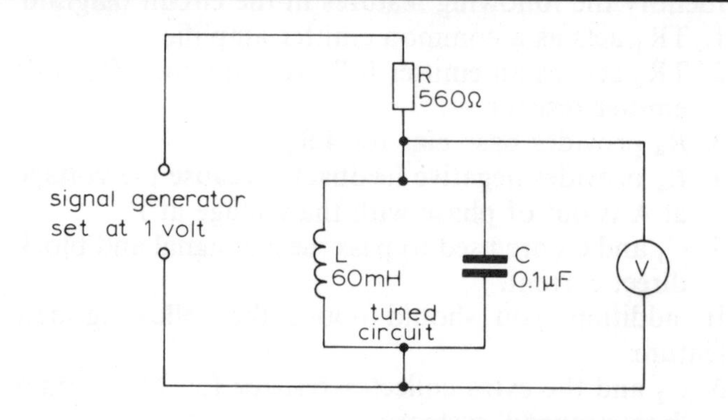

Relationship between Frequency and Voltage

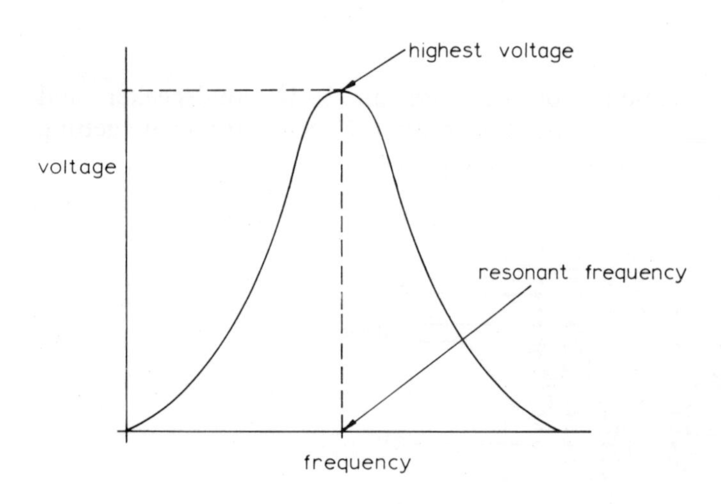

f (kHz)	*V* (volts)
1·0	
1·2	
1·4	
1·6	
1·8	
2·0	
2·2	
2·4	
2·6	
2·8	
3·0	

Instructions

In this demonstration, you are to find out what happens to the voltage across your turning circuit for changes in the frequency of the signal generator. Notice that the voltage supplied by the generator must be kept constant. Take the readings asked for and plot a graph of voltage against frequency and compare it with the one shown here.

Results

You can see that, at one frequency (about 2 kHz), the voltage reaches its maximum. At any other frequency, the voltage is less than this. This means that the tuned circuit will pass on to the rest of the radio circuit its highest voltage for this frequency. This is the frequency of the station you will hear. Any other station will generate a much smaller voltage than this, and so you will probably not hear it.

> The voltage across a tuned circuit reaches a maximum at one frequency, called the resonant frequency.

Instructions

Repeat the experiment for different values of the capacitor and find the resonant frequency for each one.

> The variable capacitor in the tuned circuit of a radio, enables different resonant frequencies to be chosen.

Tuning to a Station

These two graphs are chosen to show you what happens when you turn the variable capacitor in your radio. They are plots of voltage across the tuned circuit against the radio frequency. You can see that they are similar curves to the ones you obtained, but varying the variable capacitor has meant that the peak of the curves occurs at different frequencies.

The readings show that, when you tune to one station, then the voltages produced in the circuit by all the other stations drop considerably. Even so, you can see that it is important that stations are not allowed to use frequencies that are too close to each other.

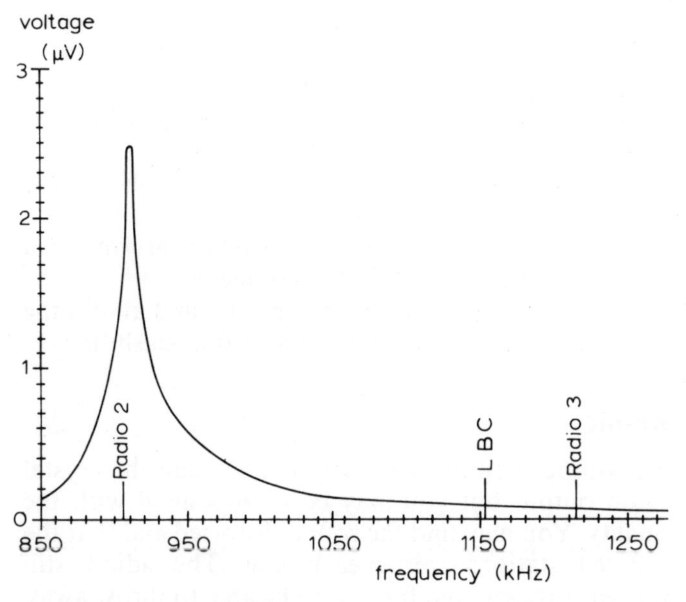

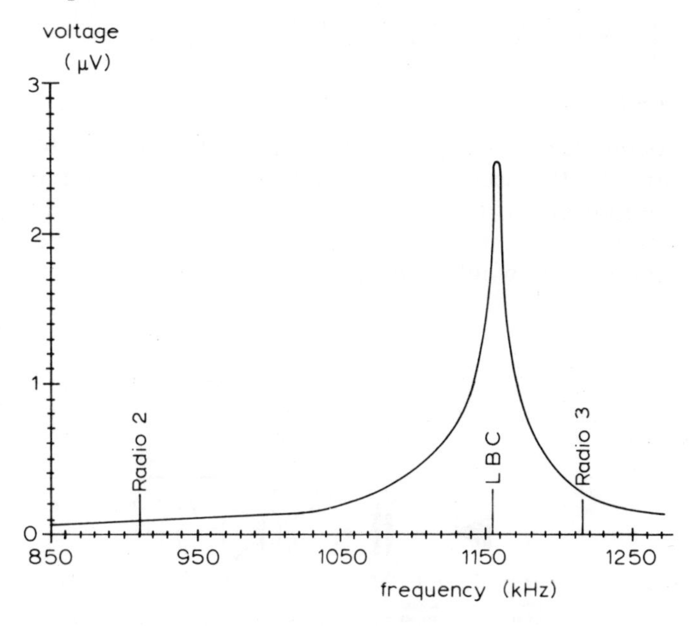

The graphs are taken from same circuit, but the value of the variable capacitor has been changed.

Circuit Tuned to Radio 2

Voltage produced by Radio 2 =
Voltage produced by Radio LBC =
Voltage produced by Radio 3 =

Circuit Tuned to Radio LBC (London Broadcasting)

Voltage produced by Radio LBC =
Voltage produced by Radio 2 =
Voltage produced by Radio 3 =

The 'Q' Factor

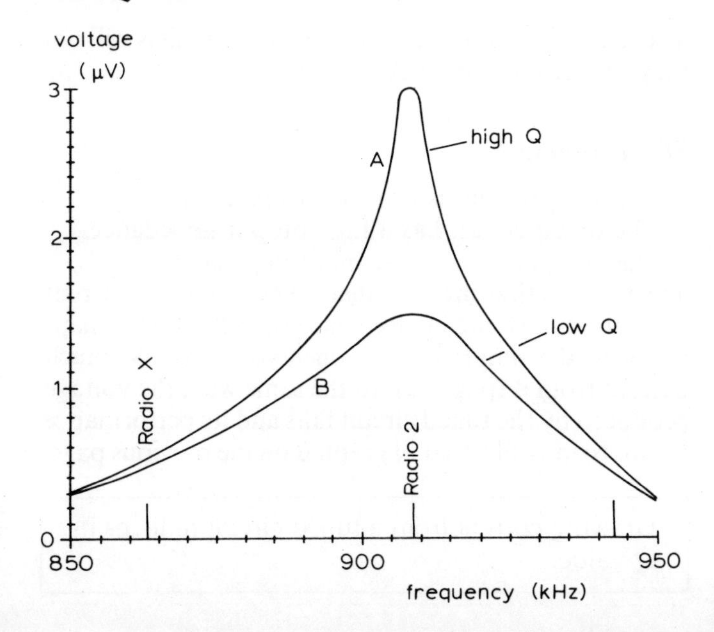

We now look at two different tuning circuits, one of which performs much better than the other. We say that the circuit that gives graph A, (which has the better performance) has a higher Q. Notice how much stronger is the signal for Radio 2 than the unwanted Radio X. This gives much better separation of stations than does the circuit of graph B. We say that graph A shows:

Greater sensitivity—gives a higher voltage on the tuned signal.

Greater selectivity—it gives a greater difference between the sizes of the signals for tuned and untuned stations.

In making radios, we must aim for a high Q.

Adding an Amplifier

The next logical step to improve your crystal radio is to add an amplifier. This will give an increased volume and should mean that you can pull in stations that were previously outside your range. It would also be convenient if you could do without the long aerial and rely on the coil to pick up the signal. This would effectively have your coil acting like half of a transformer, giving a voltage when the radio waves intersect the copper coils.

Does an Amplifier Help?

Instructions

Assemble the circuit on matrix board, allowing space to add on the second stage, as below. Use the same components as for your crystal radio and add on the first transistor stage. This single stage amplifier has the features you would expect:

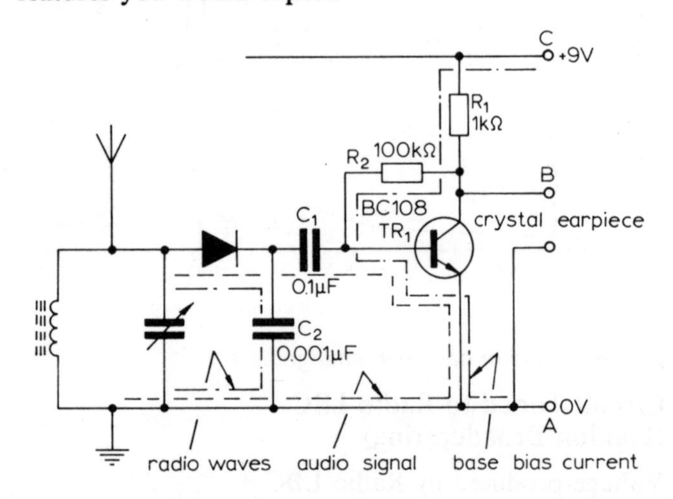

1. R_2 is a collector to base resistor and provides the bias current, shown as a dot–dash line.
2. C_1 allows the signal to pass into the base, shown as a dashed line.
3. The voltage at the collector should be nearly mid-rail and can be checked with a voltmeter.
4. C_2 acts as a radio frequency bypass and allows the radio waves to pass, shown as a dot–dash line.

Result

The output will be noticeably louder than the crystal radio output but you may be disappointed with the quality. You may find the sound distorted and that the different stations overlap each other. The radio is still not sensitive enough for you to be able to throw away the aerial and rely on the coil.

Second Stage

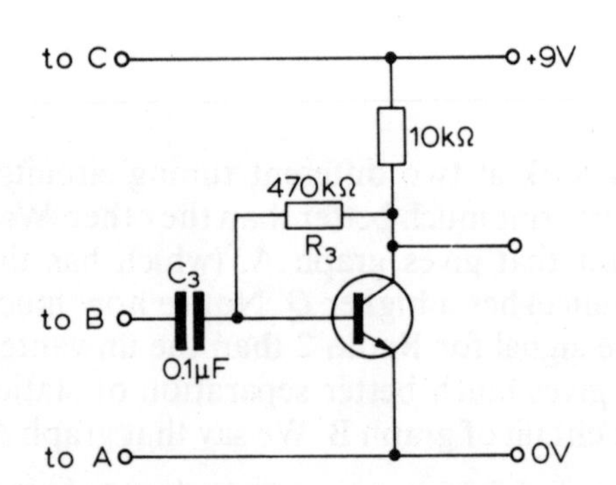

Remove the earpiece and add on the second stage. Solder the earpiece into its new position and listen to the output. The second stage has the expected features:

1. C_3 passes the signal but stops any direct current flowing into the base of TR_2.
2. R_3 passes base current since C_3 has blocked any base direct current from the first stage.

Result

Yet again, the volume increases but the circuit is still not very sensitive or selective.

The Problem

Again, the problem is one of mis-matching:

The tuned circuit has a high output impedance.

The amplifier has a low input impedance.

The result is that the amplifier takes too much current from the tuned circuit. Remind yourself of what happened to the simple battery when you took too much current from it (page 42). In the same way, the voltage produced by the tuned circuit falls and its performance begins to look like that of graph B on the previous page.

> Drawing current from a tuned circuit reduces its Q value.

Improved Amplification

We now look at two ways of adding an amplifier without drawing as much current from the tuned circuit.

Using a Transformer

Using a second coil, L_2, means that the circuit can draw the current it wants, without the tuned circuit being as badly affected as before.

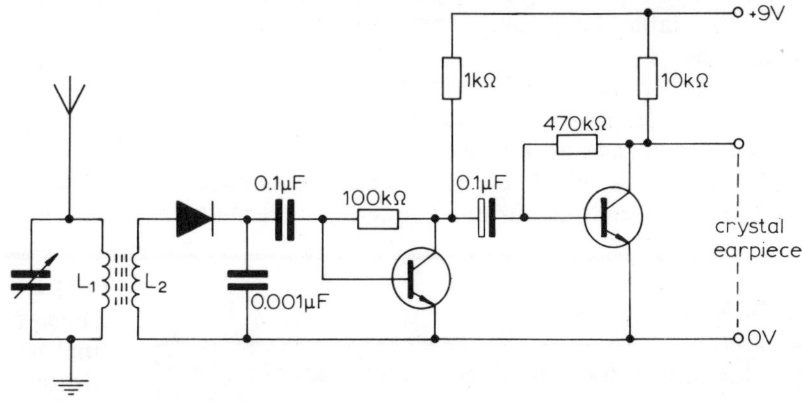

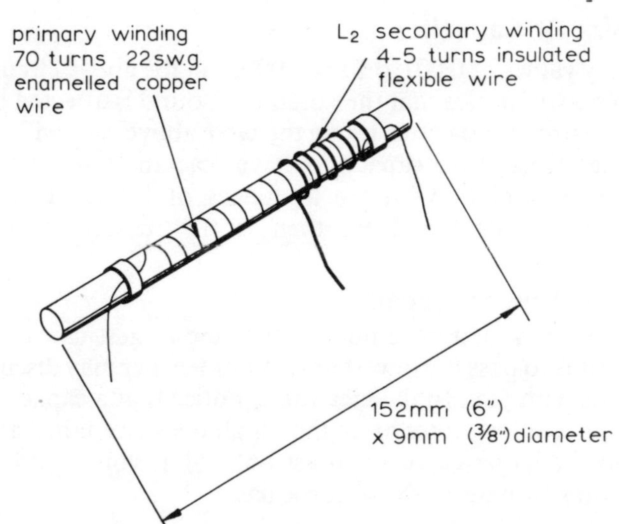

Instructions

Re-arrange your circuit to the layout shown above. The components are the same, with the addition of L_2. Notice that the tuned circuit is not electrically connected to the amplifier anymore—it is connected by transformer effect only. Try increasing the number of turns on L_2.

Result

Tune to a station again and you should find a much 'cleaner' sound and better separation between stations. Try again the reception without the long aerial, relying on the ferrite rod/coil aerial.

The tuned circuit has retained a high Q value, since little current has been drawn.

Using a High Input Impedance Amplifier

The second method relies on an amplifier with a high input impedance. It is the emitter follower which you met on page 136. This amplifier only draws a small current, because of its high impedance, and so it does not upset the tuned circuit as badly as the amplifier on the last page.

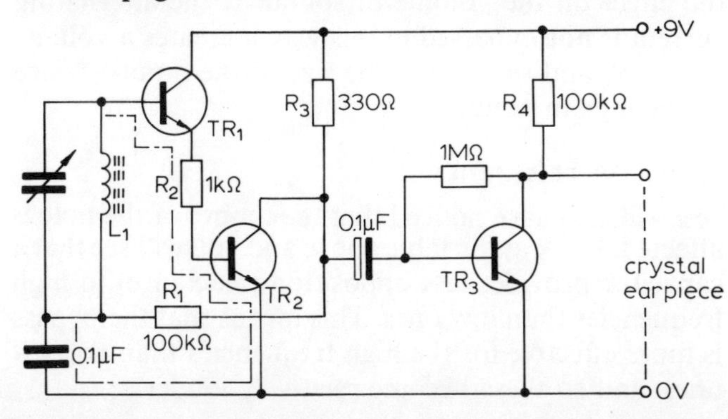

Instructions

Discard L_2 and re-arrange your layout to that shown above. Notice that a third transistor has been included to improve output.

Result

Again, tune to a station and notice the improvement in quality and volume. You should find that this circuit operates without the long aerial and so is fully portable.

1. TR_1 is the emitter-follower stage. R_2 is its emitter resistor and it has no collector resistor.
2. R_1 provides the base bias current for TR_1. Trace this current path from the 9 V rail through R_1 to TR_1.
3. The signal, shown as dot–dash line, passes through C_1 and the bases of TR_1 and TR_2.

Note: See page 152, for an explanation of how rectification is achieved by a transistor instead of a diode.

Blocking and Decoupling by Capacitors

We have put capacitors to two very good uses in this book and we shall now look more closely at these uses. They are blocking and decoupling. The reactance of a capacitor to alternating current is very important here and the table here gives some sample values of reactance.

Remember that the opposition to alternating current is called reactance and is measured in ohms. It is the equivalent of resistance in d.c. circuits.
For example, a 0·1 μF capacitor has a reactance of 1·6 kΩ to alternating current frequency 1 kHz but its reactance falls to 16 Ω at a frequency ot 100 kHz.

Frequency (kHz)	Capacitance (μF)	Reactance (Ω)
1	0·01	16 000
1	0·1	1600
1	100	1·6
100	0·1	16
0·256	22	30

Look at the two typical circuits below.

Blocking

Blocking is done by using a series capacitor.

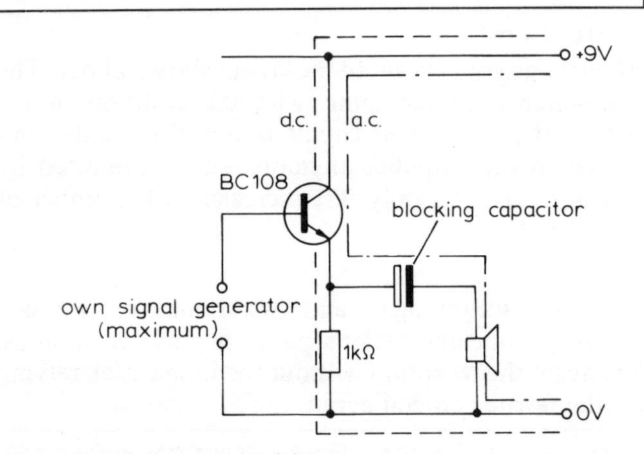

An emitter follower amplifier

Purpose

The capacitor's job is to pass the a.c. signal (dot–dash line) to the loudspeaker but block any direct current (dotted line).

Size of Capacitor

Try values of 0·001, 0·1 and 100 μF in the above circuit. You will notice that the volume of sound is affected by the size of capacitor. Using the table above, you will see that these three capacitors have reactances of 16 kΩ, 1·6 kΩ and 1·6 Ω at the frequency of 1 kHz of your signal generator. Here then is the reason for the difference in volume.

Effect on Frequency

You may also have noticed that the largest capacitor seems to pass the lowest note. Your teacher may discuss this with you. In the meantime, notice that a capacitor opposes low notes more than high ones (see table) and so the larger capacitor must be used, if you want low notes to pass to the loudspeaker.

Decoupling

Decoupling is done by using a parallel capacitor.

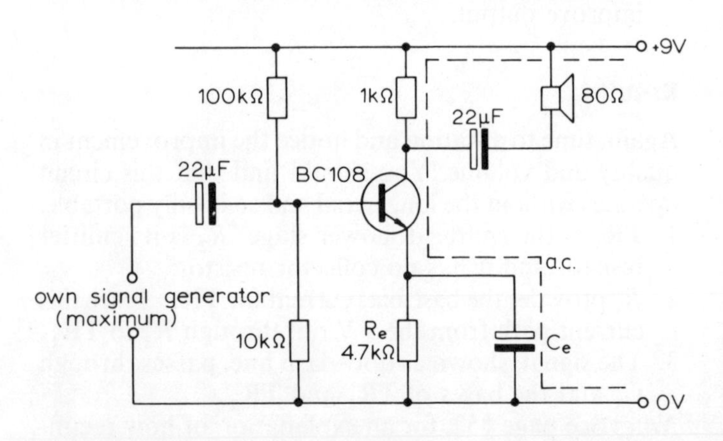

Purpose

The capacitor's job here is to provide a path of low reactance for alternating current so that it bypasses part of the circuit (R_e in this case).

Size of Capacitor

Construct the amplifier shown here. Try values of 0·01, 0·1 and 100 μF for C_e in this circuit and you will notice the effect on the volume of sound. If the alternating current is not bypassed in this way it creates a voltage across R_e and so reduces the size of the input—hence the change in volume.

Effect on Frequency

You will also have noticed that the timbre of the note is affected. Look at the table above and you will see that a capacitor provides less opposition (reactance) to high frequencies than low ones. This means that the bypass is more effective for the high frequencies than the low ones, and so these become relatively louder.

Using a Bootstrapped Amplifier

The bootstrapped amplifier, which you may have constructed already when studying page 137, is ideal for use as a radio amplifier. It has, of course, a high-impedance input, so that it only draws a small current from the tuned circuit.

Construction

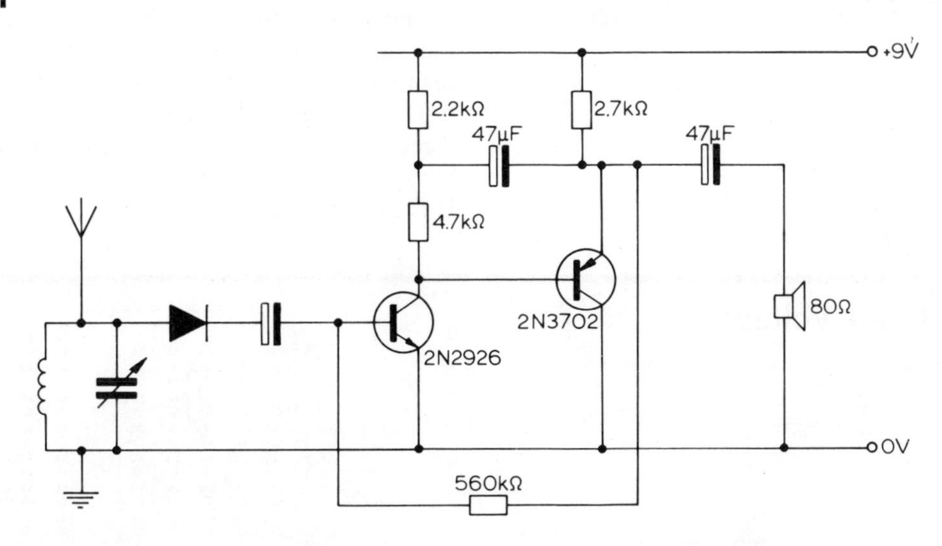

If you have already assembled this amplifier then you only need to connect your crystal radio (minus ear-piece) to the input and an 80 Ω loudspeaker to the output. If you have not made the bootstrapped amplifier yet, then assemble it from the circuit diagram above, either on Veroboard or matrix board. If you wish to add on a third transistor, as shown below, then allow more space at the input side of the amplifier.

Results

You should find that this amplifier produces enough power to give easy listening at the loudspeaker.

A Third Transistor

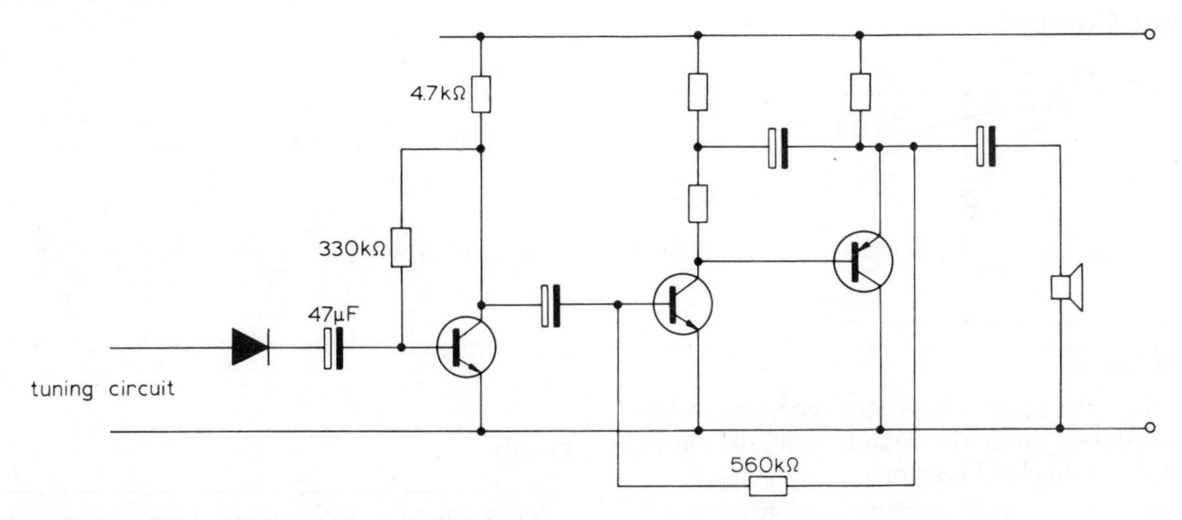

Disconnect the tuned circuit and diode from the first circuit and add a third transistor (at the input stage) as shown above. Reconnect the 'Front end' (tuning circuit) of your radio and again investigate the selection of stations that you can receive and the volume of each.

10 OTHER DEVICES

The Zener Diode

You have already met the zener diode in the power packs you have built. It is used exclusively to give a fixed voltage in a circuit. Zener diodes resemble ordinary diodes, low power ones being packaged in glass or plastic and higher power ones in metal cases.

The symbol for the zener diode

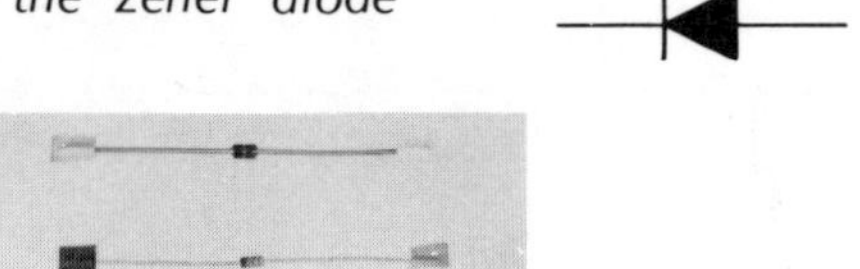

5 cm

Two zener diodes
(a) BZT88 series (b) BZX61 series

The BZY88 diode is suitable for the experiments on this page.

Use

The zener diode is so useful because it has one special property: it passes current when it is reverse biased, (if the input voltage is large enough) but, unlike the ordinary diode, it is specially designed for this and is not damaged. The voltage needed in the reverse direction for the diode to conduct is called the reverse breakdown voltage and this is a constant for any diode. It is abbreviated to V_z.

Changing the Input Voltage

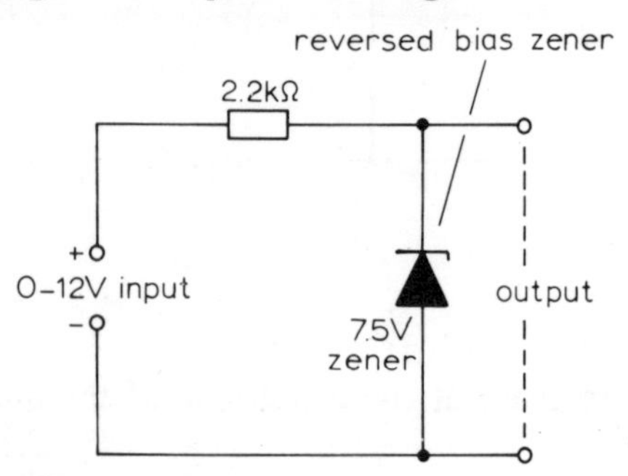

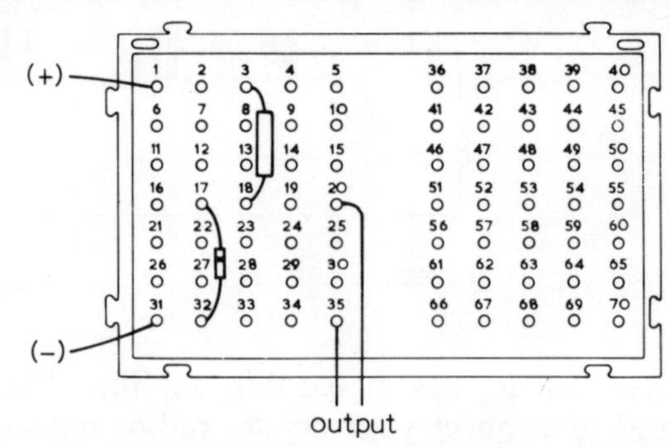

Instructions

Notice that the zener diode is reverse biased—it is going to conduct in the opposite direction to that which you would expect. Assemble the circuit as shown and connect a variable voltage supply direct current, say, 0–12 V and a voltmeter at the output. Start with 0 V input and gradually increase this up to the 12 V available. Record the output voltage that you see.

Result

When the input voltage exeeds Vz, the voltage across the zener diode stays constant at Vz.

Drawing Current

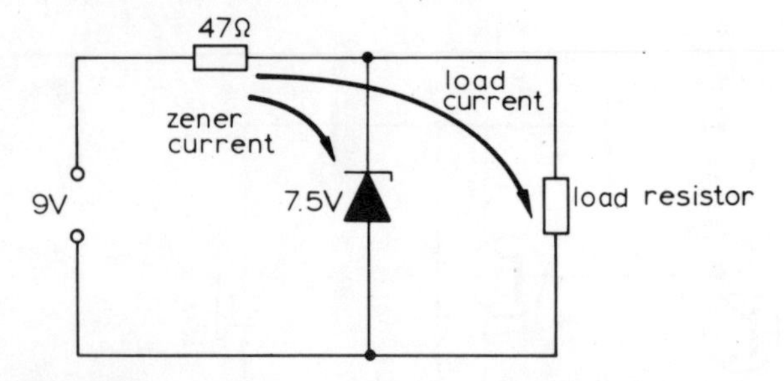

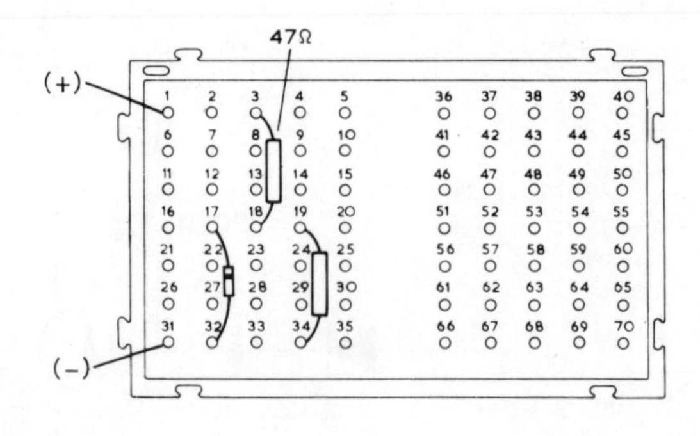

Instructions

Find whether the zener voltage stays constant as load current is drawn from the circuit. Find the output voltage for various load resistors:

Load resistor (kΩ)	100	10	1	0·1
Output voltage				

Result

As the load current drawn from across a zener diode increases, the voltage stays (nearly) constant.

The Power Transistor

So far, the heaviest d.c. load we have used with a transistor circuit is the 60 mA, 6 V light bulb. There are times when we need a transistor circuit to control a much heavier load, perhaps a heavy motor or a powerful spotlight. A transistor such as BC108 has a maximum current rating of 100 mA. Power transistors handle currents of many amperes and are used for driving these heavier loads. They are usually mounted on heat sinks to keep them cool when handling these large loads. A typical power transistor, handling up to 15 A is shown here.

The metal case is the connection to the collector terminal, so, when a heat sink is used, care has to be taken that the other two terminals cannot touch the heat sink and hence short to the collector terminal. Insulating bushes can be bought for this purpose.

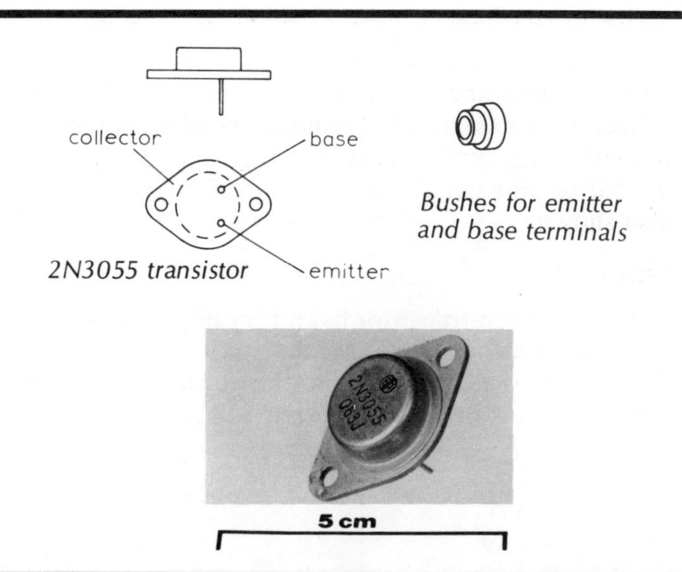

Variable Voltage Supply

Obviously, in theory, there are nearly as many uses for the power transistors as the low-power transistor. In this case, a circuit has been selected which will give a variable voltage supply, which can be used for controlling the speed of a low voltage d.c. motor or the brightness of a low voltage bulb, etc.

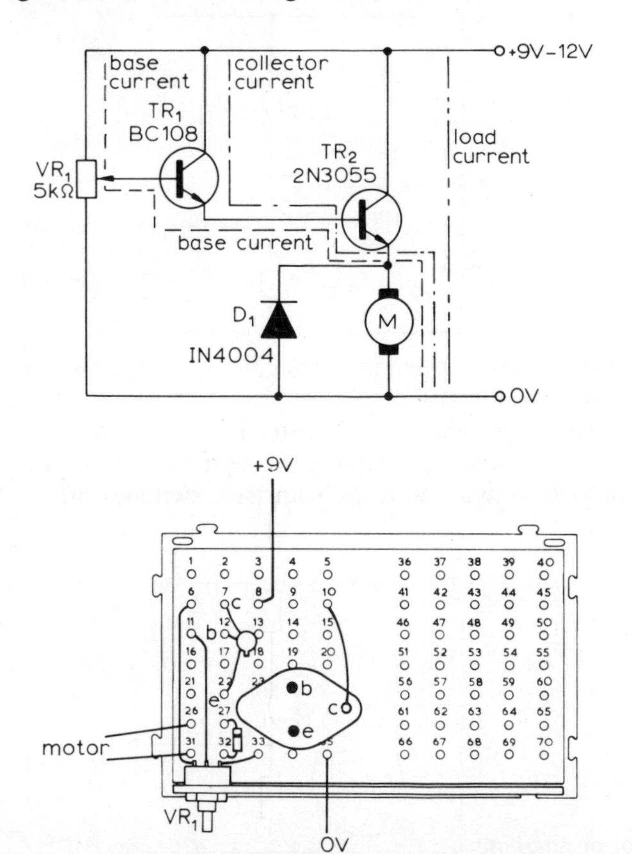

Instructions

Assemble the circuit on S-DeC, as shown. Do not use a small dry battery as these are not designed to supply these large currents. Use either a car battery or a heavy duty power supply. The diode is included to prevent high reverse voltages from the windings of the motor damaging the transistor (compare the relay).

Testing

If you are not using a heat sink, only run the circuit for a few seconds at a time, so that the transistor does not overheat. Adjust VR_1 and you should find that you can control the speed of the motor. Connect a voltmeter across the motor and you should find the voltage varying as you adjust VR_1. Take care not to exceed the maximum rated voltage of your motor.

The Circuit

The circuit contains a super alpha pair arrangement which is explained on page 154. Notice that the emitter of the TR_1 is connected to the base of TR_2. The current paths drawn in the diagram show that the base current to TR_1 switches on its collector current (as you would expect) and these two currents become the base current for TR_2, hence switching on the large current to turn the motor.

VR_1 acts as a variable voltage divider which controls the base current to TR_1 and thus the output voltage.

The Unijunction Transistor

The unijunction transistor has three terminals, called base 2, emitter and base 1, which are equivalent to the collector, base and emitter, respectively, of the common transistor. Its symbol is shown here and the figures quoted on this page refer to the 2N2646 unijunction also shown.

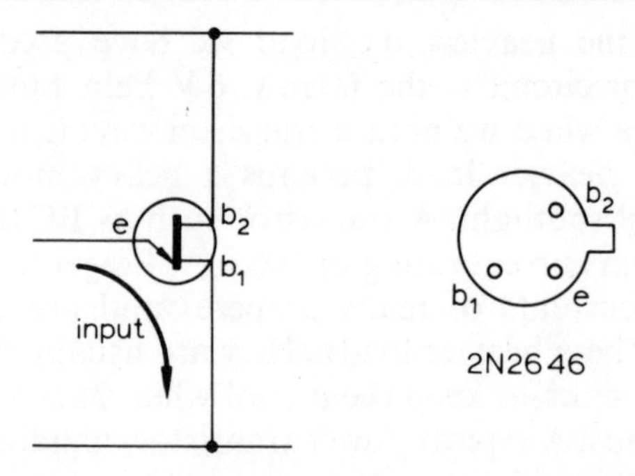

b_2 is connected to the positive rail and b_1 to the negative rail. Input is to the emitter/b_1 junction.

Operation

The unijunction transistor has a very important parameter, V_p, which is the voltage that must be supplied to the input to switch on, (trigger) the transistor. For the 2N2646 transistor, this voltage is about half the supply voltage.

Notice that, after the 'triggering', the input impedance becomes very small.

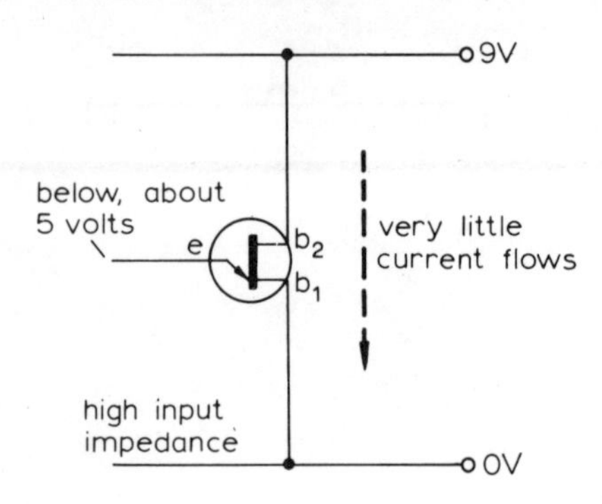

Input below V_p

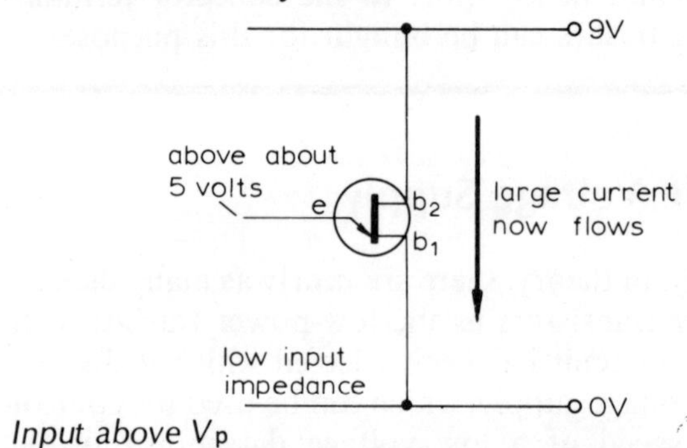

Input above V_p

A Unijunction Oscillator

This circuit gives out an audible tone from the earpiece. If a transducer, e.g. an l.d.r. or thermistor, is connected across BC then the pitch of the note will vary according to the resistance of this extra component. This gives you a unit which will tell you of changes in, for example, light or temperature by making its note change in pitch. The sensor can be mounted on a probe for remote operation in places difficult to get to.

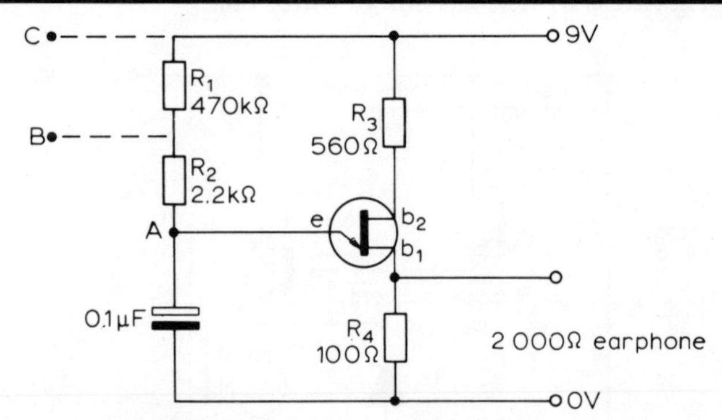

The Circuit

Remember that the current through the loudspeaker must be switched on and off for it to emit a note. Look at the part played by the capacitor in achieving this.

The voltage at A rises until, at about 5 V, the transistor switches on. This causes a large current to flow through the earphone.

However, when the transistor is on, the capacitor can now discharge very rapidly through e/b_1 so the voltage at A falls again and the transistor switches off.

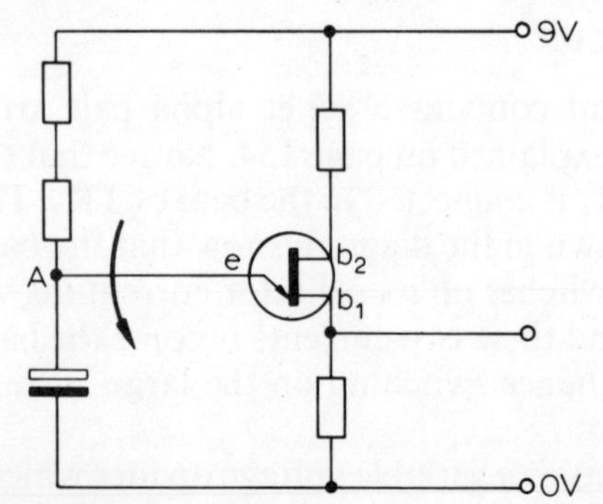

Capacitor charging

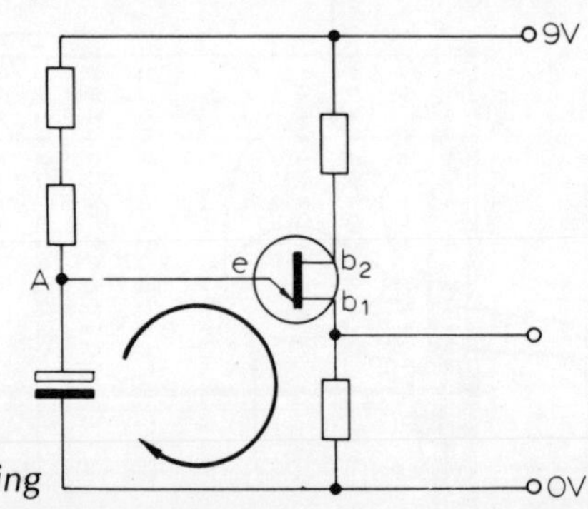

Capacitor discharging

The Thyristor or Silicon-Controlled Rectifier

The thyristor or silicon-controlled rectifier (s.c.r.) is a four-layer silicon semiconductor device with three terminals. The details are shown here.

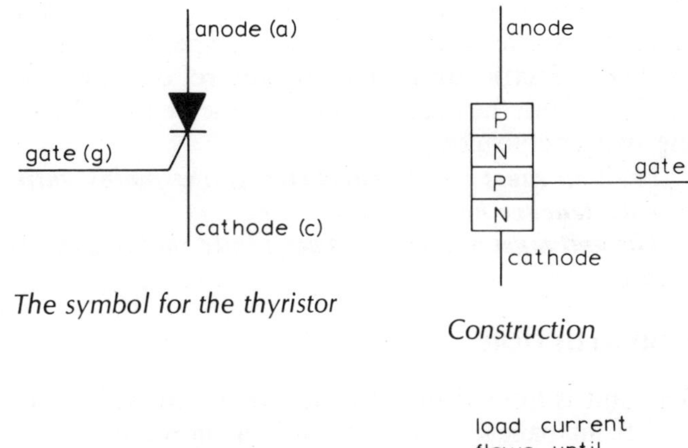

The symbol for the thyristor

Construction

Performance

As its name implies, the s.c.r. acts as a diode and passes current in one direction only, anode to cathode. The s.c.r. is switched on by a gate current of, very roughly, 20 mA, but notice that it cannot then be switched off by removing the gate current.

Because of its special properties the thyristor has some particular applications:

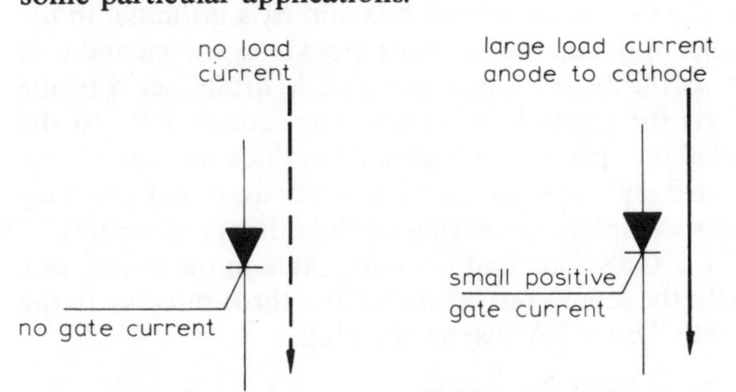

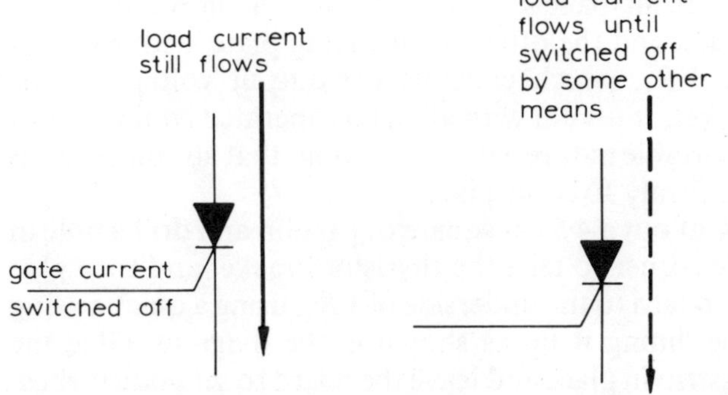

Although some s.c.r.s are given type numbers in the same way as transistors, it is quite common to find an s.c.r. with its two main specifications marked on its case. Thus a marking, THY3A/200, indicates a thyristor that will pass a current of 3 A and is safe to use up to a peak voltage of 200 V. There are, of course, other specifications, such as the gate current and voltage to trigger the device but, usually it is enough to choose a thyristor with a high enough current and voltage rating.

Simple Triggering and Latching

Instructions

Choose a heavy duty battery and load and make sure that you use a thyristor that can handle the load current. Mount the thyristor on a heat sink, if it is to take a large current, and for currents of several amperes you may need thicker cables than usual.

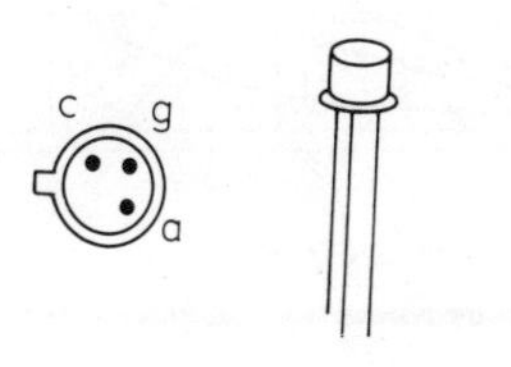

Thyristor leads

Triggering

Press S_1 for the shortest possible time and release and notice that the bulb comes on and then stays on.

Switching Off

Use either S_2 or S_3 to temporarily cut the current through the thyristor and notice that the thyristor does not come back on again when they are released.

The next project shows how this extremely useful device is employed in a drill speed controller.

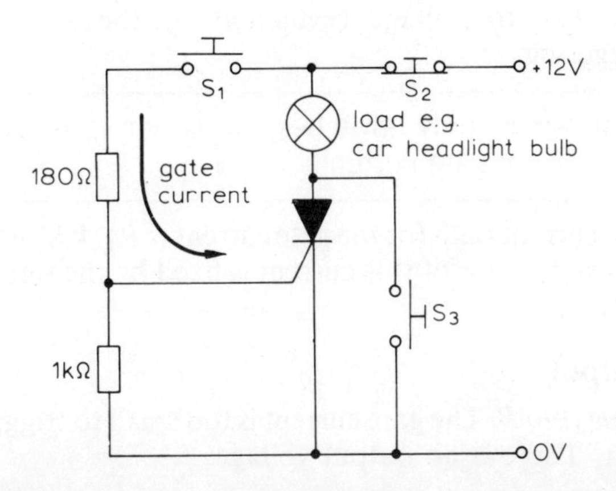

A Thyristor Application—Drill Speed Controller

This unit will give you a variable voltage supply at a three-pin socket for controlling the speed of mains powered electric drills, mixers and so on. Since it operates from the mains you must observe the following two safety rules:

The unit must not be connected to the mains until your teacher has tested it.
The unit must not be tested until fully enclosed in its box.

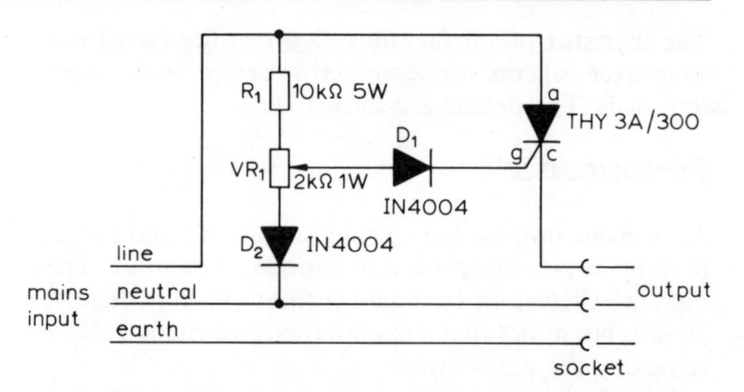

Construction

The unit is housed in a double switch-box which has two front panels. One of these is a simple three-pin socket and the other is a blanking plate, with a cutout for VR_1, which controls the output voltage at the socket. It is vital with all mains operated equipment of a portable nature, such as this unit, that all components are firmly fixed in place.

Cut out a 4·5 cm square of paxolin and drill a hole in one corner to take the thyristor bracket and then glue the board to the underside of VR_1, using a quick setting glue, lining it up as shown in the diagram. Glue the tagstrip in place and leave the board to set undisturbed.

In the centre of the blanking plate drill a hole for the spindle of VR_1 and also drill a hole for the mains cable in the side of the switch-box and fix a grommet in the hole. Make up the thyristor bracket as shown and bolt it to the board when the glue is firmly set. Fix the thyristor to its bracket and then attach VR_1 to the blanking plate and tighten the locking nut. Your board and tagstrip are now firmly fixed and you may now complete the wiring of the other components.

Fix the socket and blanking plate to the switch-box with the screws provided and fix a three-pin plug to the cable. Use a 3 A fuse in the plug.

Now hand the unit to your teacher for testing.

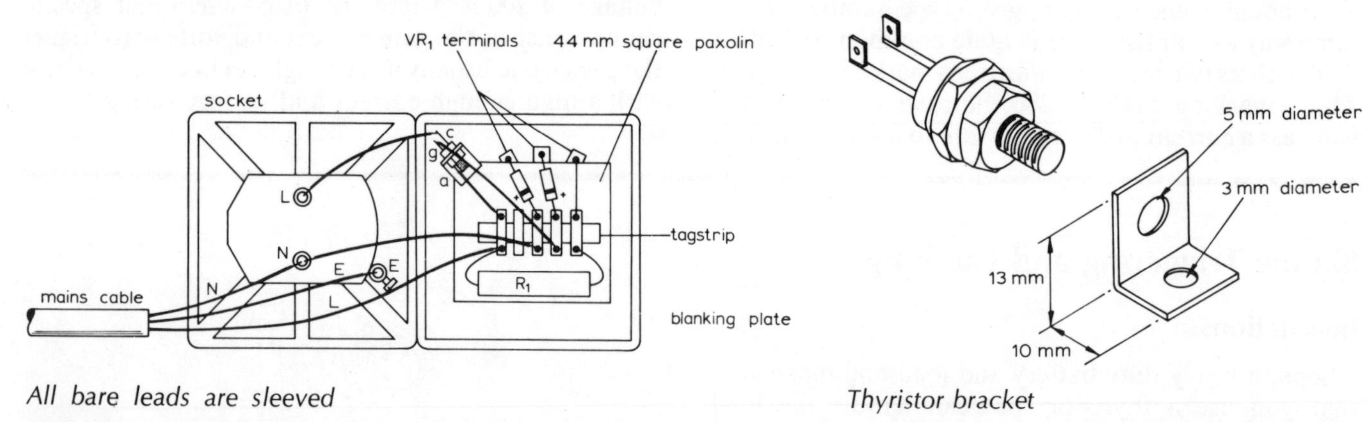

All bare leads are sleeved

Thyristor bracket

The Circuit

The key to voltage regulation by the s.c.r. is its triggering:

> An s.c.r. only switches on when it receives sufficient gate current.

The current path for the gate current is R_1, VR_1 and D_1 and so the size of this current is fixed by the setting of VR_1.

Output

Time A to B. The gate current is too small to trigger the s.c.r. There is no output voltage.

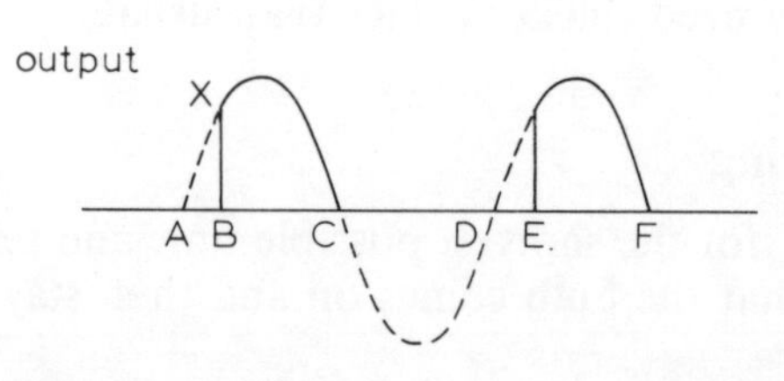

Time B. The input voltage, X, is now large enough to switch the s.c.r. on and the output is full voltage.
Time C to D. The s.c.r. behaves like an ordinary diode and blocks reverse current.
Thus the average output voltage is smaller than the input and its size is fixed by the point at which the s.c.r. triggers.

The Reed Switch

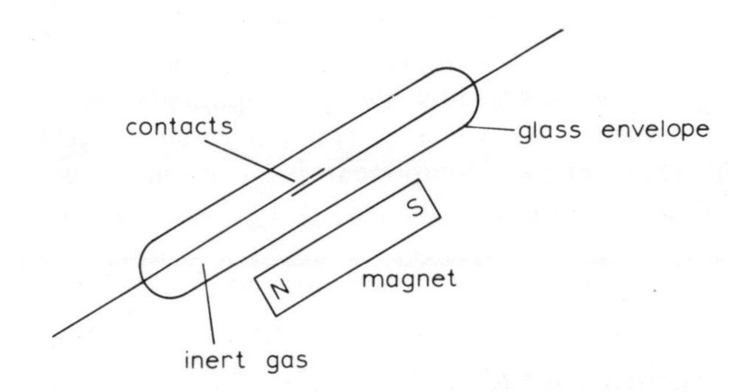

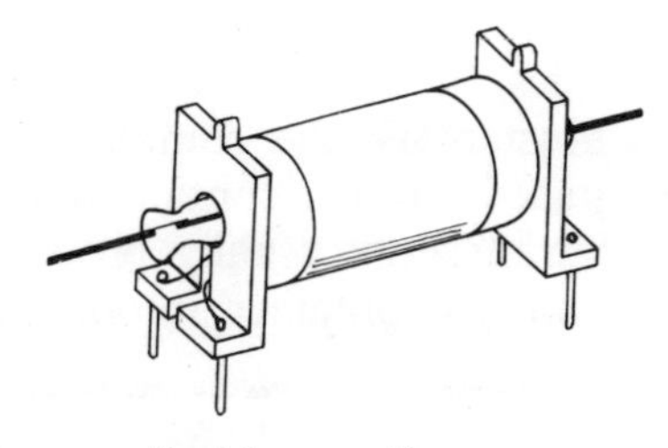

Reed switch controlled by a coil

The reed switch consists, quite simply, of two contacts inside a glass case, connected to two external terminals. When the two contacts become magnetised, they close together, thus operating as a switch. The movement of the contacts is brought about by the approach of a magnet or by an electric current through a coil.

The reed switch is operated remotely by a magnetic field.

It has a very fast switching action.

Uses

In use, some moving object will have a magnet fitted to it and the reed switch will detect its movement. It may be used to switch a circuit when an object has reached its correct position, or operate a counter to find the speed of revolution of a wheel, or simply to set off a burglar alarm when a door or window is opened.

The switch contacts are totally enclosed.

A Burglar Alarm

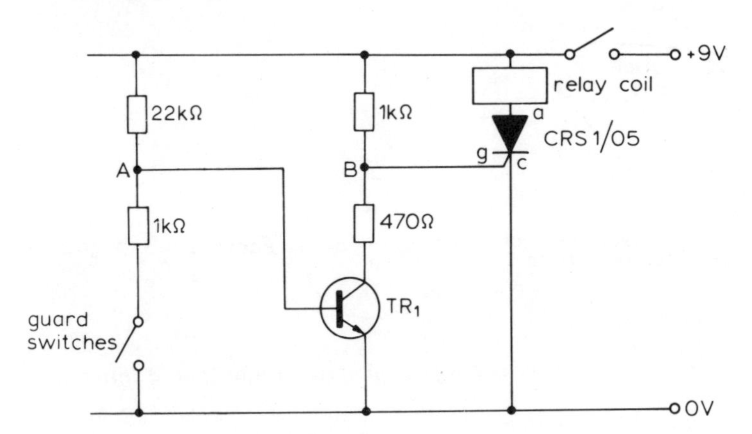

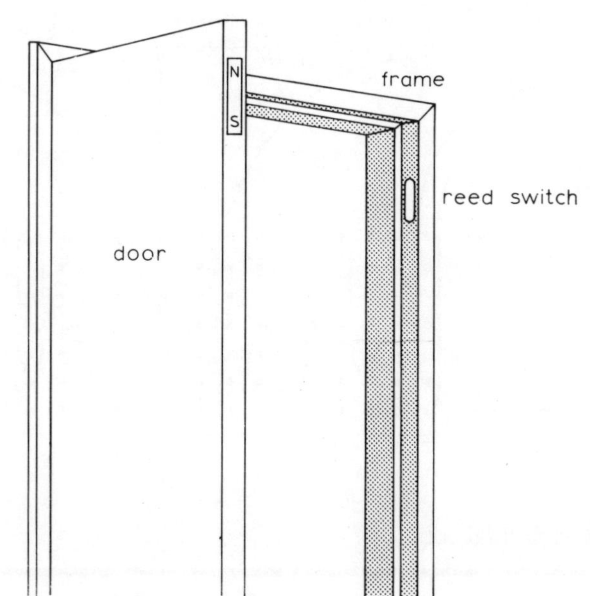

This burglar alarm system has a reed switch and magnet at every door and window to be protected. Any number can be used, placed *in series* in the position shown in the circuit diagram. When a door is opened, the reed switch opens and sets off the alarm.

Notice these features of the circuit:

1. The alarm bell is driven by a separate battery, which is switched by the relay.
2. The load current through the thyristor is switched on by a small gate current.
3. A thyristor cannot be switched off by switching off the gate current so the alarm cannot be silenced by the burglar closing the guard switch again or even ripping out the wires. The circuit is said to be *self-latching*.
4. With the guard switches closed, point A is at such a low voltage that TR_1 is off and so no gate current can flow into the thyristor.
5. Base current flows into TR_1 when a guard switch is opened and this drops point B to about 3 V, so that gate current can flow into the thyristor.

11 OTHER CIRCUITS

An Integrated Circuit Project—A Radio Receiver

If the production of the transistor marked the birth of modern electronics, then the production of integrated circuits (i.c.s) must mark the coming of age of the industry. Integrated circuits contain many resistors, capacitors and transistors usually produced on one chip of silicon, but all contained in one package.

Examine the ZN414 which you are to use and you will see that it is no larger than a single transistor and yet the diagram shows you how many components are contained within the package. This particular collection and arrangement of components is exceptionally useful as an audio amplifier.

Construction

The diagrams provide enough detail for you to be able to construct this radio.

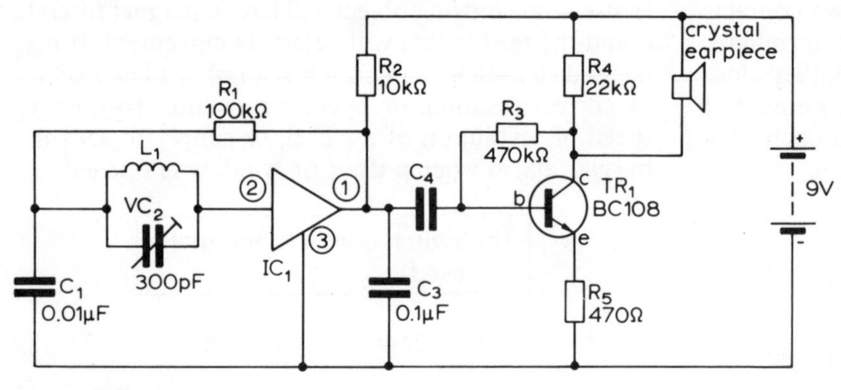

Circuit diagram
L_1 is 76 turns of 36 s.w.g. wire on a 10 mm ferrite rod

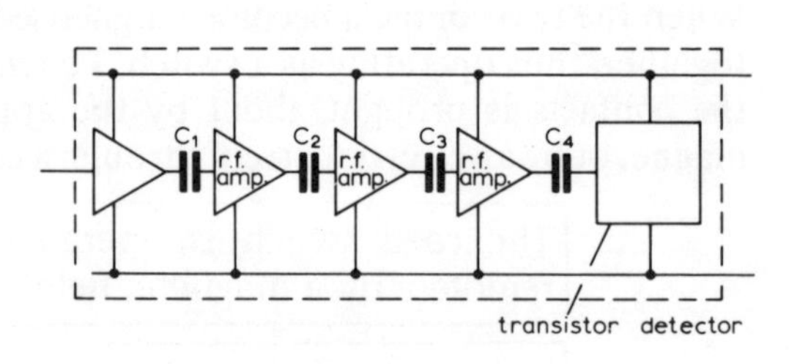

Internal circuitry of ZN414.
This integrated circuit contains 10 transistors giving three separate amplifiers of the radio signal.

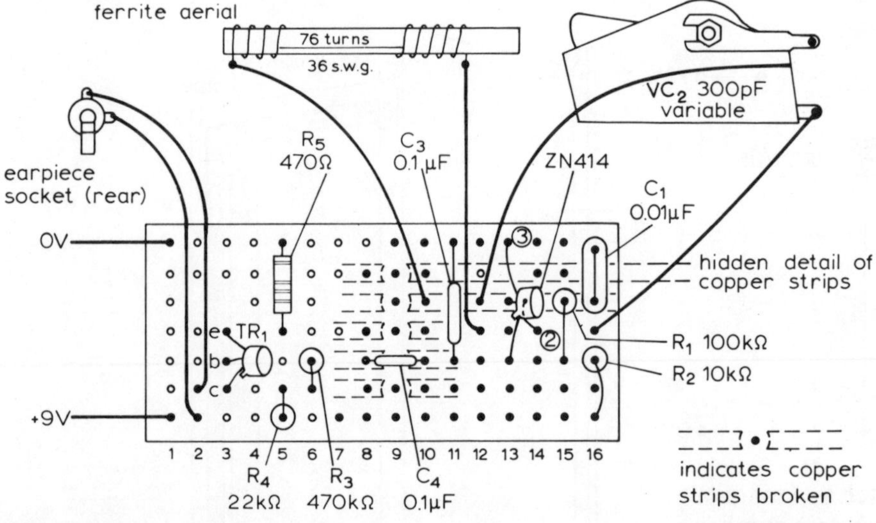

Veroboard layout

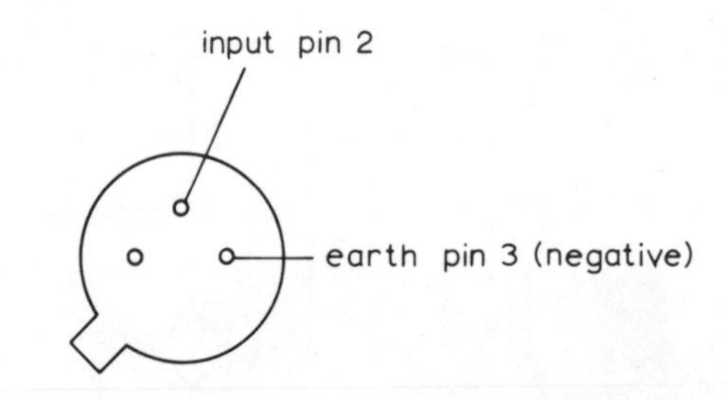

Underside view of the ZN414 integrated circuit

Note: A suitable control panel size is 125 mm × 50 mm

The Circuit

You should recognise the following features from your earlier work on radios and amplifiers.

1. L_1 and VC_2 form the tuned circuit.
2. R_1 provides base bias for IC_1.
3. C_4 blocks direct current but passes an alternating current signal to the transistor.
4. R_3 provides collector to base bias current.
5. R_4 is the collector resistor for TR_1.
6. R_5 is the emitter resistor for TR_1.
7. The output is taken from the collector of TR_1.

Testing

This circuit makes a very useful little radio and you should find that you pick up strong clear signals from quite a number of stations.

The Field Effect Transistor

The field effect transistor (f.e.t.) resembles the conventional transistor in many ways: it has three terminals, called source, gate and drain, which correspond to the emitter, base and collector, respectively, of the common transistor and is available in n-channel and p-channel versions which are, of course, the equivalent of the npn and pnp choices you have met so far. However, the f.e.t. has one important advantage over the ordinary transistor:

> The f.e.t. has a very high input impedance.
>
> The f.e.t. only draws a very small input current.

This means that the f.e.t. is ideally suited to amplifying signals from sources which cannot supply a large current. A good example of this is the tuned circuit used in radios, so now you can make a small radio using the f.e.t. at the input to the amplification section.

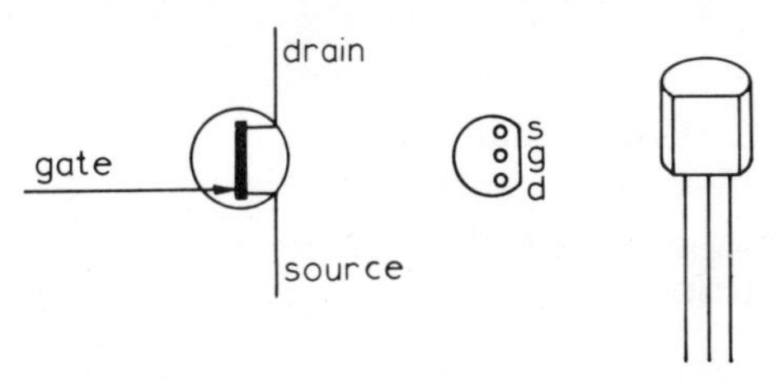

The symbol for a f.e.t. *The 2N 3819 f.e.t.*

A F.E.T. Radio

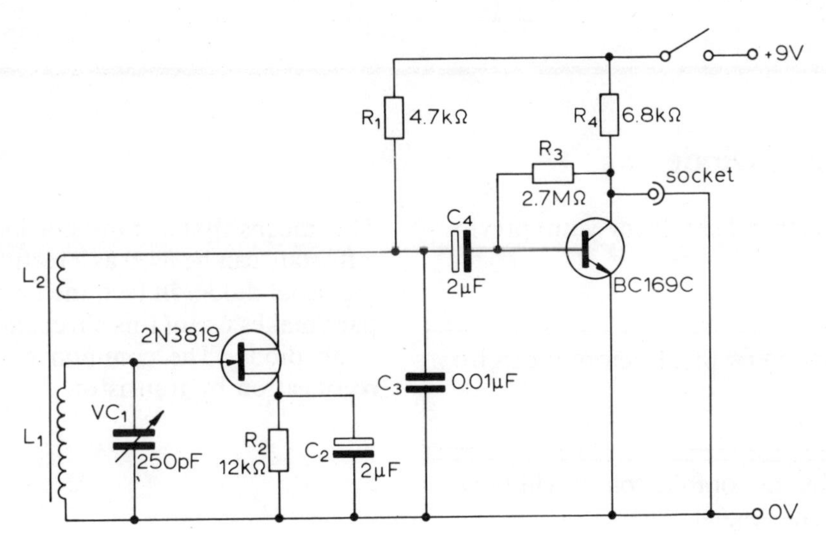

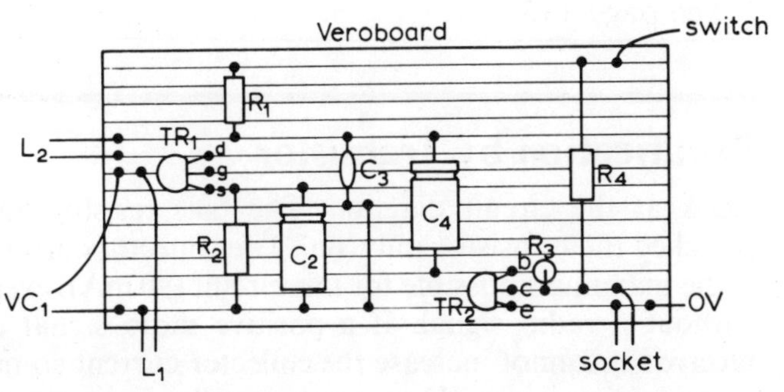

4 mm Veroboard

Construction

The components are mounted on 4 mm Veroboard, as shown, or you may design a p.c.b. if you wish. The output from the jack socket is taken to a crystal earpiece. When you are testing the set, if it gives poor reception, then L_2 has been connected the wrong way round, so wrap it the other way on the rod. Your case will need holes for the spindle of the tuning capacitor, the switch and the jack socket.

You will find that you can adjust the performance of the set by altering the position of L_2 on the ferrite rod, so choose the best position for your circuit and fix L_2 permanently in place.

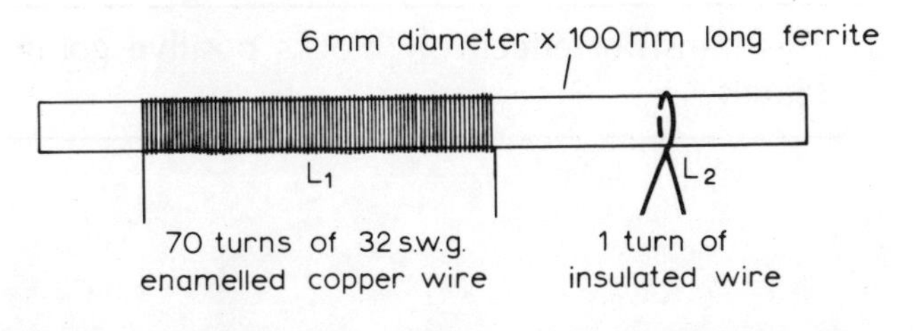

Aerial

Further Thoughts

Regenerative Feedback

The circuit on the previous page uses many of the principles which you have already met but also uses two new ideas which you can now study.

The amplified signal passing through L_2 causes a small copy to be produced in L_1 by transformer action.

This is the tuned signal, the one selected by the tuning circuit. The feedback signal adds to the original to make a larger input.

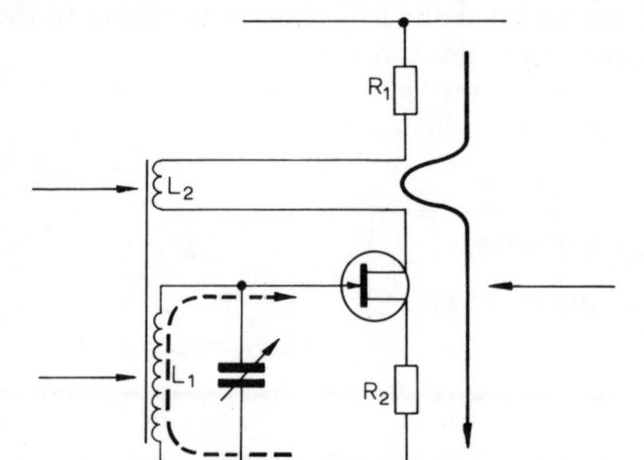

The f.e.t. produces an amplified signal in the same way as the common transistor.

The Transistor as a Diode

Remember two important facts from your previous work:

A radio signal must be rectified before it can be heard.
See page 98.

If a transistor is fully 'on' or fully 'off' it will only amplify half of an a.c. signal.
See page 115.

This means that a transistor in either the fully 'on' or 'off' state can be used as a rectifier in a radio in place of the usual diode. In fact, the f.e.t. radio on the previous page makes use of this effect and you will see that there is no diode. The common circuit below also shows rectification by transistor.

Rectification by Transistor

Look at the circuit diagram. The bias resistor has switched the transistor fully 'on'. The collector current is the maximum possible for this circuit (90 mA), even without a radio signal. If a positive radio signal is received it cannot increase the collector current so no amplification is seen. However, a negative radio signal can decrease the collector current.

This transistor effectively blocks positive going signals.

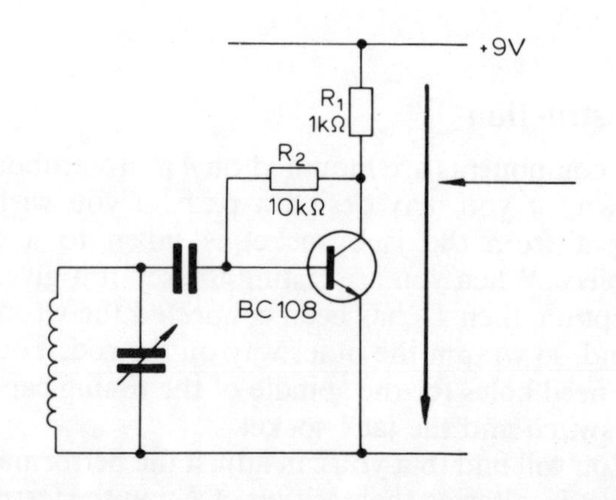

Nine Volt Power Supply

This circuit provides a 9·4 V supply, with a maximum current of 150 mA.

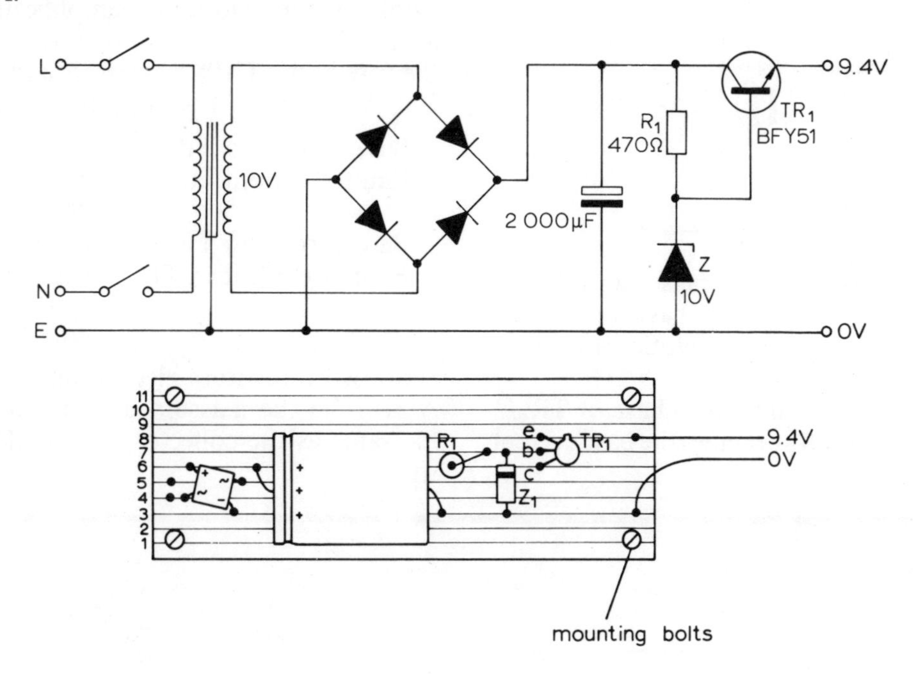

Construction

This circuit is a modification of the circuit shown on page 97. You only need to change the value of R_1 and add on TR_1, a BFY 51 type. In fact R_1 and Z_1 need not be the high power types used previously, but it is easier to leave Z_1 in place if you have already built the earlier circuit. TR_1 may be fitted with a push-on heatsink to keep it cool.

Follow the instructions given on pages 13 and 148 for construction of the mains equipment.

The Circuit

Within limits, the zener diode always has a voltage drop of 10 V across it, so, after allowing for the 0·6 volt drop across the base–emitter junction of the transistor, this leaves 9·4 V across the output. Now check the circuit in use (see table) and compare it with the simpler circuit which you built.

Performance

By connecting different load resistors across the output, find what happens to the output voltage as you draw different currents from the unit. Calculate what current is being taken and put your results in table form.

Load Resistor (kΩ)	Load current	Output voltage
10		
1		
0·1 (1 W)		

With a load resistor of 100 Ω connect an oscilloscope across the output and examine the smoothing. Record the size of the ripple shown on the oscilloscope.

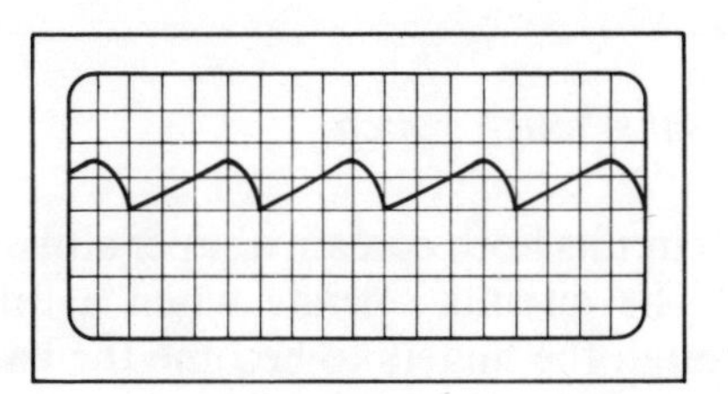

Ripple with voltage setting at . . . V/cm.

Super Alpha Pair

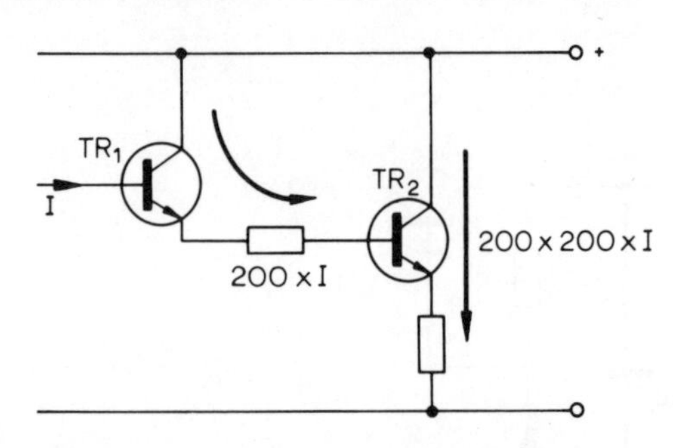

A super alpha pair consists of two transistors wired as emitter followers—this means they have an emitter resistor but no collector resistor. Notice two very important facts:

1. The emitter of TR_1 is connected to base of TR_2.
2. The emitter current of TR_1 becomes the base current of TR_2.

Amplification

Look at the enormous amplification that can be achieved by using two transistors in this way. Assume they are both BC108 transistors and each has a gain of 200. A current of I is put into the base of TR_1:

Base current to TR_1 $= I$
Emitter current in TR_1 = gain × base current
$= 200 \times I$
Base current to TR_2 $= 200 \times I$
Emitter current in TR_2 = gain × base current
$= 200 \times 200 \times I$
$= 40\,000\,I$

So two transistors give a gain of 40 000!

Note: It has been assumed here that emitter current is the same as the collector current. This is very nearly true.

Touch Switches

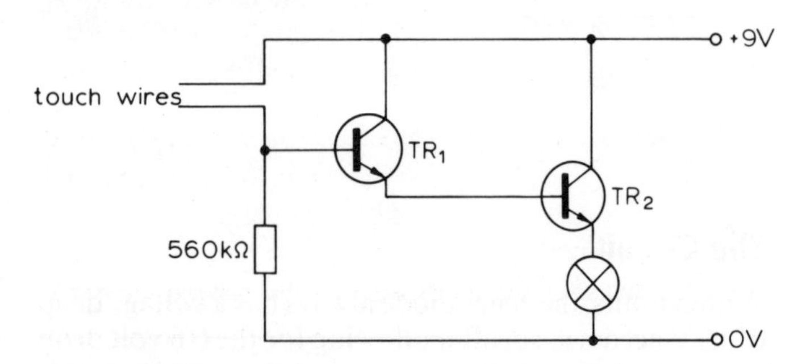

A touch switch

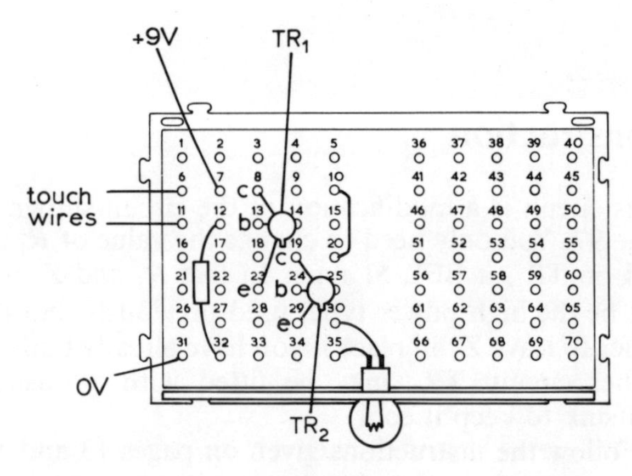

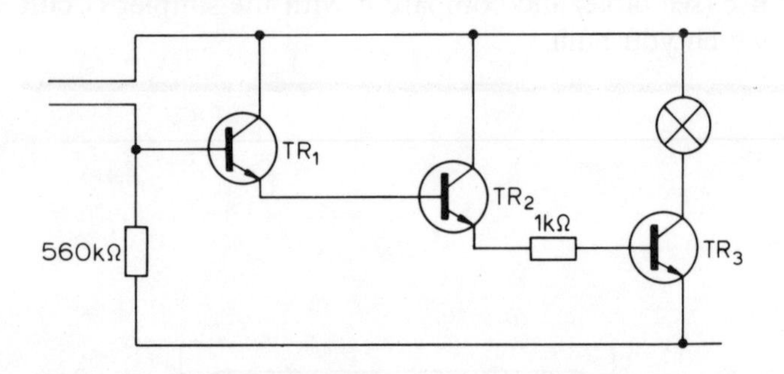

A more sensitive touch switch

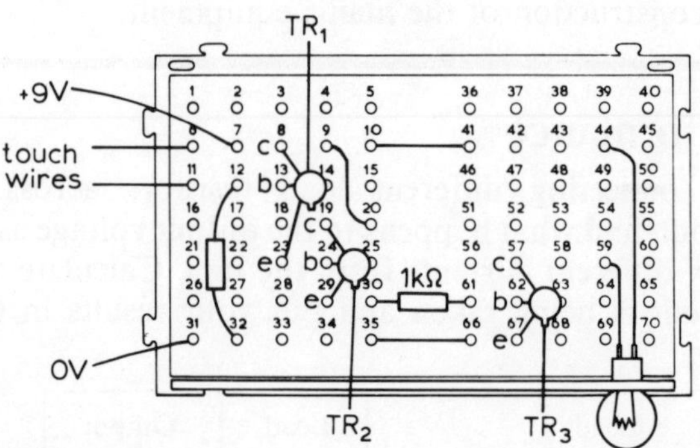

These two circuits both contain a super alpha pair, TR_1 and TR_2. The circuits operate when a tiny current passes through the fingers to become the base current to TR_1. The bulb brightness is affected by the resistance of the fingers (and their surface) so that the bulb brightness can be affected by dampness of the fingers, finger pressure and so on. This can lead to many modifications.

Modifications

Using a p.c.b., you may design a touch switch to suit your needs or the circuit may be modified to be a moisture meter, rain indicator, water level indicator and so on. The bulb may be replaced by a relay for switching another circuit and the sensitivity may be changed by altering the values of the resistors.

Relay Delay Switch

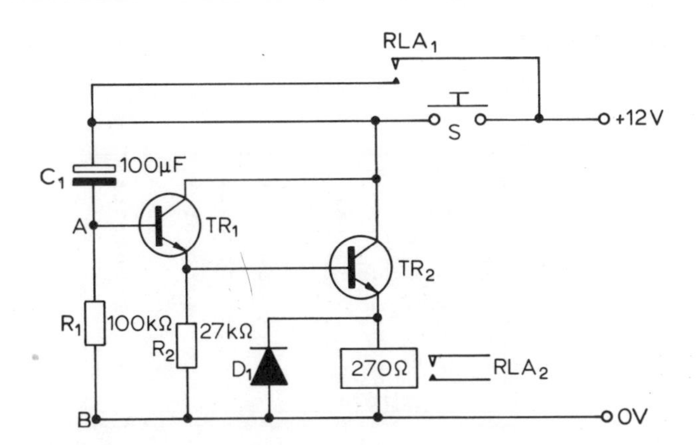

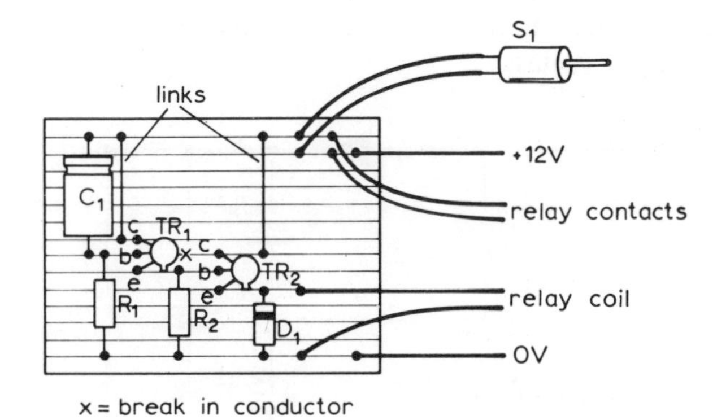

x = break in conductor

This is a very useful circuit for the motorist returning home after dark. It will keep a light switched on for about a minute, giving him time to get indoors, and then it will switch the light off. The car light it controls could be, for instance, one of its spotlights but it is advisable not to use the car's headlights or sidelights on which the motorist's safety depends, just in case the circuit should cause any malfunction.

Construction

The components are mounted on Veroboard, as shown. The relay must have two sets of normally open (n.o.) contacts, RLA_1 and RLA_2. One set, RLA_2, is connected in parallel to the switch of the light to be controlled. RLA_1 is connected as shown. This relay should be capable of switching a large current at its contacts. The push switch, S_1, should be mounted in an accessible place in the car. The circuit board and relay should be mounted in a small case, bolted in place in the car, perhaps under the bonnet.

Testing

Before it is mounted in the car, connect the unit to a battery and press switch S_1 momentarily. The two sets of contacts, RLA_1 and RLA_2, should close and remain closed for about a minute, then open again. The circuit is now ready for mounting in the car and connecting to the car battery and the light switch. Remember that your unit will suffer a lot of vibration, so all parts must be firmly fixed in place.

The Circuit

You see here two familiar part circuits—an *RC* series combination, (R_1 and C_1) and a super alpha pair (TR_1 and TR_2). (Strictly speaking, they are not a super alpha pair, since some of TR_1's current can bypass TR_2 via R_2. However, with R_2 at 27 kΩ this current is small.)

Series *RC*

When a voltage is supplied to the R_1 and C_1 combination, then the capacitor begins to charge up. At first, the voltage at point A will be high. As the capacitor charges up, this voltage will gradually fall. It is this voltage at A that controls the two transistors. Connect a voltmeter across AB and record the voltage change and time. Maximum voltage at A = ——— V Minimum voltage at A = ——— V Time that the relay stays on = ——— s.

Temporarily connect another 100 µF capacitor in parallel with C_1 and repeat these measurements.

Super Alpha Pair

Remember the useful property of the super alpha pair, that its amplification is so high that it does not draw very much current for the base of TR_1. This means that it does not upset the series *RC* combination.

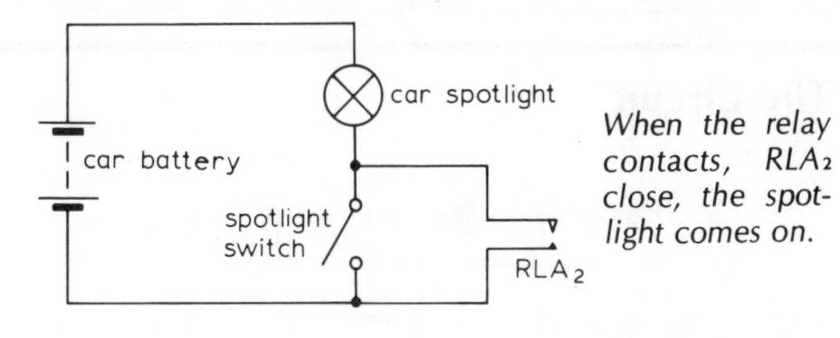

When the relay contacts, RLA_2 close, the spotlight comes on.

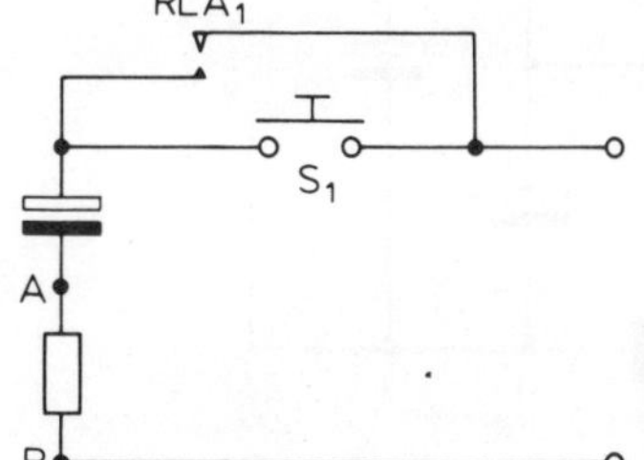

As soon as the relay contacts, RLA_1 close, they take over from S_1. When the relay switches off, RLA_1 opens and the unit is disconnected from the battery, until S_1 is again pressed.

The diagrams show the parts played in the circuit by the relay contacts. It is assumed that the capacitor discharges itself by leakage so it needs a long time to discharge every time it is used.

Voltage Doubler

As its name implies, the voltage doubler enables you to produce an output voltage which is approximately double the input voltage.

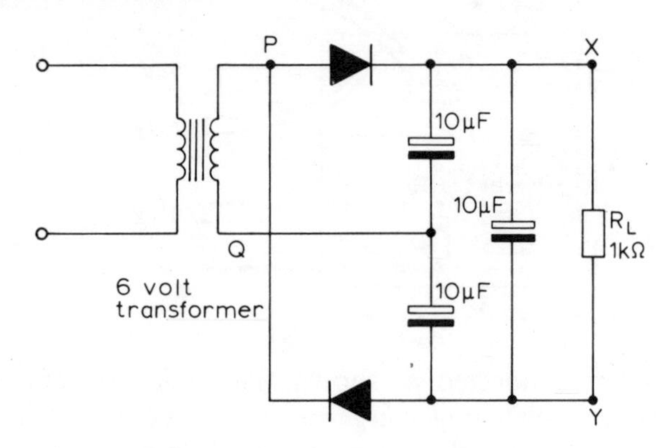

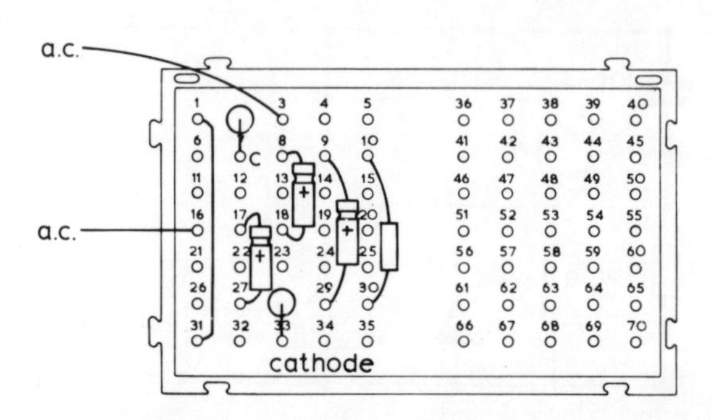

Construction

Assemble the circuit on S-DeC, using the values shown. The experiment can be done with alternating voltage connected to P and Q, instead of the transformer. (This must, of course, be a low voltage a.c. source and not the mains!) Connect either a d.c. voltmeter or oscilloscope across XY, and take the measurements shown below.

Result

Input voltage = 6 V   Output voltage = ———

Performance

Like all voltage supplies, the circuit's performance changes when you start to use it and draw current.

With capacitors of 10 μF

Load resistor (Ω)	Output voltage (V)
10 000	
1000	
100	

With capacitors of 1000 μF

Load resistor (Ω)	Output voltage (V)
10 000	
1000	
100	

Examine its performance by changing the load current drawn from the circuit, using different values for the load resistor, R_L. Also, you can examine the effect of using larger capacitors.

Result

The output voltage drops as the load current increases.

The output voltage is more stable with larger capacitors.

The Circuit

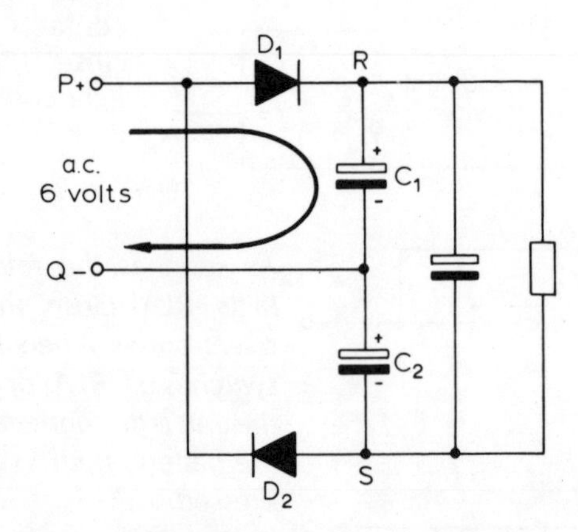

P positive

When P goes positive, current flows through D_1, and capacitor C_1 charges up to the supply voltage (approximately).

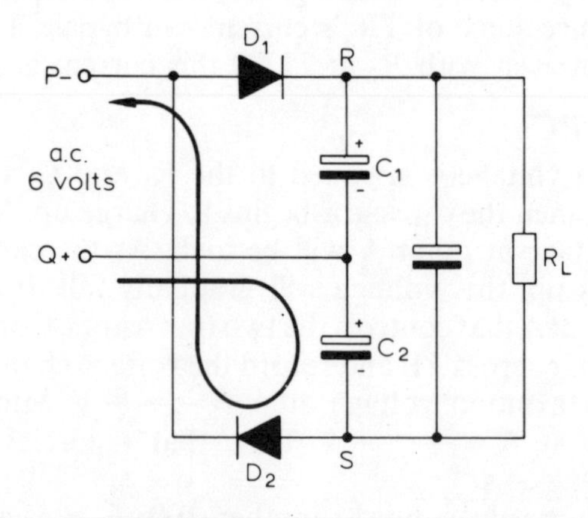

P negative

When P goes negative, current flows through D_2, and capacitor C_2 now charges up to supply voltage. The voltage across RS is now equal to the sum of the voltages across C_1 and C_2.

The Schmitt Trigger— Used As A Light Switch

This is a circuit with quite general uses in switching. In this case, it is to be used as a car parking-light switch, controlled by a light dependent resistor, making it fully automatic in operation. The circuit is adjusted so that the relay switches on the parking light when dusk approaches.

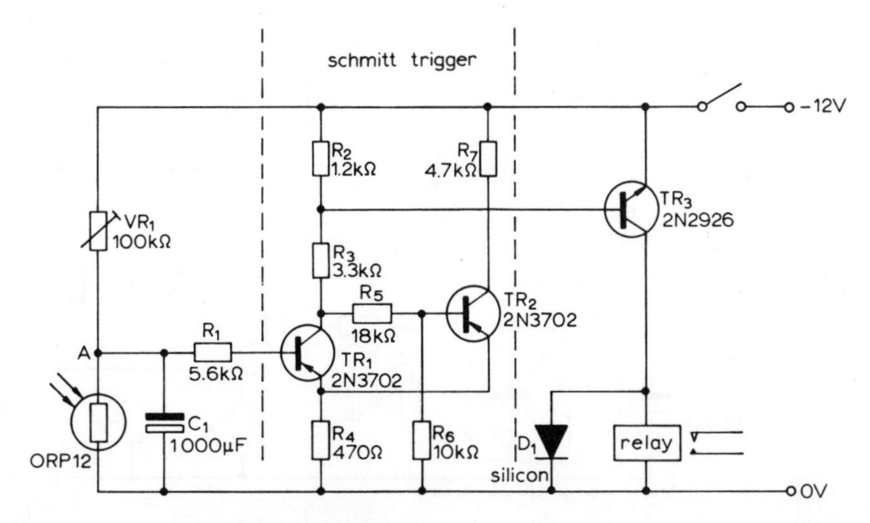

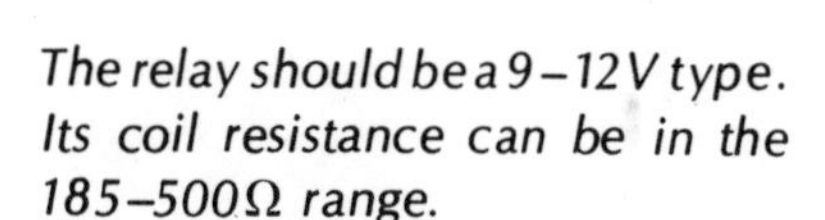
The relay should be a 9–12V type. Its coil resistance can be in the 185–500Ω range.

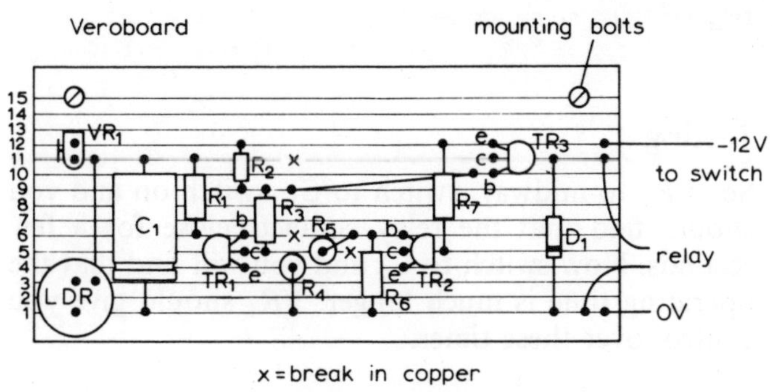

Construction

Assemble the circuit on Veroboard, as shown. The layout is quite spacious and may be reduced in size, if you wish. Arrangements must be made for the l.d.r. to 'see' the level of light, and so you may wish to mount this in the car, separate from the board. The relay contacts can control a simple series circuit with a parking light and battery. The unit includes a switch so that it may be disconnected when the car is not parked on the street.

Testing and Adjusting

You will find that the unit takes a few seconds to adjust to any change, so allow these few seconds when taking readings from the circuit. (This is quite deliberate in the design, so that the parking light isn't switched off every time the headlights of a passing car fall on the l.d.r. Perhaps you can work out how C_1 achieves this.)

You should find the following results:

VR_1 set at minimum: relay is always switched on.
VR_1 set a maximum: relay is always switched off.

Now adjust VR_1 so that the relay switches on as the level of light falls to that of dusk. The unit is now ready for boxing up and installing in the car. Take the following voltage measurements:

Voltage at point A in total darkness = ______
Voltage at point A in bright light = ______
Voltage at point A at 'switch-on' = ______
Voltage across relay in darkness = ______
Voltage across relay in bright light = ______
Also, time taken to switch from on to off = ______

The Circuit

VR_1 and the l.d.r. form a potential divider which controls TR_1. With TR_1 switched on, TR_2 is off and TR_3 is on.

When the voltage at A starts to switch TR_1 on, this causes changes in the circuit which make TR_1 switch on even harder. This is caused by positive feedback—the first voltage change causes a second voltage change which is fed back to make the first voltage even larger.

It is positive feedback that causes a Schmitt trigger to be either fully on or fully off, but never to remain in between.

The Schmitt Trigger—Used As A Time Switch

You should recognise the Schmitt trigger and relay driver parts of the circuit from the last page. The light sensing section, however, has been replaced by a timing section. This gives a circuit which has timing periods from about 1 s to 100 s, operating via a relay.

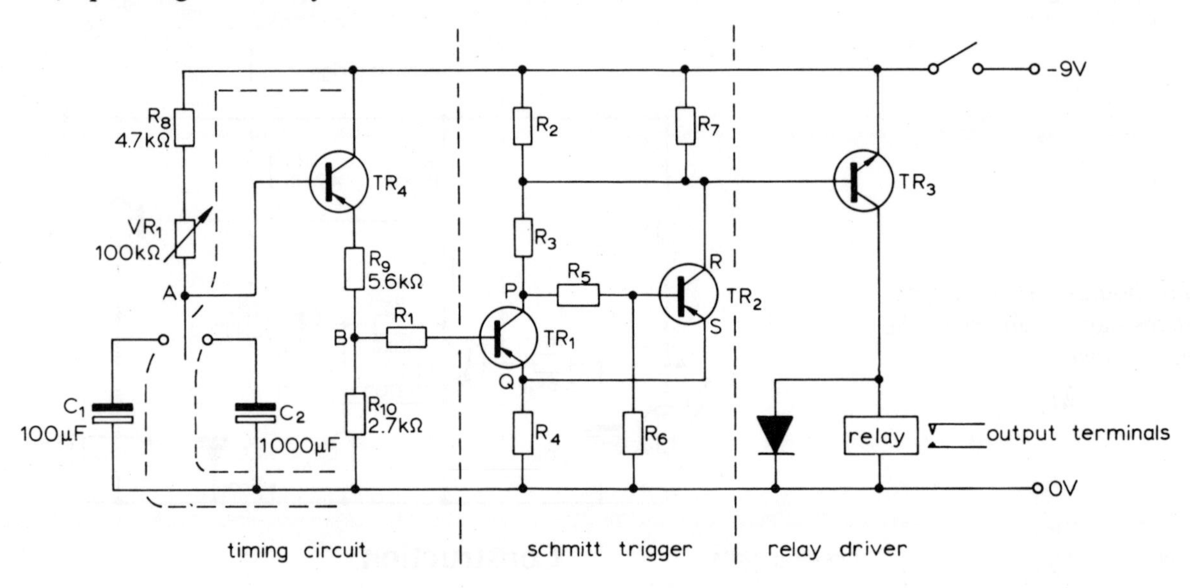

Construction

Use a Veroboard layout, as on the last page, and design the extra part for the timing circuit yourself. If you box up this project, your front panel will need:

1. A switch to select either C_1 or C_2, to give the two different timing ranges, 0·5 s–10 s or 5 s–100 s.
2. A knob for VR_1 to select the desired time.
3. A graduated scale for the time.
4. Two output terminals for the circuit that is switched by the relay.
5. An on/off switch for the unit.

Testing

Set VR_1 to midway, switch to C_1, switch on and you should find that the relay contacts close for a few seconds. Now switch to C_2 and you will find that the operating time is much longer. VR_1 should give you control over these times.

The Circuit

TR_4 is part of the timing circuit which switches on TR_1.

TR_1 switches on TR_3, the relay driver.

TR_2 provides positive feedback which makes the switching rapid and complete.

1. Timing Circuit

This is merely a capacitor and a resistor, (R_8 and VR_1, in this case) in series. As the capacitor charges up, the voltage at A changes, as you found. Draw a part circuit showing the current path for this timing section.

Set the timer to about 20 s and switch on. Follow the voltages at A and B.

Record the voltages when the relay comes on.

Voltage at A = ——. Voltage at B = ——. (When relay comes on.)

2. Schmitt Trigger

In the simplest transistor circuits, the collector voltage can be made to vary between 0 V and the supply voltage. This is not so with the Schmitt trigger.

Set the timer again to about 20 s and switch on. Take the collector to emitter voltages across PQ and RS.

> In a Schmitt trigger, the collector/emitter voltages remain near to 0 V or near to the supply voltage, but never remain in between.

This is the usual requirement of a switch—that it should rapidly change from on to off, as needed, but it should not linger in between.

The Darlington Pair

The Darlington Pair is similar in some ways to the super alpha pair, but it is normally wired with the load to the collector terminal and not the emitter.

The Darlington pair circuit gives very large values of amplification.

Now compare a Darlington pair with a single transistor. Assume you are using BC108 transistors which have a gain of, say, 200, and you are trying to switch on a 60 mA bulb.

Single Transistor

Required collector current = 60 mA

$$\text{Gain} = \frac{\text{collector current } (I_c)}{\text{base current } (I_b)}$$

$$200 = \frac{60}{I_b} \qquad I_b = \frac{60}{200} = 0{\cdot}3 \text{ mA}$$

So a base current of 0·3 mA will switch on a collector current of 60 mA

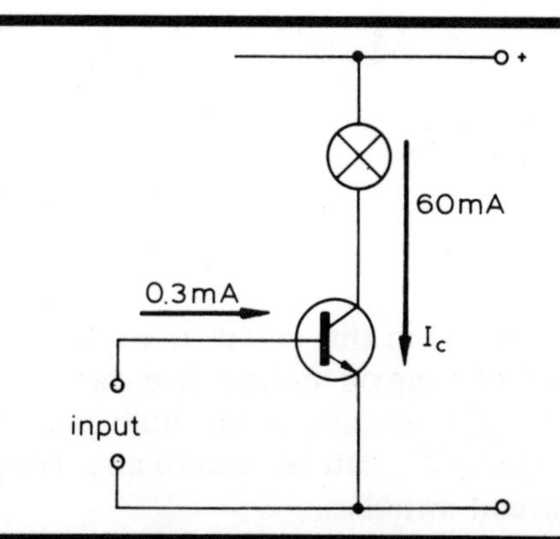

Darlington Pair

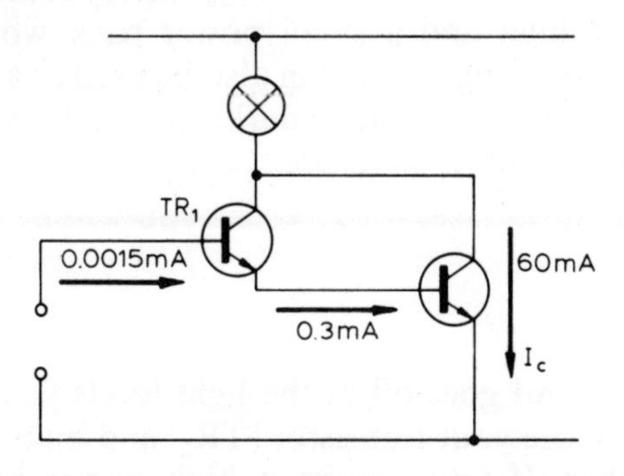

As before, for a collector current of 60 mA (through TR_2) a base current (into TR_2) of 0·3 mA is needed. So TR_1 only has to pass a collector/emitter current of 0·3 mA. Now calculate the base current required by TR_1 to do this:

$$\text{Gain} = \frac{I_c}{I_b}$$

$$200 = \frac{0{\cdot}3}{I_b} \qquad I_b = \frac{0{\cdot}3}{200} = 0{\cdot}0015 \text{ mA.}$$

Thus a base current of 0·0015 mA into the base of TR_1 will switch on a collector current of 60 mA in TR_2.

The Darlington pair achieves its high amplification because the emitter current of TR_1 becomes the base current of TR_2.

Sensitive Switch

This is, by now, a familiar sort of arrangement. A voltage divider at the base of a transistor controls its switching action. The transducer can be an l.d.r., a thermistor, resistor etc.

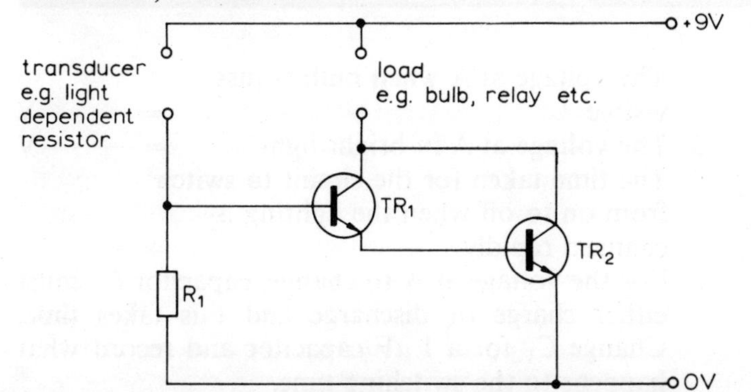

Amplifier

This circuit will give very good amplification and, if TR_2 is a power transistor, then a high volume can be expected. The loudspeaker must be chosen to match the power wanted.

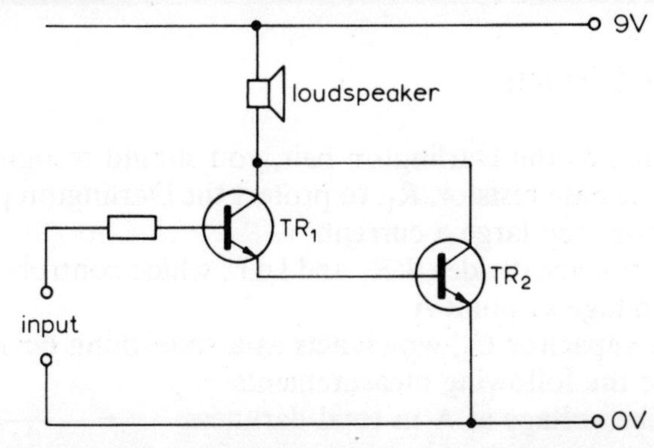

A Light Operated Switch

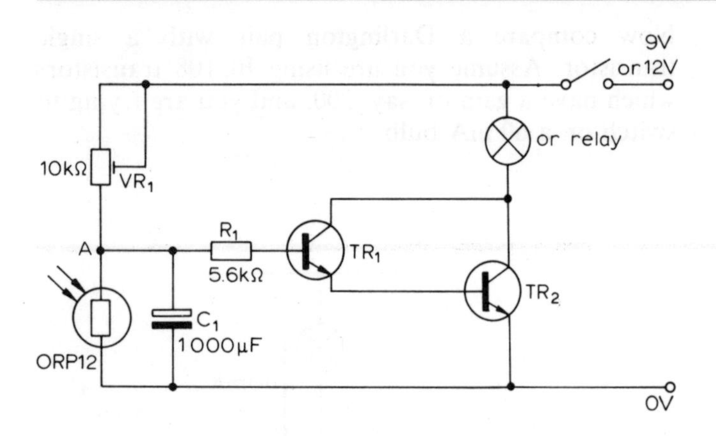

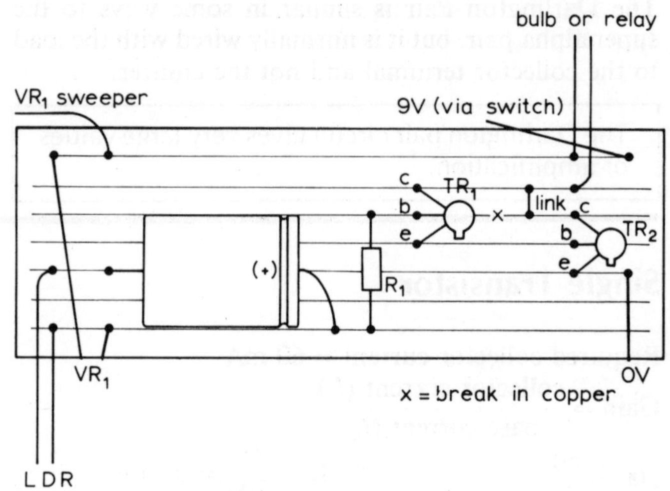

The bulb (or relay) in this circuit is controlled by the amount of light falling on the l.d.r. In darkness, the bulb is on and in light conditions the bulb goes off. The variable resistor, VR_1, can be used to alter the point at which the circuit switches.

Uses

An obvious use for the circuit is a car parking light, with the unit wired into the car battery circuit. If the bulb is replaced by a relay, then the circuit could switch on, for instance, an outside (mains) light, after dark. See your teacher about the dangers of connecting your relay to the mains supply.

If the circuit is used in this way, battery consumption would be high and a small power pack would save money in the long run. It can also be used as a warning light, for instance, outside a darkroom. If the bulb is on, then the darkroom is in use.

Construction

The circuit is shown assembled on Veroboard. You may test the circuit first on S-DeC, if you wish. Leads attached to the l.d.r. would enable it to be mounted independently of the circuit board, and you may have to consider shielding the l.d.r. so that it is not affected by any stray light.

Testing

Set VR_1 to its midpoint and cover the l.d.r. The bulb should light up and, if the l.d.r. is now illuminated, the bulb should go out. Now set VR_1 so that the bulb comes on and goes off at the light levels you wish.

Note: Choose your transistor (TR_2) and bulb to match each other. If you require a high power bulb then choose a power transistor that will take the necessary current. A BFY51, for instance, will take a current of 1 A and handle 800 mW of power, whilst a 2N3055 will handle 15 A. If you only need the 60 mA bulb commonly used in this course, then a BC108 will do.

Make allowance for the fact that, at switch on, bulbs draw a much larger current than their rated value until the bulb becomes hot.

The Circuit

As well as the Darlington pair, you should recognise:

1. The base resistor, R_1, to protect the Darlington pair from too large a current.
2. A voltage divider, VR_1 and l.d.r., which controls the voltage at point A.
3. A capacitor C_1, which acts as a smoothing device.

Take the following measurements:

1. The voltage at A in total darkness = ——— V
2. The voltage at A when bulb is just visible = ——— V
3. The voltage at A in bright light = ——— V
4. The time taken for the circuit to switch from on to off when the lighting is changed rapidly = ———
5. For the voltage at A to change, capacitor C_1 must either charge or discharge and this takes time. Change C_1 for a 1 μF capacitor and record what happens to the switching time.

Oscillators

Oscillator is a term that refers to many different kinds of circuit that all have one thing in common—they oscillate Put simply, this means that the circuit moves from one extreme to another and back again, continuously, without any outside signal. In practical terms, oscillators are producers of a.c. waveforms.

A Phase Shift Oscillator

Instructions

Construct the oscillator shown below, which produces alternating current in the audible range. By suitable use of the contacts the unit can either be a continuity tester, morse code practice device, warning bell and so on. Note that it only contains one transistor so it will only power high resistance earphones.

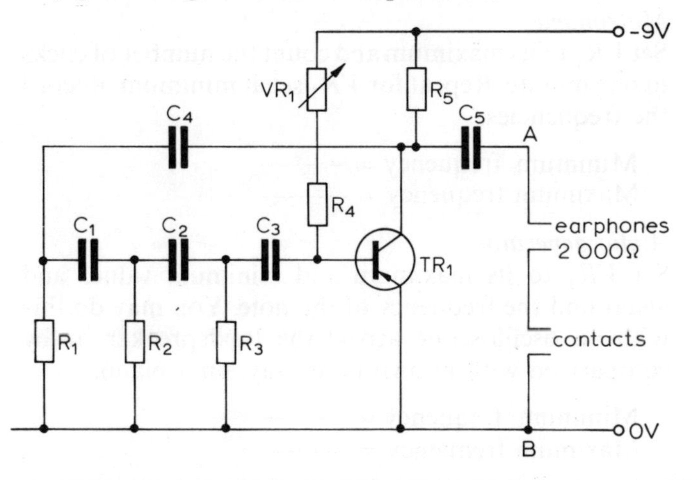

R_1, R_2, R_3, R_5 2.2 kΩ C_1, C_2, C_3, C_4, C_5 0.1 μF
R_4 220 kΩ TR_1 2N3702
VR_1 220 kΩ

Testing

With the battery connected, a note should be heard in the earphones when the contacts are joined. Adjustment of VR_1 should change the pitch of the note. Examine the shape of the output across AB on an oscilloscope.

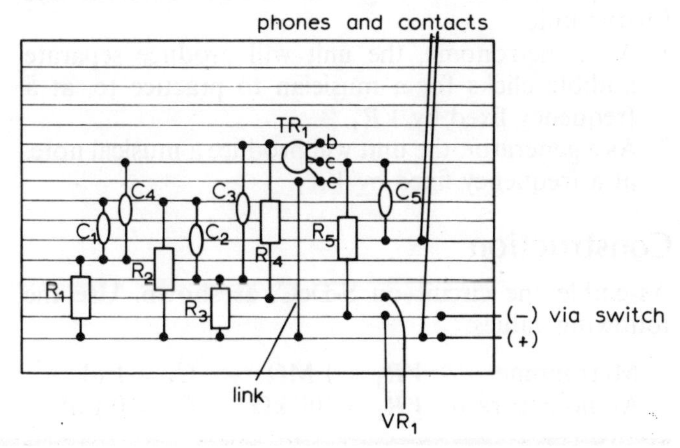

How Does Oscillation Occur?

There are two main processes to be understood in the operation of a simple one-transistor oscillator and neither of these will be completely new to you.

Feedback

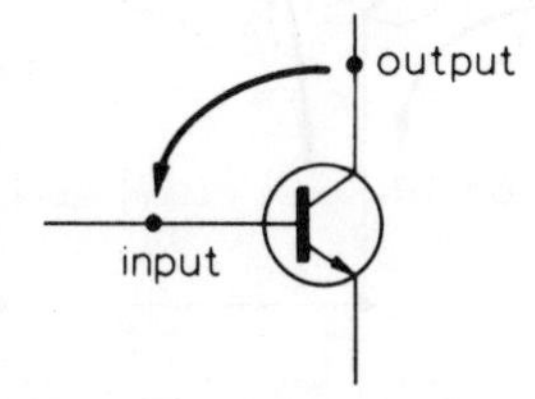

Let us imagine that, somehow, the circuit has managed to produce its first cycle of alternating current at the output. Now examine how the circuit could produce the second cycle, then the third and so on. The answer is by feedback. A fraction of the voltage of the first output cycle becomes the input for the second cycle. The feedback path (in the oscillator) is via C_4 and the combination C_1, C_2 and C_3.

The circuit acts like an amplifier which provides its own signal; every output wave is an amplified version of a fraction of the previous one.

Phase Shift

When this feedback occurs, the timing must be just right. The feedback wave must arrive at the input when the first output has just finished. Imagine what would happen if it arrived too early or too late. This means that you need a delay in the feedback of one cycle. This one cycle delay means that you have a continuous train of waves arriving at the input.

Every feedback wave starts just as the previous one finished. A one cycle delay is called a 360° phase shift.

Your teacher may discuss with you how the circuit achieves the 360° phase shift. In actual fact, the transistor achieves half of it and the capacitors achieve the other half.

Another Audio Oscillator

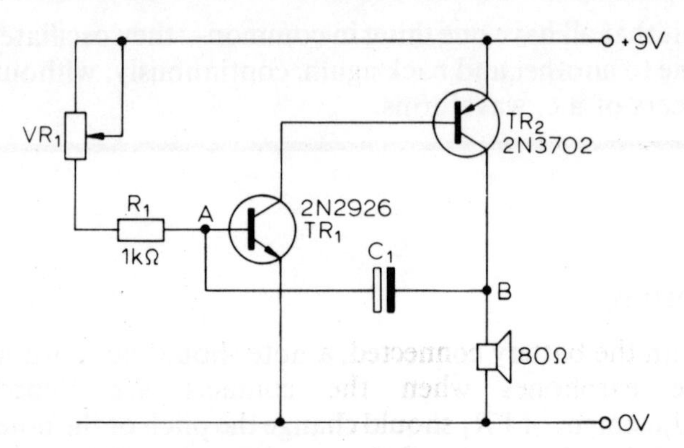

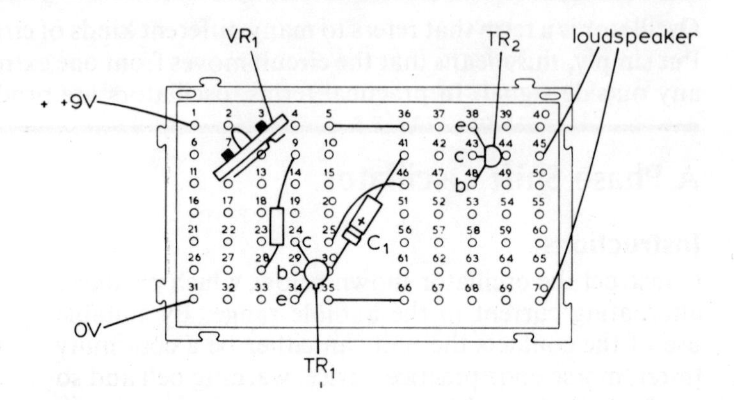

TR₁ 2N2926; TR₂ 2N3702

You are to use different values of components for (VR_1 and C_1) in this circuit to give at least two different uses for the unit:

1. As a metronome, the unit will produce separate audible clicks for a musician to practice to, at a frequency fixed by VR_1.
2. As a generator, the unit will produce a musical note, at a frequency fixed by VR_1.

Construction

Assemble the circuit on S-DeC, as shown. Use the following values:

Metronome:	$VR_1 = 1\ \text{M}\Omega$	$C_1 = 1\ \mu\text{F}$
Audio generator:	$VR_1 = 100\ \text{k}\Omega$	$C_1 = 0{\cdot}1\ \mu\text{F}$

Testing

Metronome

Set VR_1 to its maximum and count the number of clicks in one minute. Repeat for VR_1 set at minimum. Record the frequencies:

Minimum frequency = ——
Maximum frequency = ——

Audio generator

Set VR_1 to its maximum and minimum values and again find the frequency of the note. You may do this with an oscilloscope across the loudspeaker or by comparison with known notes, say, on a piano.

Minimum frequency = ——
Maximum frequency = ——

The Circuit

Look at two part circuits:

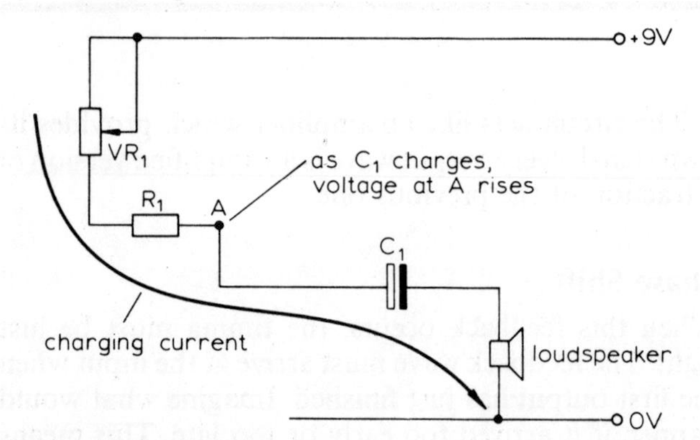

Capacitor charging circuit

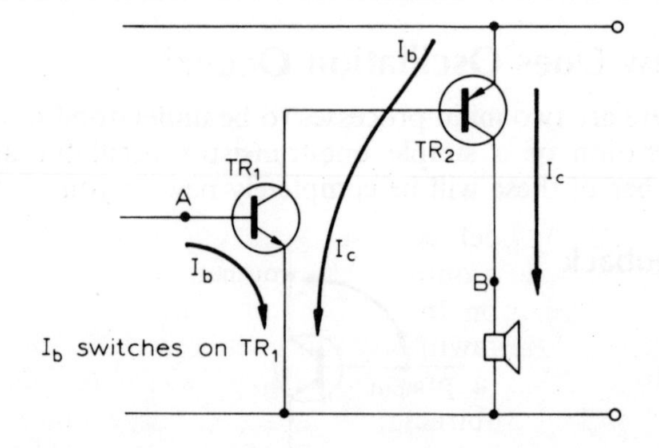

Transistor currents

The current for the loudspeaker is the collector current of TR_2. Remember, however, that a constant current through the loudspeaker produces no sound. Only a varying current can produce a sound. The capacitor, by charging and discharging, causes a varying current.

1. As the capacitor charges up, the voltage at A rises and the two transistors switch on, causing an increasing current in the loudspeaker.
2. However, with the transistors now on, C_1 can discharge, causing the transistors to switch off and the loudspeaker current to fall.

As this process is repeated, clicks can be heard in the loudspeaker.

12 THE COMPUTER

Introduction

We have now become accustomed to the computer being able to perform a vast range of tasks, from the simple arithmetic needed to work out a bank balance to the complex task of guiding a spacecraft to the planets. But what is this machine and how does it manage to carry out such a variety of jobs? Obviously we use different computers to carry out different tasks but they all operate on similar principles. Look at the basic layout of a very simple computer.

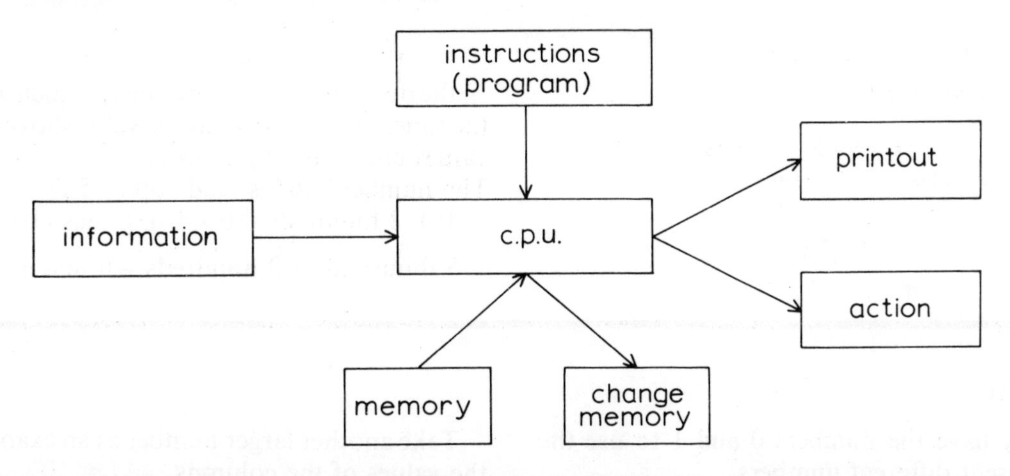

The Process

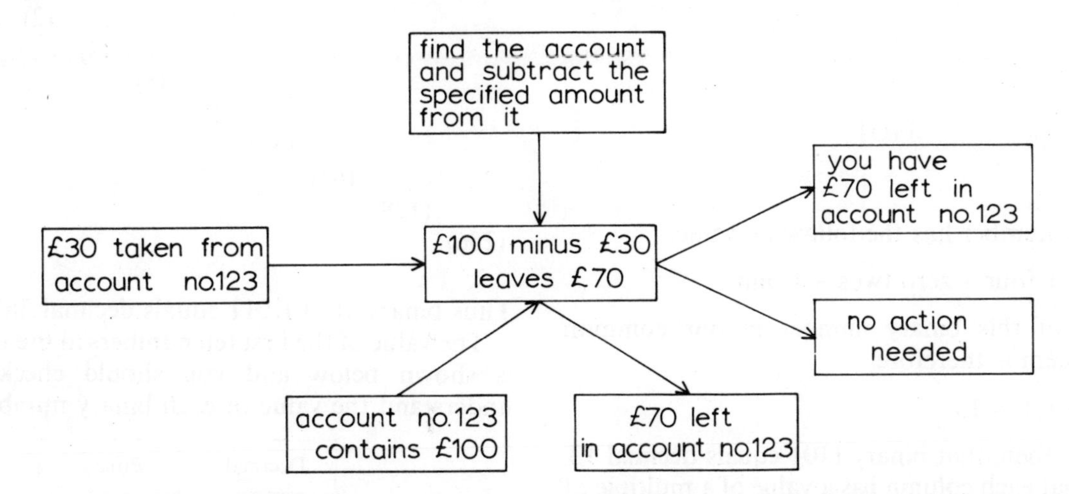

Using the model, consider what happens when, say, money is taken out of a computerised bank account. The information (number of the bank account and amount withdrawn) is fed into the central processing unit (C.P.U.), a programme gives the computer its instructions (subtract the amount withdrawn from the amount in the account), the C.P.U. then 'goes into' the memory (to find how much money was in the account) and then the C.P.U. does the subtraction sum.

Of course, if the account contained only £20 then the computer would have to take some action, perhaps by ordering the sending of a warning letter to the account holder. This block diagram gives some idea of how many activities are involved and how they are interconnected.

However, it is still hard to imagine how electronic circuits can perform some of these processes and, in particular, there are four processes which are at the heart of these operations:

Storing numbers and words	Arithmetical processes
Steering numbers through the computer	Giving instructions to the computer

This chapter will attempt to solve some of these mysteries for you.

Decimal and Binary Numbers

To carry out all its tasks, the computer uses just two numbers, 0 and 1, instead of our familiar 0, 1, 2, . . . , 9. This system of two numbers is called the binary system.

To understand how to use the binary system, it is necessary to examine how we use our more common decimal system.

Decimal System

Look at the number 5263:

hundreds units
thousands tens
5263

In the decimal system of numbers, which we use most of the time, each column has the value shown. The column values are all multiples of ten.

The number 5263 is made up of 5 thousands ($10 \times 10 \times 10$), 2 hundreds (10×10), 6 tens (10) and 3 units.

5 thousands + 2 hundreds + 6 tens + 3 units.

Binary System

When we only have the numbers 0 and 1 to use the columns represent different numbers.

fours units
eights twos
1101

This binary number has the following value:

1 eight + 1 four + zero twos + 1 unit.

The value of this binary number in our common decimal system is therefore:

$8 + 4 + 0 + 1 = 13$

We can say, then, that binary 1101 equals decimal 13.

Notice that each column has a value of a multiple of 2. Obviously, the binary number system requires the use of more digits than the decimal system but its advantages will become clear shortly.

Take another larger number as an example and learn the values of the columns.

1 0 1 1 0 1 0 1 1

Column value	Value
(1)	1
(2)	+2
(4)	+0
(8)	+8
(16)	+0
(32)	+32
(64)	+64
(128)	+0
(256)	+256
	363

Thus binary 101101011 equals decimal 363

The value of the first ten numbers in the two systems is shown below and you should check that you understand the value of each binary number

Decimal	Binary
1	1
2	10
3	11
4	100
5	101
6	110
7	111
8	1000
9	1001
10	1010

Why Use Binary?

The answer to the question 'Why use binary?' lies in the nature of transistor operation. You have seen the transistor switching in two states which you called on and off, producing high and low voltages at output. This means that we have a way of representing the numbers 0 and 1.

Binary 0 and 1 are represented by the on/off states of a transistor

Every number is written in the computer as 0's and 1's by putting transistors in on and off states. When a computer does an addition sum, it combines the voltages of the two numbers in such a way that the voltages produced represent the answer. Obviously, binary is used because two is the highest number of different states into which a transistor can be put.

Addition in Binary

Look at the rules for addition in binary. The largest digit available is 1, so, if addition produces a higher number than 1, a carry into the next column is needed. For example:

```
  1        1
 +1       +1
 --       +1
 10       --
          11
```

Now look at a larger addition sum, but remember that the rules are always the same. The first three columns have been explained for you.

So addition in binary, in the computer, can be done by combining the high and low voltages which represent the 0's and 1's.

```
   10110
 +101100
 -------
 1000010
```

gives 0 and carries 1

gives 1

obviously gives 0

Display

Simple binary numbers can be displayed in the experiments on the next few pages by the use of indicator bulbs or l.e.d.'s. The bistable you built on page 127 contains a bulb which indicates whether a transistor is on or off and so can represent a binary digit. An array of bulbs, as shown here, can represent a complete binary number:

represents binary number

1 1 0 0 1

A Binary Counter

The bistable circuits you constructed in Chapter 8 can be combined into a binary counter which will count the number of pulses which you insert. The counting is, of course, done in binary and the answer is displayed by the bulbs of the bistables.

A Simple Adder

Instructions

Arrange four bistables, as constructed on page 127, in a row. The first one should include the fourth transistor stage (drawn in a pale line) whilst the other three need only have the bistable and indicator stages (heavy line). Connect the output of the first bistable to the input lead of the second, then the output of the second to the input of the third and so on. Comnect their top and bottom rails together.

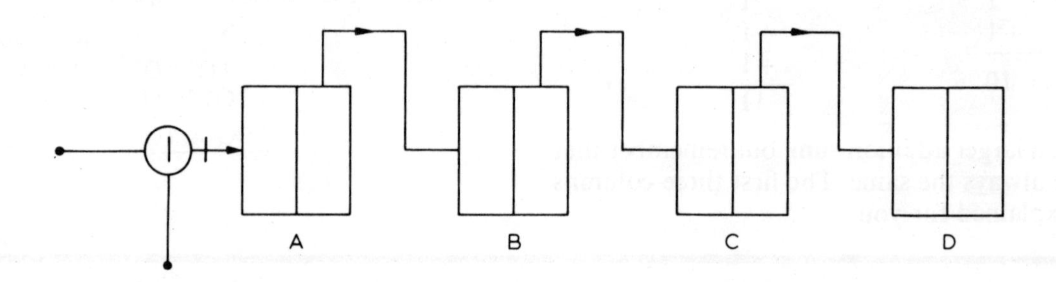

Operation

Start with all four bistables off (all bulbs off) and then insert pulses at the gate input. As each pulse is entered, note the number indicated by the bulbs and the total number of pulses used. Check against the table shown here, noting that the results have been written from the right.

Number of pulses entered	Bulb display			
	D	C	B	A
0	0	0	0	0
1	0	0	0	1
2	0	0	1	0
3	0	0	1	1
4	0	1	0	0
5	0	1	0	1
6	0	1	1	0
7	0	1	1	1
8	1	0	0	0
9	1	0	0	1
10	1	0	1	0
11	1	0	1	1
12	1	1	0	0
13	1	1	0	1
14	1	1	1	0
15	1	1	1	1
16	0	0	0	0

Result

It should be obvious that the unit has done a binary count. Check against your binary table and you should see that the numbers indicated by the bulbs are the binary equivalent of the decimal number inserted.

The Use of Integrated Circuits

The binary counter you made is considered to be clumsy by the standards of today's electronic industry and units to do the same task as your circuit would be made at a fraction of the size of the one you built. Computers are made from components called integrated circuits and microprocessors. These contain large numbers (in some cases, thousands) of transistors, diodes, resistors and capacitors all in one package, but of the same sort of size of just one component of the types you have used so far. An extremely useful set of integrated circuits is the 74 series and you can now use some of these in your computer experiments.

Using the 74 Series

These integrated circuits all have 14 terminals and can simply be pushed into a standard 14 pin socket. This means that once the socket has been suitably mounted then any of this series can be connected immediately, without any further soldering.

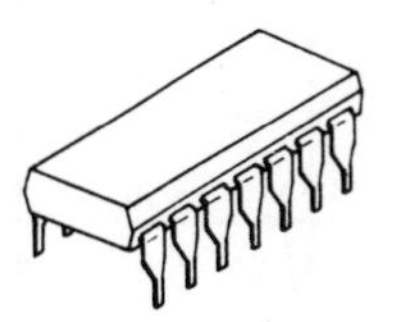

Integrated circuits

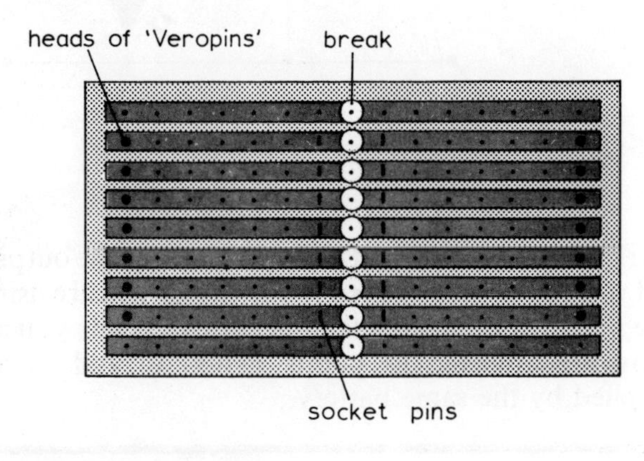

Veroboard—copper side

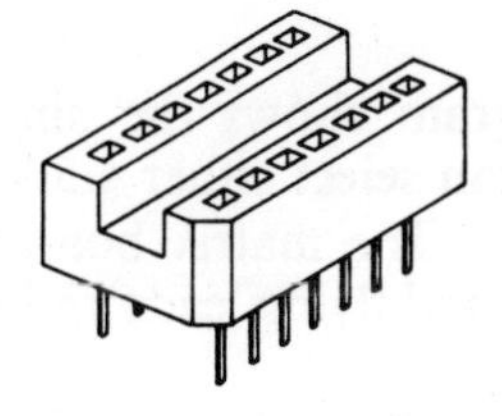

14 pin socket

Instructions

Use 2·5 mm Veroboard to hold the 14 pin socket and break all the copper strips at hole 8, as shown. Press 14 Veropins through, from the copper side, so that you can solder their heads to the copper strip. Press a socket through, from the non-copper side, so that you can also solder its pins to the copper strip. This means that the Veropins are now connected to the socket pins.

Connect a length of insulated wire to pin 14 (labelled positive) and pin 7 (labelled negative). These are your supply terminals.

Any 74 series integrated circuit can now be pushed into this holder and connected by the Veropins to any circuit.

Identification

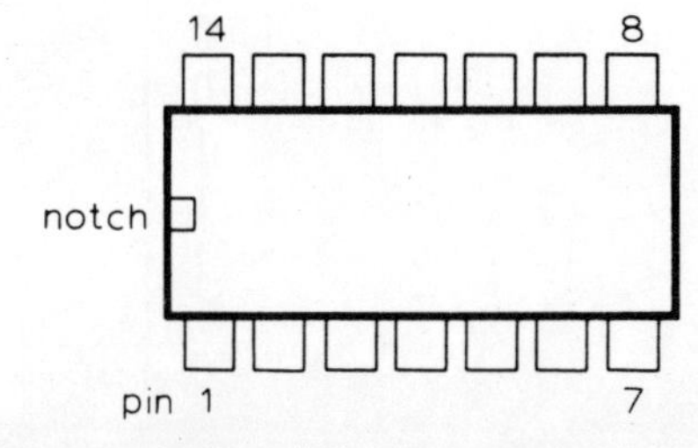

Use this diagram to identify the pin numbers of the SN7400 series. Hold the integrated circuit with the pins pointing away from you, and you will notice a notch on the upper surface. Hold the integrated circuit with the notch on the left, and the pins are numbered as shown.

Integrated Circuits—Input and Output

To use these circuits to perform operations with numbers, you have to be able to supply them with the numbers 0 and 1. We use the supply voltage of + 4·5 V to represent the number 1 and this is called logical one voltage whilst zero voltage represents the number zero and is called logical zero voltage.

Simple switches, which can select either the high or low voltage, are used to input the numbers, as you need them.

Selecting 0 and 1

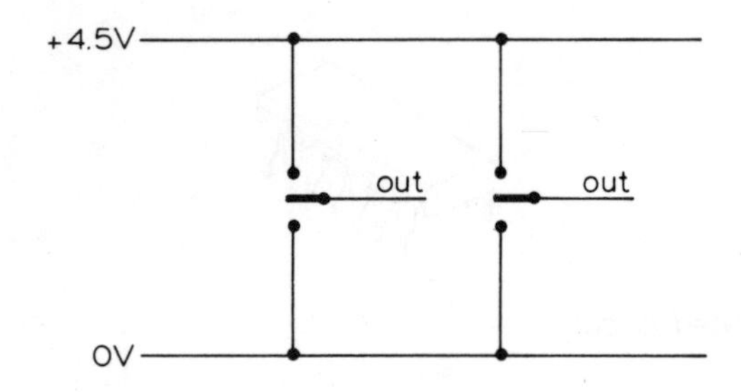

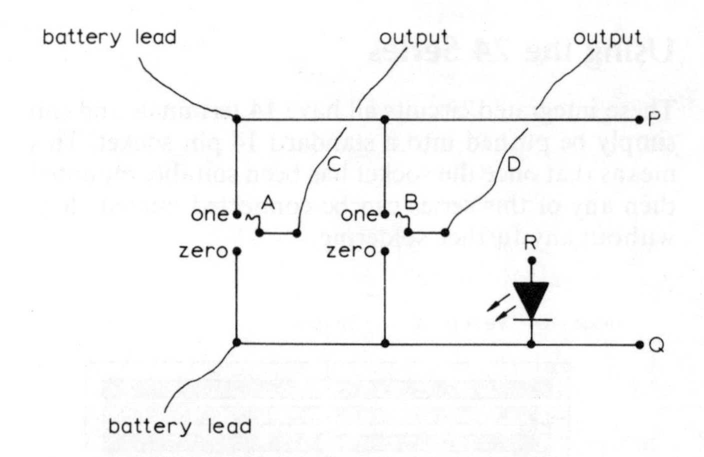

Instructions

The circuit diagram shows the simple change-over switches which can select either 4·5 V (logical one) or 0 V (logical zero). The matrix board diagram shows positions of pins and flying leads and also an l.e.d. (see below).

A, B, C and D are flying leads and should have small crocodile clips attached. A and B can be clipped to either the upper or lower pin to select either one or zero.

Mark the pins as shown. Leads C and D are the outputs and will supply one or zero to the circuit you are using. The supply terminals of the integrated circuits you are using are attached to points P and Q and so are supplied by the same battery.

Displaying 0 and 1

To display the numbers 0 and 1 the l.e.d. shown above is used.

l.e.d. on represents one
l.e.d. off represents zero

A circuit may be connected to point R and the l.e.d. will show whether a one or zero is indicated.

To demonstrate the use of the switches, one of the leads, C or D, may be connected to R; but the l.e.d. must be protected by a 220 Ω resistor in series with it.

Operate your switches to their 0 and 1 positions and verify that the l.e.d. displays in the manner stated above.

The 74 series of integrated circuits may have their outputs connected directly to point R, without the need for the resistor so the l.e.d. will tell you directly whether your integrated circuit is registering a 0 or 1 at its output.

Note: If an input lead is not connected, it is at logic 1. This is just a peculiarity of the way the SN7400 integrated circuits are made.

Some Computer Operations—Logic Gates

To achieve processes more complicated than the simple binary addition you have already seen, a computer has to be able to combine and transport the logical one and zero voltages (numbers 1 and 0) in a large variety of ways. These operations are carried out by gates, the first of which is the invert gate. It is not possible in a book of this nature to be able to show how the use of these gates can achieve complicated computer operations, so your work will end with a study of the simpler gates.

The INVERT Gate

An INVERT gate will produce the opposite binary number to the input. This is a very useful function in a computer and the integrated circuit SN7404 contains six of them. Test the operation of one of these INVERT gates in the apparatus you have assembled already.

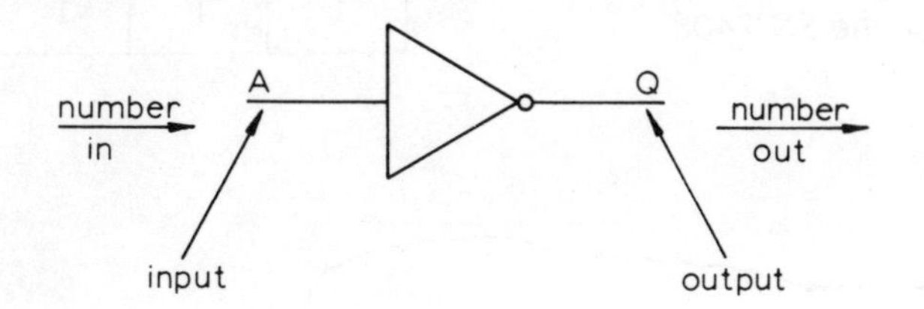

Symbol for the INVERT gate

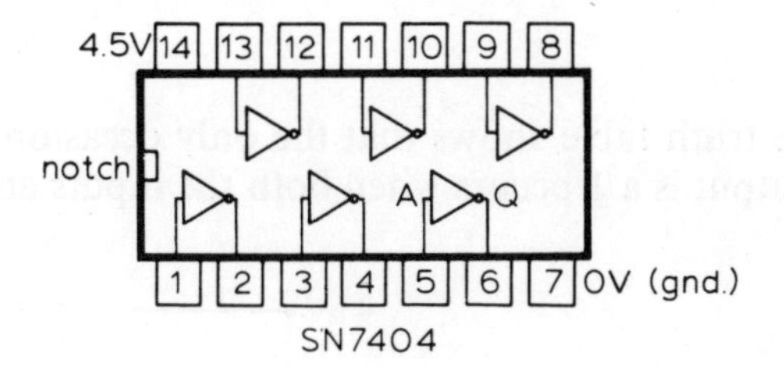

Internal circuit for the SN7404

The internal circuit diagram shows the six INVERT gates in the SN7404 and the terminals involved. For example, the inputs to the gates are the terminals numbered 1, 3, 5, 9, 11 and 13 and any one of these can be used in your experiment.

A *truth table* tells you what the output from a gate will be for all possible inputs. The table here shows that for the possible inputs (0 and 1) the output will be 1 and 0 respectively.

Truth table

Input (A)	Output (Q)
0	1
1	0

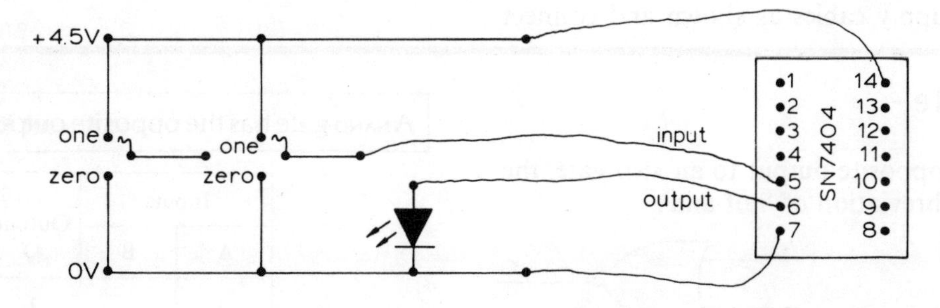

Instructions

Place an SN7404 in the socket board and connect up the two supply cables, as shown, to terminals 7 and 14. Connect one of your switches to the input of an INVERT gate, say terminal 5, and connect its output, terminal 6, to the l.e.d. Now use your switch to input a 1 to the gate and check that the output is zero (l.e.d. off) and repeat for a zero input.

Check these results against the truth table.

Construction of INVERT Gate

You have already met a circuit which will act as an INVERT gate. The simplest transistor circuit will output a low voltage at the collector (logical zero) when a high voltage (logical one) is input at the base. This is invert operation. It is, however, far more bulky than the integrated circuit above.

Continued on p. 170

The AND Gate

The AND gate operates with at least two numbers. It has, therefore, at least two inputs, labelled A, B . . . . . . For a two input AND gate:

An AND gate will output a 1 when both A and B are at 1.

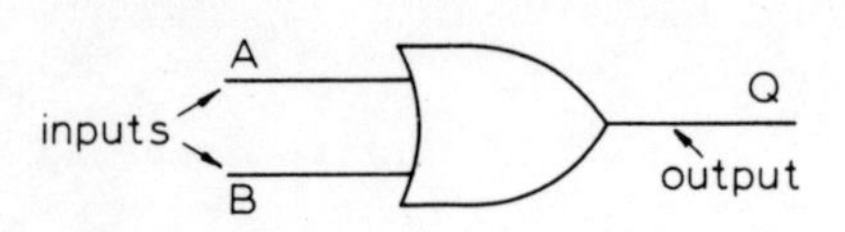

Symbol for the AND gate

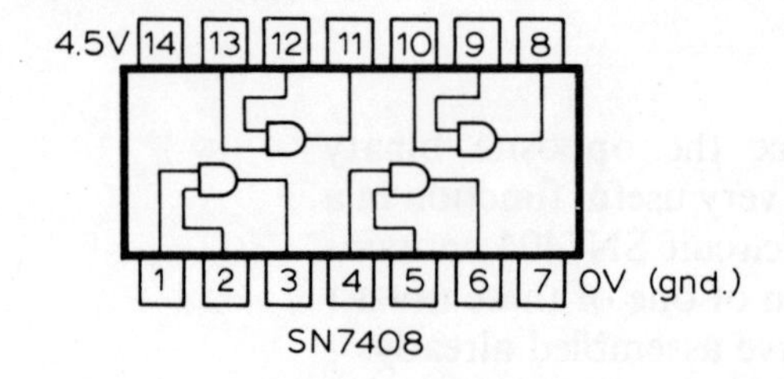

Internal circuit for the SN 7408

Truth table

Inputs		Output
A	B	Q
0	0	0
0	1	0
1	0	0
1	1	1

The truth table shows that the only occasion when the output is a 1 occurs when both the inputs are at 1.

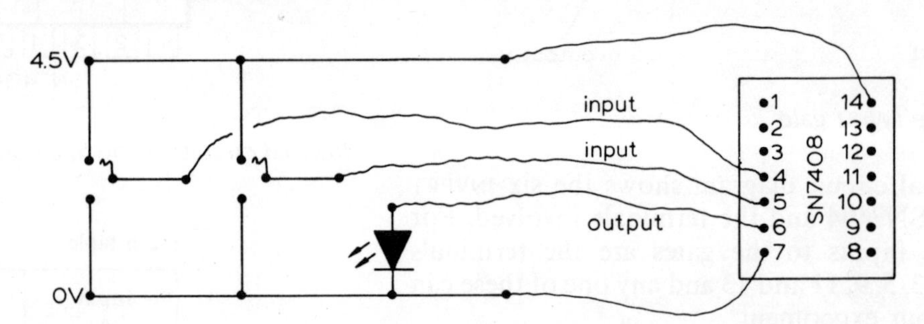

Instructions

Place an SN7408 in the socket holder. The diagram shows that it contains 4 AND gates, terminals 4 and 5 being the input to one of these and 6 being its output. Connect up the supply cables as shown and connect each of your switches to one of the input terminals. Work through the truth table for all possible inputs and check the output against the table.

The NAND Gate

A NAND gate has the opposite output to an AND gate.

This gate has the opposite output to an AND gate, the name being an abbreviation of 'not and'.

Symbol for the NAND gate

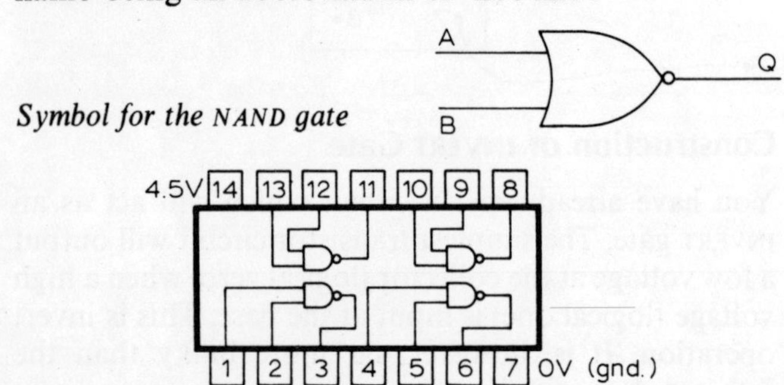

Internal circuit for the SN7400

Inputs		Output
A	B	Q
0	0	1
1	0	1
0	1	1
1	1	0

Instructions

Replace the SN7408 with an SN7400 which contains 4 NAND gates. Test the truth table in the same way as before.

Continued on p. 171

The OR Gate

The OR gate is another gate with at least two inputs (A and B). It performs as follows:

An OR gate will output a 1 when either A or B is at 1

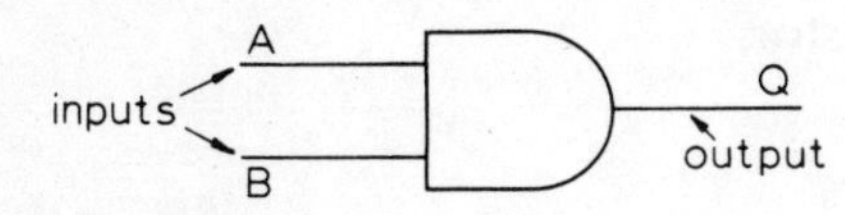

Symbol for the OR gate

Truth table

Inputs		Output
A	B	Q
0	0	0
0	1	1
1	0	1
1	1	1

Unfortunately, there is no SN7400 series integrated circuit available which contains the OR gate. However, a method of assembling an OR gate from two integrated circuits is shown later.

The NOR Gate

The word NOR is a compression of 'not or' so naturally:

The NOR gate has the opposite output to an OR gate.

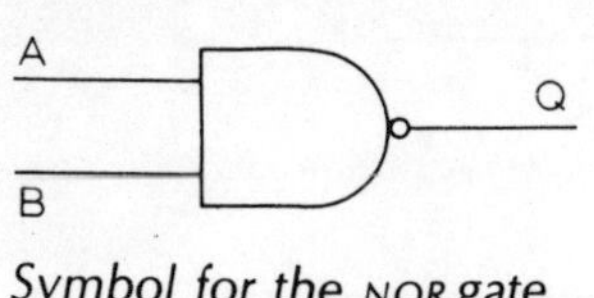

Symbol for the NOR gate

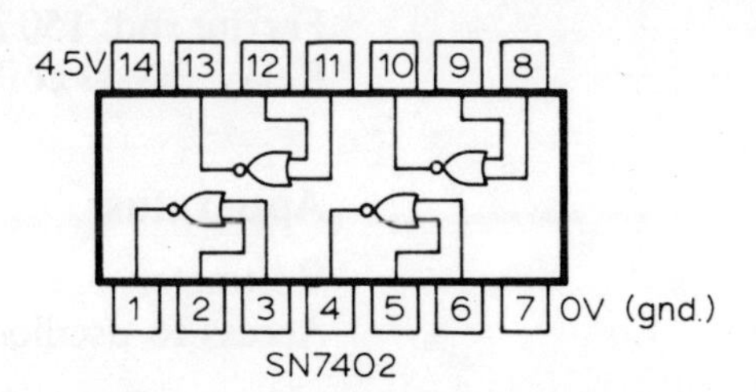

Internal circuit for the SN7402

Truth table

Inputs		Output
A	B	Q
0	0	1
0	1	0.
1	0	0
1	1	0

The SN7402 has four NOR gates, as shown above, but notice that the input and output connections for each gate are at different pin numbers from the previous gates.

Instructions

Place the SN7402 integrated circuit in the socket and connect the input and output terminals of one of the NOR gates in the usual manner. Use the two switches to input 0's and 1's to test the above truth table. You should see that this table has opposite output to the OR table at the top of the page.

Constructing an OR Gate

It should be obvious from examining the above two tables that if you invert the output from the NOR gate you obtain the output shown in the OR table. Use an SN7402 NOR gate with an SN7404 INVERT gate to construct the logic circuit shown below and check its performance.

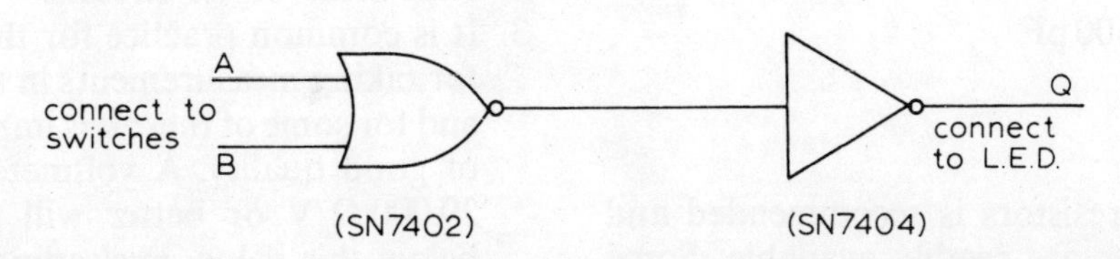

This concludes the study of the basic gates.

Components List

This book contains 35 constructional projects and it is not envisaged that readers would make all of them. The amount of time available for the study of this course will probably preclude that amount of practical work and so a components list for these projects is not included. The book also contains some 70 teaching circuits or sets of experiments, in which the assembly of components is not permanent and the components would be re-used.

Transistors

Most circuits use the BC108
Also 2N3702
2N2926
2N3055
BFY51

Diodes

OA91
1N4001
1N4004

Zener Diode

7·5 V

Other Semiconductors

Light emitting diode
Thyristor
Unijunction transistor 2N2646

Capacitors

It is usually economic to buy capacitors in bulk packs, specifying the type and range of values required. Two useful types are the Mullard miniature electrolytic (values of 1 μF upwards) and the C280 type (below 1 μF). Many mail order firms offer these types in different combinations of values:

(in microfarads) 2000, 1000, 470, 250, 100, 47, 10, 1, 0·1, 0·01, 0·001,

Note: Most values are not critical and substitution of values which approximate can usually be tried.

Variable capacitor: 500 pF

Resistors

Again bulk buying of resistors is recommended and packs of the E12 series are readily available. Some resistors are required far more frequently than others (e.g. 1 kΩ and 10 kΩ) and familiarity with the text will soon indicate which should take priority.

Special values: 22 Ω, 33 Ω, 47 Ω, (all 5W) and 10 Ω(2W)

Variable resistors: 100 kΩ, 10 kΩ, 5 kΩ, 1 kΩ, (2W) 100 Ω, (1W)

Light dependent resistor: ORP12

Thermistor.

Inductors

60 mH

Miscellaneous

Switches
Bulbs
Earphone
Loudspeaker: 80 Ω
Battery: PP6 or PP9
Ferrite rod: 150 mm × 10 mm
Relay: 300 Ω coil

Apparatus

Voltmeter
Access to oscilloscope and signal generator

Integrated Circuts

SN7404
SN7408
SN7400
SN7402

Notes

1. The majority of the text can be covered using a very restricted quantity of components, so it is advisable for the reader to examine the text to decide which parts he or she wishes to study before buying the components.
2. Many schools and institutions have low voltage supplies, powered by the mains, to replace batteries but these must be used with care because many do not give a smoothed output and will cause confusion with some of the circuits.
3. It is common practice for the voltmeter to be used for taking measurements in the circuits in this book and for some of these it is important that the meter is of good quality. A voltmeter with a sensitivity of 20,000 Ω/V or better will give good results but below this value, misleading readings may be obtained with a few circuits.